RADIATION HEAT TRANSFER

RADIATION HEAT TRANSFER

A Statistical Approach

J. Robert Mahan

Department of Mechanical Engineering
Virginia Polytechnic Institute and State University

JOHN WILEY & SONS, INC.

This publication is designed to provide accurate and authoritative information in regard to the subject matter covered. It is sold with the understanding that the publisher is not engaged in rendering professional services. If professional advice or other expert assistance is required, the services of a competent professional person should be sought.

Wiley also publishes its books in a variety of electronic formats. Some content that appears in print may not be available in electronic books. For more information about Wiley products, visit our web site at www.wiley.com.

Library of Congress Cataloging-in-Publication Data:

Mahan, J. R.
 Radiation heat transfer: a statistical approach / James Robert Mahan.
  p. cm.
 ISBN 0-471-21270-9 (cloth : alk. paper)
  1. Heat—Radiation and absorption. I. Title.

 QC331 .M28 2001
 536'.3—dc21

                                                    2001046951
Printed in the United States of America.

10 9 8 7 6 5 4 3 2 1

To Bea

CONTENTS

13   The Distribution Factor for Nondiffuse, Nongray, Surface-to-Surface Radiation   371

14   The MCRT Method Applied to Radiation in a Participating Medium   390

15   Statistical Estimation of Uncertainty in the MCRT Method   413

PREFACE

This book is a result of the author's thirty years of experience teaching and directing research in radiation heat transfer at Virginia Tech. As is often the case, the book evolved from class notes distributed to critical graduate students. Therefore, it bears the brand not only of the author but also of a generation of bright young scholars who continuously challenged the author to get it right and make it relevant. For better or for worse, the result is a book written for students rather than for professors.

The material in this book is divided into three parts:

Part I: Fundamentals of Thermal Radiation
Part II: Traditional Methods of Radiation Heat Transfer Analysis
Part III: The Monte Carlo Ray-Trace Method

If the book is to be used in a one-semester course it is recommended that one of the two options indicated in the figure on page xvi be followed. Both options would use the first six chapters, which present the fundamentals of thermal radiation. A one-semester course emphasizing the traditional methods of radiation heat transfer, which includes the net-exchange formulation, would be based on the first six chapters plus Chapters 7, 8, 9, 10, and 11; while a one-semester course emphasizing the statistical formulation (the Monte Carlo ray-trace method) would use the first six chapters plus Chapters 11, 12, 13, 14, and 15. Chapter 11, "Introduction to the Monte Carlo Ray-Trace Method," is included in both of these options. In addition to these two options, the book is ideally suited for a two-semester (or three-quarter) sequence that covers all of the material.

While authors of recent radiation heat transfer textbooks have included the MCRT method as a viable option, it has usually been presented as an option of secondary importance. In this textbook the method has been promoted to its rightful position as an equal partner in radiation heat transfer modeling. The goal of this book is to present the subject at a level of detail and nuance that will allow the uninitiated practitioner to begin formulating accurate models of complex radiative systems without first assuming away all of the complexity.

If the MCRT method has been criticized in the past for its excessive demand on computer resources, such criticism stands without merit today in a world inundated by a virtual tidal wave of inexpensive computing power. Software tools such as the MCRT-based Program FELIX, the student version of which is packaged with this

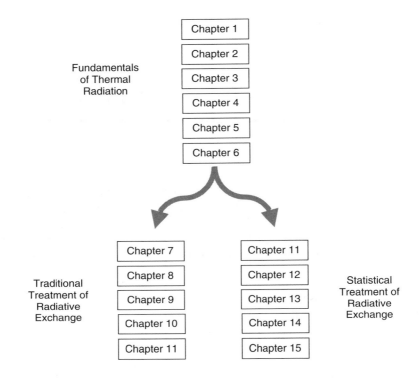

book, are now fully capable of free interaction with a wide range of CAD and spreadsheet systems. It seems that this trend must inevitably lead to a future that places increased value on the material in Part III of this book at the expense of the material in Part II. Still, change comes slowly. The traditional methods of Part II are well entrenched in our technical culture and are likely to remain influential in the foreseeable future.

J. R. MAHAN, PhD, PE
Blacksburg, Virginia

ACKNOWLEDGMENTS

In the mid 1960s one of the most remarkable heat transfer faculties ever assembled came together at the University of Kentucky under the leadership of Dean R. M. Drake, Jr. As a result, from September 1966 through December 1970 I had the privilege of studying heat transfer under the tutelage of Professors Drake, R. C. Birkebac, C. J. Cremers, Roger Eichhorn, and J. H. Lienhard IV. The education I received from these outstanding contributors to heat transfer knowledge, and especially the guidance I obtained from my advisor, Cliff Cremers, largely account for any success I have enjoyed in my subsequent career as an engineering educator.

My good fortune continued in the fall of 1970 when I began my teaching career at Virginia Tech under the direction of J. B. Jones, to whom I am greatly indebted for any acquired teaching skills I may now possess. After my first year in Blacksburg Professor Jones packed me off to NASA's Langley Research Center, where I spent the summer of 1971 under the mentorship of the late George E. Sweet, a founding father of earth radiation budget research. It was this experience as a NASA/ASEE Summer Faculty Research Fellow that determined the course of my future career in radiation heat transfer. In recent years my NASA-sponsored research has been ably monitored by Robert B. Lee III, whose support and encouragement have made it possible for me to work with a steady stream of outstanding graduate students. Over the course of a long career it has been my privilege to direct the doctoral research of a number of exceptional young engineers and scientists. Farshad Kowsary, Thomas H. Fronk, Nour E. Tira, Douglas A. Wirth, Edward L. Nelson, Jeffrey A. Turk, Pierre V. Villeneuve, Martial P. A. Haeffelin, Kory J. Priestley, Félix J. Nevárez, María Cristina Sánchez, Ira J. Sorensen, Dwight E. Smith, and Amie S. Nester have all made important contributions to this book. In addition, the book has benefited from the thesis research of more than forty master-of-science students, including Leo D. Eskin, who in 1980 became the first of my students to study the Monte Carlo ray-trace method.

The high standards of intellectual pursuit and accomplishment set by my heat transfer colleagues at Virginia Tech—Tom Diller, Doug Nelson, Elaine Scott, Karen Thole, J. R. Thomas, W. C. Thomas, and Brian Vick—have been a constant source of inspiration.

A recent semester spent as a visiting professor at the United States Naval Academy permitted me the time to pull the final manuscript together and send it off to my publisher, while at the same time exposing me to a whole other way of doing things. I owe a debt of gratitude to Patrick Moran, chair of mechanical engineering at the Academy, for arranging my visit there, and especially to my department head at Virginia Tech, Walter O'Brien, for arranging my leave. While at the Academy I had the privilege of collaborating with three outstanding heat transfer colleagues there: Matthew Carr, Karen Flack, and Ralph Volino.

Finally, and most of all, I owe more to my wife, Bea, than I will ever be able to adequately express or repay. Her proofreading, encouragement, support, companionship, and love have made all the difference.

J. R. MAHAN, PhD, PE
Blacksburg, Virginia

PART I

FUNDAMENTALS OF THERMAL RADIATION

INTRODUCTION TO THERMAL RADIATION

In this chapter radiation heat transfer is defined as the heat interaction between a system and its surroundings that occurs in the absence of an intervening medium. Thermal radiation is then identified as electromagnetic radiation emitted solely due to the temperature of the emitter, and the dual wave–particle description of electromagnetic radiation is introduced. The chapter concludes with an atomic-scale description of emission of thermal radiation by a solid.

1.1 THE MODES OF HEAT TRANSFER

Students of thermodynamics learn that *heat* is the energy interaction that occurs solely due to the temperature difference between a system and its surroundings. Engineers commonly apply the somewhat redundant term "heat transfer" to this same phenomenon. The study of heat transfer is traditionally divided into three categories, or "modes": conduction, convection, and radiation. Some authorities include heat transfer with phase change, or "boiling heat transfer," as a fourth mode.

In electrical nonconductors *conduction heat transfer* occurs when thermal energy stored in vibratory modes of atoms is passed from atom to atom through the interatomic forces that maintain atomic spacing. The process is similar for electrical conductors except that now the migration of free electrons contributes significantly to or even dominates heat diffusion. A simple model for heat conduction in electrical insulators is illustrated in Figure 1.1. In this model the atoms are represented as point masses of mass M and the interatomic forces are represented as linear springs of modulus K. If one of the point masses (the "source") is set into vibration along one or more of its three independent axes, or degrees of freedom, its neighbors will be jostled. This will produce a wave that propagates through the lattice according to the

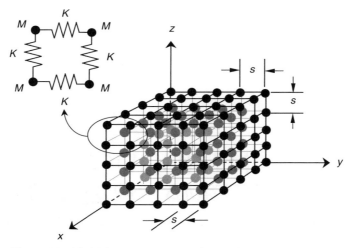

Figure 1.1   Model for heat conduction; an idealized crystal structure

laws of classical mechanics. The atoms behind the spreading wave front will be left to vibrate with amplitude that depends on the distance from the source, the effective mass of the atoms, the stiffness of the intermolecular bonds, and the manner in which the array of atoms is constrained at its boundaries. The amplitude and frequency of the vibrations are related to the physical quantity *temperature* such that heat is transferred from regions of higher temperature to regions of lower temperature in accordance with the second law of thermodynamics.

Convection heat transfer also requires the presence of an intervening physical medium between the system and its surroundings. The essential difference between convection and conduction is that in the case of the former the intervening medium must be a fluid that flows between the system being heated or cooled and its surroundings. Consider the heated flat plate shown in Figure 1.2. As the fluid flows over the plate it picks up "heat" (a student of thermodynamics would say "thermal en-

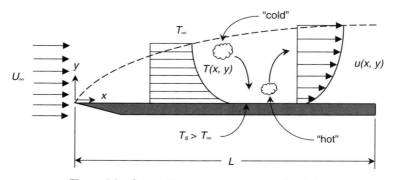

Figure 1.2   Convection heat transfer from a flat plate

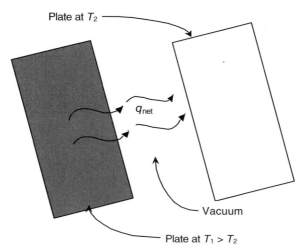

Plate at T_2

q_{net}

Vacuum

Plate at $T_1 > T_2$

Figure 1.3   Radiation heat transfer between two flat plates

ergy") and stores it in its thermal capacity. It then carries this stored thermal energy to the cooler surroundings where some of it is deposited. Note that conduction heat transfer is the actual mechanism by which thermal energy is picked up from the heated plate and deposited in the cooler surroundings. Thus, heat conduction is an essential element of heat convection.

In contrast to conduction and convection heat transfer, *radiation heat transfer* does not require the presence of an intervening medium. Rather heat is transferred from the warmer body to the cooler body by *electromagnetic radiation,* as indicated in Figure 1.3. The nature of electromagnetic radiation and its emission and absorption by a material substance is the subject of this book. However, for the moment let us turn our attention to the distinguishing characteristics of the three modes of heat transfer.

1.2   CONDUCTION HEAT TRANSFER

The model describing conduction heat transfer is *Fourier's law of heat diffusion,* which states that the local heat flux $\mathbf{q}$ (W/m^2) is proportional to the local temperature gradient ∇T (°C/m); that is,

$$\mathbf{q} = -k\,\nabla T \tag{1.1}$$

The "constant" of proportionality k (W/m·°C) in Equation 1.1 is the *thermal conductivity,* and the minus sign is a consequence of the second law of thermodynamics, which requires that spontaneous heat transfer be directed from higher to lower temperatures, or down the temperature gradient. Originally the thermal conductivity was assumed to be a scalar quantity; however, the more modern view is that it is a tensor k_{ij} whose off-diagonal elements are usually but not always equal to zero.

Equation 1.1 may be thought of as the definition of the thermal conductivity. Then, if in a particular situation the relationship between the local heat flux and the local temperature gradient is observed to deviate from Fourier's law, this deviation may be attributed to a variation in the thermal conductivity with position, direction, or temperature, or with some combination of these three quantities. In other words, for a given local component of the heat flux vector, say q_x, the negative of the ratio of the heat flux component to the local temperature gradient in the same direction, $\partial T/\partial x$, is taken as the definition of the local value of the element k_{xx} of the thermal conductivity tensor. In Cartesian coordinates the thermal conductivity tensor is written

$$k_{ij} = \begin{vmatrix} k_{xx} & k_{xy} & k_{xz} \\ k_{yx} & k_{yy} & k_{yz} \\ k_{zx} & k_{zy} & k_{zz} \end{vmatrix} \tag{1.2}$$

where, for example,

$$k_{xy} \equiv -\frac{q_x}{\partial T/\partial y} \tag{1.3}$$

The elements k_{xx}, k_{xy}, ... of the thermal conductivity tensor are in general a function of both position and temperature, and the values of the off-diagonal elements of the thermal conductivity tensor are in most cases so small that they can be treated as zero.

When the thermal conductivity is independent of temperature it can be said that the local conduction heat transfer varies as the local temperature, assuming the surrounding temperature remains fixed. However, because of the actual temperature dependence of the thermal conductivity, the local heat transfer may vary as the local temperature to a power of as much as 1.1; that is,

$$q_{\text{cond}} \propto T^{1.0-1.1} \tag{1.4}$$

1.3 CONVECTION HEAT TRANSFER

The model for convection heat transfer is *Newton's law of cooling*, which, for external flow, states that the heat flux from a surface at temperature T_S to a surrounding fluid, whose temperature far from the surface is T_∞, is proportional to the temperature difference $T_S - T_\infty$; that is,

$$q_{\text{conv}} = h\,(T_S - T_\infty) \tag{1.5}$$

The *heat transfer coefficient h* ($\text{W/m}^2\cdot{}^\circ\text{C}$), defined by Equation 1.5, is an aerodynamic quantity. This means that its value depends on the details of the boundary layer flow across the surface being heated or cooled (laminar or turbulent, attached or separated, etc.). Now, the flow field surrounding a solid body is coupled to the temperature field through the kinematic viscosity and the mass density, whose values are generally temperature dependent. As a consequence of this and of the fact that the specific heat and thermal conductivity of the fluid are also temperature dependent, the heat transfer coefficient h in Newton's law of cooling is a function of temperature. When this is taken into account convection heat transfer from a surface may vary with surface temperature to a power as high as two; that is,

$$q_{\text{conv}} \propto T_S^{1-2} \tag{1.6}$$

1.4 RADIATION HEAT TRANSFER

We have seen that conduction and convection heat transfer are described by convenient *models* that define the thermal conductivity in the case of conduction and the heat transfer coefficient in the case of convection. The two models are referred to as "laws" because they are based on observation rather than derived from first principles. Their success is based on the wide latitude permitted by the adjustability of the coefficients involved. Radiation heat transfer, on the other hand, is described by a model subject to rigorous derivation, at least for one important special case (that of the blackbody, introduced in Chapter 2). For this reason it is perhaps inaccurate to refer to the model describing radiation heat transfer as a law. However, following common usage we will refer to this model as the *Stefan–Boltzmann law of radiation heat transfer,* which states that the heat flux emitted from a blackbody at absolute temperature T_b is

$$q_{\text{rad}} = \sigma T_b^4 \qquad \text{(blackbody)} \tag{1.7}$$

The coefficient σ in Equation 1.7 is the *Stefan–Boltzmann constant.*

The German physicist Josef Stefan (1835–1893) first suggested the form of Equation 1.7 in 1879 on the basis of data already available in the literature [1]. Stefan deduced that a straight line would result if the instantaneous cooling rate of a body suspended in a vacuum were plotted against the difference between its absolute temperature to the fourth power and that of its surroundings to the fourth power. Five years later Stefan's student, the Austrian physicist Ludwig Boltzmann (1844–1906), derived the form of Equation 1.7 on the basis of classical thermodynamics [2]. Boltzmann's development is described in Chapter 2. In contrast to Fourier's law and Newton's law of cooling, the coefficient in the Stefan–Boltzmann law is a physical constant whose value can be obtained exactly from well-established laws of physics.

Equation 1.7 must be modified when "real" surfaces and bodies, rather than the highly idealized blackbody, are considered. Under certain circumstances the modification may take the form of a factor ε, called the *emissivity,* which multiplies the Stefan–Boltzmann constant. Then under these circumstances the net heat flux from surface 1 to surface 2 for the arrangement of parallel flat plates shown in Figure 1.3 is well approximated by

$$q_{\text{rad, net}} = \frac{\varepsilon}{2 - \varepsilon} \sigma(T_1^4 - T_2^4) \tag{1.8}$$

for sufficiently small spacing between the plates compared to their size. In writing Equation 1.8, the emissivity ε has been assumed to be the same for both surfaces.

We will learn in Chapter 3 that the emissivity is introduced to account for the fact that real surfaces generally emit and absorb less heat than a blackbody at the same temperature. In fact, under certain circumstances Equation 1.8 can be used to define the emissivity in much the same way that Fourier's law defines the thermal conductivity and Newton's law of cooling defines the heat transfer coefficient. As in the case of the thermal conductivity k and the heat transfer coefficient h, the emissivity can vary with temperature. Because of this variation the net radiation heat flux from a surface varies with the surface temperature to a power ranging between four and five; that is,

$$q_{\text{rad, net}} \propto T_S^{4-5} \tag{1.9}$$

The relations discussed in this section are summarized in Table 1.1. It can be concluded that radiation heat transfer becomes increasingly more important as other modes of heat transfer are suppressed (cryogenics, spacecraft thermal control, radiometry, thermal imaging) and when temperatures become high (combustion, electric arcs and thermal plasmas, atmospheric reentry).

1.5 THE ELECTROMAGNETIC SPECTRUM

It has already been stated that thermal radiation is an electromagnetic phenomenon. Electromagnetic waves, whose properties are explored elsewhere in this book, are capable of carrying energy from one location to another, even—indeed, especially—in

Table 1.1 Summary of the relationship between temperature and heat flux for the three modes of heat transfer

Heat Transfer Mode	Model, or "Law"	Coefficient Defined by Model	Range of Temperature Exponent
Conduction	Fourier's law $\mathbf{q} = -k \, \nabla T$	Thermal conductivity k	1.0–1.1
Convection	Newton's law of cooling $q = h \, \Delta T$	Heat transfer coefficient h	1.0–2.0
Radiation	The Stefan–Boltzmann law $q_e = \varepsilon \sigma T^4$	Emissivity ε	4.0–5.0

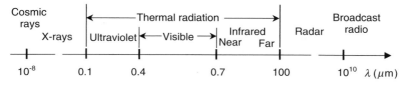

Figure 1.4   The electromagnetic spectrum

a vacuum. A familiar example is solar energy, which travels about 150 million kilo-meters through the vacuum of space to earth, where it is stockpiled in growing plants and acts as the power source that drives the earth's weather and climate. Other ex-amples of electromagnetic energy carried by waves are radar and broadcast radio, mi-crowaves, X-rays, cosmic rays, and the "light" signals received by our eyes. In fact, thermal radiation is but a very small segment of a broad electromagnetic spectrum that includes all of these phenomena and more. Some essential features of the elec-tromagnetic spectrum are illustrated in Figure 1.4.

The frequency ν (Hz) of an electromagnetic wave is related to its wavelength λ (m) according to

$$\nu = \frac{c}{\lambda} = \frac{c_0}{n\lambda} \qquad (1.10)$$

where c (m/s) is the speed of light in the medium through which the radiation is pass-ing, c_0 ($\approx 2.9979 \times 10^8$ m/s) is the speed of light in a vacuum, and $n \equiv c_0/c$ ($-$) is the *index of refraction*.

Thermal radiation is electromagnetic radiation emitted by a material substance solely due to its temperature. The part of the electromagnetic spectrum considered to be "thermal" is not cleanly bounded; however, for the purposes of this book we will say that it extends from the short-wavelength limit of the ultraviolet ($\approx 0.1~\mu$m) to the long-wavelength limit of the far infrared ($\approx 100~\mu$m). Note that the part of the electromagnetic spectrum visible to the human eye occupies only about 0.3 μm out of a spectrum that ranges (for most practical purposes) over more than seventeen or-ders of magnitude!

In this book we will use the adjective "spectral" to indicate any quantity whose value varies with wavelength. Then the spectral distribution of radiation emitted from a source is the variation with wavelength (per unit wavelength) of the source strength. It is interesting to note that the spectral distribution of radiation emitted by the sun peaks at about 0.55 μm, or in the center of the visible spectrum. The human eye has evidently evolved to be optimally suited to exploit the earth's natural light source.

1.6   THE DUAL WAVE–PARTICLE NATURE OF THERMAL RADIATION

A young French doctoral student, Louis de Broglie (1892–1987), published a disser-tation in 1924 that included an idea for which he was accorded the Nobel Prize in

1929. In his dissertation de Broglie (pronounced duh-BROY-ee) hypothesized that atomic particles could exhibit wave properties just as a photon can be assigned mass, thereby forever smearing the line separating waves and particles [3]. The de Broglie wavelength of a particle whose momentum is $p = mc$ is given by

$$\lambda = \frac{h}{p} \tag{1.11}$$

where h is *Planck's constant* ($h = 6.6237 \times 10^{-34}$ J·s). A discussion of the origin and significance of Planck's constant is deferred to Chapter 2. The German (later American) physicist Albert Einstein (1879–1955) had already established in 1905 the two principles underlying de Broglie's startling result: (1) the celebrated equivalence of mass and energy,

$$e = mc^2 \tag{1.12}$$

(special relativity) and (2) the photon view of light,

$$e = h\nu \tag{1.13}$$

(the photoelectric effect). It is left to the student to derive Equation 1.11 from Equations 1.10, 1.12, and 1.13.

The significance of Equation 1.11 is that it establishes the dual wave–particle nature of electromagnetic radiation. In some cases it is convenient to consider that light is emitted and propagates as a stream of energy packets called *photons.* Electrons can occupy only certain discrete energy levels within an atom. Thus, when an electron moves from one energy level to another within an atom, the principle of conservation of energy requires that a corresponding amount, or *quantum,* of energy be absorbed by or emitted from the atom. If the atomic transition occurs from energy level E_a to a lower energy level E_b, then, according to Equation 1.13, the frequency of the light emitted by the atom for this *bound–bound transition* is

$$\nu = \frac{E_a - E_b}{h} \tag{1.14}$$

In some cases a free electron with speed s will be captured by an atom and come to rest at some energy level, say E_a. In the case of such a *free–bound transition* the frequency of the light emitted will be

$$\nu = \frac{1}{h}\left(\frac{1}{2}m_e s^2 - E_a\right) \tag{1.15}$$

Because all speeds s are available to a free electron, Equation 1.15 permits a *continuum* of frequencies (or wavelengths), whereas Equation 1.14 permits emission only

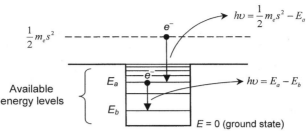

Figure 1.5 The atomic potential well illustrating a bound–bound and a free–bound transition

at discrete frequencies. In the case of high-temperature combustion processes, thermal plasmas, and electric arcs, a continuum is also produced by the acceleration of electrons as they follow curved paths. This latter phenomenon is sometimes referred to as *bremsstrahlung* (from the German "burning radiation").

Equations 1.14 and 1.15 are the basis of the science of *spectroscopy,* which permits an atom or molecule to be identified by the spectral line structure of light emitted when it is excited. The transitions represented by Equations 1.14 and 1.15 are illustrated schematically in Figure 1.5. These transitions are reversible, with the reverse transitions corresponding to absorption rather than emission. Figure 1.6 shows most of the visible emission spectrum at the exit of an electric-arc-powered plasma generator using methane as the body gas. The CH and C_2 bands are associated with rotational and vibrational energy storage modes of these diatomic molecules (see Section 6.8), while the hydrogen line is due to a bound–bound electronic transition.

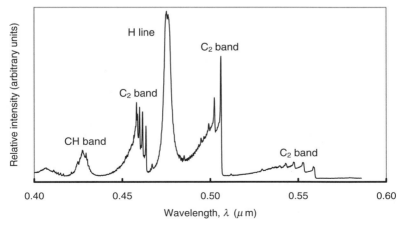

Figure 1.6 Visible portion of the emission spectrum from ionized methane [4]

The photon view of thermal radiation is the basis of the probablistic description of radiation heat transfer presented in Part III of this book, the *Monte Carlo ray-trace* (MCRT) method. In the MCRT method thermal radiation is described in terms of the mean behavior of a large number of energy bundles, or packets of photons, whose individual lives are governed by the laws of chance. Chapter 5 establishes a criterion for when we can model thermal radiation as a "raylike" phenomenon, and when we must resort to solution of the equations that describe its "wavelike" nature.

1.7   WAVE DESCRIPTION OF THERMAL RADIATION

In many cases of engineering interest it is more convenient to treat thermal radiation as a propagating wave than as a packet of photons. This is especially true when the effects of polarization, diffraction, and interference must be considered, as in certain optics and instrumentation applications. This view is also useful for modeling and understanding surface radiation properties and scattering from particles.

In 1864 the British physicist and mathematician James Clerk Maxwell (1831–1879) tied electricity and magnetism together and established that electromagnetic energy propagates through space as a wave at the speed of light [5]. Maxwell's equations for a homogeneous, isotropic medium in the absence of sources and sinks of electrical charge may be written

$$\nabla \times \mathbf{H} = \varepsilon \frac{\partial \mathbf{E}}{\partial t} + \frac{\mathbf{E}}{r_e} \tag{1.16a}$$

$$\nabla \times \mathbf{E} = -\mu \frac{\partial \mathbf{H}}{\partial t} \tag{1.16b}$$

$$\nabla \cdot \mathbf{E} = \frac{\rho_e}{\varepsilon} \tag{1.16c}$$

and

$$\nabla \cdot \mathbf{H} = 0 \tag{1.16d}$$

In Equations 1.16, $\mathbf{H}$ (A/m) is the *magnetic field strength* (or *"intensity"*), $\mathbf{E}$ (V/m) is the *electric field strength,* ε (C^2/N·m^2) is the *permittivity,* or *dielectric constant,* of the medium, r_e (Ω·m) is the *electrical resistivity* of the medium, μ (N/A^2) is the *permeability* of the medium, and ρ_e (C/m^3) is the *electric charge density.*

1.8   SOLUTION TO MAXWELL'S EQUATIONS FOR AN ELECTRICAL INSULATOR

For the case of an electrical insulator ($r_e \to \infty$, $\rho_e = 0$), Maxwell's equations describe an electromagnetic wave traveling through the medium at the speed of light. When the electromagnetic wave is propagating in the x direction, Maxwell's equations lead to the *wave equations*

$$\mu\varepsilon \frac{\partial^2 E_y}{\partial t^2} = \frac{\partial^2 E_y}{\partial x^2} \tag{1.17a}$$

$$\mu\varepsilon \frac{\partial^2 E_z}{\partial t^2} = \frac{\partial^2 E_z}{\partial x^2} \tag{1.17b}$$

$$\mu\varepsilon \frac{\partial^2 H_y}{\partial t^2} = \frac{\partial^2 H_y}{\partial x^2} \tag{1.17c}$$

and

$$\mu\varepsilon \frac{\partial^2 H_z}{\partial t^2} = \frac{\partial^2 H_z}{\partial x^2} \tag{1.17d}$$

where E_y, E_z, H_y, and H_z are the y and z components of the electromagnetic wave. In the special case where the magnitudes of the two components of the electric field and the two components of the magnetic field are equal, Equations 1.17 describe an *unpolarized,* or *randomly polarized,* wave.

1.9 POLARIZATION AND POWER FLUX

We say that an electromagnetic wave is *linearly polarized* if the electrical component is oriented in only a single direction. It can be shown that, in the free field far from the source, the magnetic component of an electromagnetic wave is in phase with and oriented in a direction at right angles to the electric component for linearly polarized radiation. A y-polarized electromagnetic wave freely propagating in the x direction is depicted in Figure 1.7. The power flux carried by an electromagnetic wave is given by the *Poynting vector,*

$$\mathbf{P} = \mathbf{E} \times \mathbf{H} \quad (\text{W/m}^2) \tag{1.18}$$

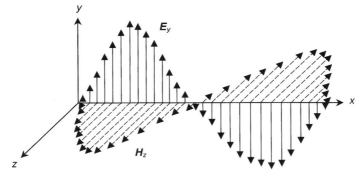

Figure 1.7 A *y*-polarized electromagnetic wave propagating in the *x* direction

In the case of the y-polarized wave depicted in Figure 1.7 we have

$$\mathbf{P} = P_x \mathbf{i} = E_y H_z \mathbf{i} \tag{1.19}$$

The quantity E_y in Figure 1.7 is the electric component of a *transverse-magnetic-polarized* electromagnetic wave, whose most general form is

$$E_y = f(x\sqrt{\mu\varepsilon} - t) + g(x\sqrt{\mu\varepsilon} + t) \tag{1.20}$$

where f and g are any single-valued, twice-differentiable functions of x and t. The function f describes a wave traveling in the positive x direction, and the function g describes a wave traveling in the negative x direction. The general right-running function f is shown at two different instants in time in Figure 1.8. From the figure it is clear that

$$f(x_1\sqrt{\mu\varepsilon} - t_1) = f[(x_1 + dx)\sqrt{\mu\varepsilon} - (t_1 + dt)] \tag{1.21}$$

Thus, because the function f is single valued it must be true that

$$x_1\sqrt{\mu\varepsilon} - t_1 = (x_1 + dx)\sqrt{\mu\varepsilon} - (t_1 + dt) \tag{1.22}$$

or

$$\frac{dx}{dt} \equiv c = \frac{1}{\sqrt{\mu\varepsilon}} \tag{1.23}$$

where c is the *speed of light* in the medium. To an accuracy of seven significant figures the speed of light in a vacuum is

$$c_0 = \frac{1}{\sqrt{\mu_0\varepsilon_0}} = 2.997925 \times 10^8 \text{ m/s} \tag{1.24}$$

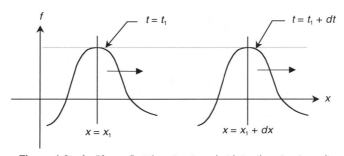

Figure 1.8   An "f wave" at time $t = t_1$ and at later time $t = t_1 + dt$

Now suppose E_y is a periodic electric wave whose Fourier series representation is

$$E_y\left(\frac{x}{c} - t\right) = E_0 + \sum_{m=1}^{\infty} |E_{y,\,m}| e^{2\pi i m\,(x/c - t)/T} \tag{1.25}$$

where $T = 1/v$, the period of the fundamental harmonic of $E_y(x/c - t)$. Then the mth harmonic of E_y is

$$E_{y,\,m} = |E_{y,\,m}|\,e^{i\omega_m\,(x/c - t)} \tag{1.26}$$

where

$$\omega_m = 2\pi v_m = \frac{2\pi m}{T} \quad \text{(r/s)} \tag{1.27}$$

In terms of the index of refraction, $n \equiv c_0/c$, Equation 1.26 may be written

$$E_{y,m} = |E_{y,m}|\,e^{i\omega_m(nx/c_0 - t)} \tag{1.28}$$

Equation 1.28 describes a harmonic wave propagating without attenuation in the positive x direction. A version of this result will be used in Chapter 4 in an analysis aimed at understanding the phenomena of reflection and refraction that occur when radiation is incident to a plane interface separating two dielectrics having different indices of refraction. The mathematical description of the propagation of electromagnetic waves through electrically conducting media (finite r_e) is deferred until Chapter 4, when consideration of the interaction of electromagnetic waves with an interface between a vacuum (or air) and an electrically conducting medium is used as the basis for predicting trends in the surface radiation properties of metals.

1.10 DIFFRACTION AND INTERFERENCE

When two or more harmonic propagating electromagnetic waves share the same space, their instantaneous local electric fields are presumed to superimpose, i.e., to linearly combine. For example, if two harmonic waves having the same frequency are propagating in opposite directions along the same axis, as in the case of the f and g waves in Equation 1.20, a standing wave will result. However, in most radiation heat transfer applications a given space will be filled with randomly polarized waves arriving from all directions and at a wide range of frequencies. Furthermore, the components that do happen to have the same frequency are assumed to have random directions, polarizations, and phase relationships. Under these hypothetical circumstances it is reasonable to assume that the resulting radiation field will be "well mixed" so that the effects of diffraction and interference can be ignored. The justification for ignoring these effects is that if interference does occur at a given location and instant in time, on average both constructive and destructive interference will occur in equal amounts and so their effects will cancel. While this is a reasonable assumption in the usual radiation heat transfer context, it may lead to serious errors in certain optics and instrumentation applications. We return to this question in Chapter 5.

1.11 PHYSICS OF EMISSION AND ABSORPTION OF THERMAL RADIATION

Emission and absorption of photons by individual atoms is described above in terms of electronic transitions between energy levels within the atom or between the energy of a free electron and an atomic energy level. The probability that such a transition will lead to the emission or absorption of a photon by an atom is favored by high temperatures in gases but is relatively small in solids. For this reason when spectroscopy is used to identify the chemical makeup of a solid material, the sample is either burned or subjected to an electric arc to vaporize it and excite its atoms. While electronic transitions in atoms account for the absorption of relatively high-energy (short wavelength) radiation by cold gases and by cold electrically conducting solids (the photoelectric effect), the importance of this phenomenon as a heat transfer mechanism is negligible in most applications of engineering interest. We therefore seek a more efficient mechanism to explain emission from solids.

All matter is composed of atoms. Sometimes these atoms are organized into molecules, and often the resulting matter is a liquid or gas or an amorphous solid. However, for the purposes of the following discussion we will consider the idealized crystal structure of Figure 1.1 in which interatomic bonding forces maintain a regular array of atoms. The simplifications inherent in this model do not in any way limit the validity of the result it produces.

1.12 ELECTRICAL DIPOLE MOMENT

Each of the atoms in the crystal structure of Figure 1.1 acts as an oscillator whose stored energy is determined by its state of thermal equilibrium and certain laws of physics. Atoms are composed of an equal number of positively (protons) and negatively (electrons) charged particles, and, so, are electrically neutral. However, thermal "jostling" of the atom can lead to its mechanical distortion such that at a given instant it may have an *electrical dipole moment,* as shown in Figure 1.9. An atom having an electrical dipole moment, when placed in an electric field, will be subject to a torque that will tend to align its $\pm$ axis with the electric field.

The electron "cloud" (in this simple model) is assumed to have a uniform charge density, $-\rho_e$ (C/m^3), where C is the abbreviation for the unit of electrical charge, the *coulomb.* Thus, the charge q (C) of the electron cloud is given by

$$q = -\tfrac{4}{3}\pi R^3 \rho_e \qquad (1.29)$$

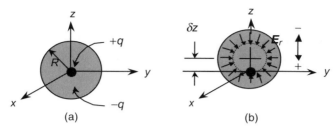

Figure 1.9 Atomic model (a) at equilibrium and (b) perturbed along the z axis

The local electric field E (V/m) on an imaginary closed surface S (m^2) is related to the electric charge Q (C) enclosed by the surface by the *Gauss flux theorem,*

$$\int_S \mathbf{E} \cdot \mathbf{n}\, dS = \frac{Q}{\varepsilon_0} \tag{1.30}$$

In Equation 1.30, $\mathbf{n}$ is the local outward-directed surface normal on surface S, and ε_0 ($= 8.854 \times 10^{-12}$ C^2/N·m^2) is the *permittivity of free space.*

For the idealized model of the jostled atom, the electric field $E_r = |\mathbf{E}_r|$, indicated by the radially inward-directed arrows in Figure 1.9b, is a constant on the spherical surface S of radius δz. The charge enclosed by S is

$$Q = \tfrac{4}{3}\pi\, (\delta z)^3\, \rho_e \tag{1.31}$$

Thus, applying the Gauss theorem, Equation 1.30,

$$4\pi(\delta z)^2\, E_r = -\frac{\tfrac{4}{3}\pi\, (\delta z)^3\, \rho_e}{\varepsilon_0} \tag{1.32}$$

or

$$E_r = -\frac{\rho_e}{3\varepsilon_0}\delta z \tag{1.33}$$

Eliminating the charge density ρ_e between Equations 1.29 and 1.33 yields

$$E_r = \frac{q}{4\pi\varepsilon_0 R^3}\delta z \tag{1.34}$$

We see from Figure 1.9b that the nucleus of the atom, whose charge is $+q$, lies in the spherical surface of radius δz where the magnitude of the radially inward-directed electric field is E_r. Therefore, the restoring force that tries to reestablish the equilibrium geometry of Figure 1.9a is

$$F = qE_r = \frac{q^2}{4\pi\varepsilon_0 R^3}\delta z \tag{1.35}$$

1.13 THE ATOMIC OSCILLATOR

For a linear spring–mass system such as the one shown in Figure 1.10, the restoring force of the spring is

$$F = K\, \delta z \tag{1.36}$$

Then, by analogy with the spring–mass system, the atom, once perturbed, acts as an oscillator with spring constant

$$K = \frac{q^2}{4\pi\varepsilon_0 R^3} \tag{1.37}$$

and mass

$$M = N m_e \tag{1.38}$$

where m_e is the mass of an individual electron and N is the number of electrons. The energy stored in each vibrational mode of the atomic oscillator, whose mechanical analog is shown in Figure 1.10, is

$$W = \frac{\mathscr{P}^2}{8\pi\varepsilon_0 R^3} \tag{1.39}$$

where $\mathscr{P} = q\, \delta z_{max}$ is the *electric dipole moment* of the distorted atom. According to classical *equipartition of energy* applied to an atomic oscillator in thermal equilibrium with its surroundings at temperature T (K),

$$W = kT \qquad \text{(per mode)} \tag{1.40}$$

where $k\,(= 1.380 \times 10^{-23}\,\text{J/K})$ is *Boltzmann's constant,* which may be thought of as the gas constant per molecule, or per atom in this case.

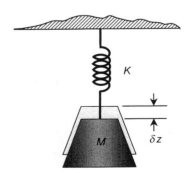

Figure 1.10 The atomic oscillator as a simple spring–mass system

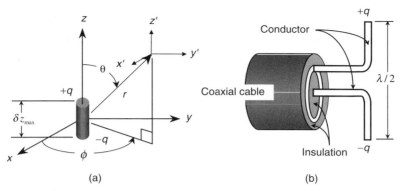

Figure 1.11 (a) The atomic oscillator as a dipole antenna and (b) a practical dipole antenna

1.14 THE ATOMIC OSCILLATOR AS A DIPOLE ANTENNA

The atomic oscillator can now be modeled as a *dipole antenna* carrying an alternating current

$$I = \frac{dQ}{dt} = \frac{d}{dt}(qe^{i\omega t}) = i\omega qe^{i\omega t} \tag{1.41}$$

where $i = \sqrt{-1}$ and $\omega = 2\pi\nu = 2\pi c_0/\lambda$. An elemental dipole antenna and its practical implementation are shown in Figure 1.11. The actual atomic oscillator has three vibrational degrees of freedom, one each along the x, y, and z axes. It is established in Appendix A that the total power radiated from three orthogonal, co-located dipole antenna elements like the one shown in Figure 1.11a is

$$P_r(\lambda, T) = \frac{\mathcal{P}^2}{\varepsilon_0} 6\pi^3 c_0 \lambda^{-4} \quad \text{(W)} \tag{1.42}$$

where now $\mathcal{P} = q\delta$ with $\delta = \delta x_{\max} = \delta y_{\max} = \delta z_{\max}$. Equating the two expressions for W, Equations 1.39 and 1.40, and rearranging, there results

$$\frac{\mathcal{P}^2}{\varepsilon_0} = 8\pi R^3 kT \tag{1.43}$$

Thus,

$$P_r(\lambda, T) = 48\pi^4 R^3 kTc_0\lambda^{-4} \quad \text{(W)} \tag{1.44}$$

1.15 RADIATION DISTRIBUTION FUNCTION

The *radiation distribution function* of a source is defined such that, if it is multiplied by $d\lambda$ and the result integrated over some wavelength interval $[\lambda_1, \lambda_2]$, the total power per unit surface area emitted by the source over that wavelength interval is obtained. Thus, the power in watts radiated from our atomic oscillator in the wavelength interval $\Delta\lambda = \lambda_2 - \lambda_1$ is

$$P_r(\Delta\lambda, T) = \int_S \int_{\lambda_1}^{\lambda_2} p_{r\lambda}(\lambda, T) \, d\lambda \, dS \qquad (1.45)$$

where $p_{r\lambda}(\lambda, T)$ is the radiation distribution function (W/m$^2\cdot\mu$m). Bearing in mind the rather gross assumptions already made about our oscillator (arrangement in a crystal matrix, uniform charge distribution of the "electron cloud," etc.), we are justified in dividing Equation 1.44 by a dimensional constant whose dimensions are length cubed, whose units are m$^2\cdot\mu$m, and whose order of magnitude is the same as $d\lambda \, dS$. Note that due to the small size of the quantities involved this is equivalent to taking the partial derivatives of Equation 1.45 with respect to λ and S. When we do this we obtain for our atomic oscillator,

$$p_{r\lambda}(\lambda, T) = 8\pi \, CkTc_0\lambda^{-4} \quad (\text{W/m}^2\cdot\mu\text{m}) \qquad (1.46)$$

The factor C in Equation 1.46 accounts for all of the differences, including the step taking us from Equation 1.45 to Equation 1.46, between our rather simplified model of the atomic oscillator and reality. However, we shall see in Chapter 2 that if $C = 1$ Equation 1.46 is in agreement with a result for the same quantity derived directly from classical mechanics.

The analysis in the final five sections of this chapter verifies a principle often referred to as *Prevost's law,* which states that all material substances at a temperature above absolute zero continuously emit radiation, even when they are in thermal equilibrium with their surroundings.

Team Projects

TP 1.1 The following is a partial list of serials with which radiation heat transfer students and practitioners should be familiar:

1. *Journal of Heat Transfer* (ASME)
2. *International Journal of Heat and Mass Transfer*
3. *Journal of Quantitative Spectroscopy and Radiative Transfer*
4. *Applied Optics*
5. *Optical Engineering*
6. *AIAA Journal*
7. *Journal of the Optical Society of America* (Parts A and B)
8. *Numerical Heat Transfer*

Visit the library and find each of these journals on the shelves. You will find them in two locations: the recent numbers will be found unbound and stacked on shelves in one area of the library, and older numbers will be bound and in the stacks in another part of the library. (Note that very old numbers of some serials may be stored at an off-site location.)

1. Find and write down the call number for each of these journals.
2. Make a photocopy of the table of contents of the most recent unbound number for each of these journals. Use a marker to highlight articles treating radiation heat transfer.
3. Explore the stacks near the journals listed above. Are there other journals of a similar vein? If so, add them to the list and complete parts 1 and 2 above for them also. Submit a comprehensive word-processed list, with call numbers, of all journals you found.
4. Browse through the most recent numbers of these journals and find an article that you think you might like to read and share with the class.

TP 1.2   Technical papers reporting more recent developments in radiation heat transfer are published in *conference proceedings.* Conference proceedings are also typically archived in university libraries. Visit the library and locate the conference proceedings that include papers on radiation heat transfer. Such conferences are sponsored by the American Society of Mechanical Engineers (ASME), the American Institute of Aeronautics and Astronautics (AIAA), the Society of Photo-Optical Instrumentation Engineers (SPIE), the Optical Society of America (OSA), the American Society for Photogrammetry and Remote Sensing (ASPRS), the Resource Technology Institute (RTI), the National Geophysical Union (NGU), the American Meteorological Society (AMS), and the International Radiation Symposium (IRS), to name but a few.

1. Compile a complete list of the conference proceedings you were able to locate (bound and unbound) that treat aspects of radiation heat transfer. Limit your search to conference proceedings published in the last five years. Note the call number in each case.
2. Make a photocopy of the table of contents of the most recent of these conference proceedings.
3. Browse through the conference proceedings that you have found and find a paper that you think you might like to read and share with the class.

Discussion Points

DP 1.1   Figure 1.2 illustrates forced convection heat transfer from a flat plate. What can we say about the value of the Prandtl number in this case? Explain your answer.

DP 1.2   Equation 1.3 purports to define an off-axis element of the thermal conductivity matrix. Discuss the possibility of k_{xy} being nonzero in the context of the second law of thermodynamics and the definition of "heat" as the energy interaction between a body and its surroundings that occurs because of a temperature difference. You may also want to consider the idealized crystal structure model in Figure 1.1. Related topics: nonequilibrium thermodynamics and Onsager's relations.

DP 1.3   Consider Equation 1.8. What does the spacing between the two plates in Figure 1.3 have to do with the degree of approximation of the equation? How would Equation 1.8 be modified if the surroundings of the two plates were maintained at a temperature of absolute zero and the spacing between the two parallel plates was arbitrary?

DP 1.4   Referring to Equation 1.9, explain how the exponent can exceed 4 without violating the principle discovered by Stefan and explained by Boltzmann. Does this mean that a body can emit more radiation than a blackbody at the same temperature?

DP 1.5   In the last paragraph of Section 1.4 it is stated that radiation heat transfer becomes dominant, among other situations, in two extreme cases: very low temperatures (cryogenics) and very high temperatures (combustion). Is there a contradiction here? Please explain.

DP 1.6   Most vegetation appears green to the eye. We know that the solar spectrum peaks in the green part of the visible spectrum. Is this a coincidence? Speculate on how natural selection might have played a role in the evolution of chlorophyll.

DP 1.7   Show formally that the magnetic component of a linearly polarized electromagnetic wave is oriented at right angles to the electric component for free-field radiation.

DP 1.8   How are microwave antennas and infrared emitters alike and how are they different? In the context of this question, consider the difference in the way a microwave oven and a "convection" oven work.

DP 1.9   Two students should be designated to lead a debate of the proposition, "Resolved: It is important for an engineering graduate student to understand the material in Appendix A."

DP 1.10   Suppose you and your team set out to derive a *radiation distribution function,* that is, a mathematical relation that predicts the amount of radiation emitted from an ideal emitter as a function of wavelength and temperature. (No emitter can emit more radiation at a given wavelength than an ideal emitter at the same temperature.) How would you know if your eventual distribution function was correct? Among other things (such as experimental data) you would want to know in advance the asymptotic values of the function as temperature and wavelength approach the physical limits of zero and infinity. Turn-of-the-century physicists, when faced with this problem, immediately came up with the four asymptotic conditions—two in temperature and two in wavelength—to which their candidate distribution functions would have to adhere. Can you? In this context please criticize the radiation distribution function represented by Equation 1.46. Does it conform to your expectations? Explain.

DP 1.11   Discuss the atomic model of Figure 1.9b. Would not the electric field $\mathbf{E}_r$ be distorted by the presence of the positively charged nucleus at the outer rim of the electron cloud of radius δz? [*Hint:* What adjectives would you use to describe the equation identified as the Gauss flux theorem, Equation 1.30?]

DP 1.12   Criticize the model that is the basis of the last five sections of this chapter. Which assumptions and simplifications might intervene in determining the value of the constant C in Equation 1.46?

Problems

P 1.1   The data in Table 1.2 are from a classical article published by Dulong and Petit in 1817 [6]. In the experiment described in the article, bodies were suspended in a vacuum jar whose outer walls were surrounded by melting ice. The table gives the initial temperature of the bodies and the decrease in temperature after one minute. Stefan used these same data in 1879 (62 years later!) to establish the absolute-temperature-to-the-fourth-power dependence of radiative heat emission from a solid body. Can you reproduce Stefan's reasoning?

Table 1.2   Data from Dulong and Petit [6] used by Stefan [1] to establish the absolute temperature-to-the-fourth-power dependence of radiative heat emission

Initial Temperature of Body (°C)	Temperature Decrease After 1 min (°C)
240	10.69
220	8.92
200	7.40
180	6.10
160	4.89
140	3.88
120	3.02
100	2.30
80	1.74

P 1.2 Derive Equation 1.11, the expression for the de Broglie wavelength of a particle of mass m traveling at speed c with momentum $p = mc$, from Equations 1.10, 1.12 and 1.13.

P 1.3 Derive the wave equation, Equation 1.17a, from Maxwell's equations, Equations 1.16, for the case of an electrical insulator ($r_e \to \infty$, $\rho_e = 0$).

P 1.4 Demonstrate by direct substitution that Equation 1.20 is the most general solution to the wave equation for E_y, Equation 1.17a.

P 1.5 Resolve the units in Equation 1.30. That is, show that the units are the same on both sides of the equation.

P 1.6 Suppose that in the simple atomic model of Figure 1.9a the electric charge density ρ_e varies as the inverse square of the radial distance r from the center of the atom; that is,

$$\rho_e(r) = \rho_{e0} \left[1 - \left(\frac{r}{R} \right)^n \right] \tag{1.47}$$

where the radial position r is measured from the center of the electron cloud. What would be the charge q of the electron cloud in terms of ρ_{e0}, R, and n?

P 1.7 For sufficiently small perturbations ($\delta z \to 0$) it may be assumed that the electron cloud in Figure 1.9b remains spherical. Then for the electric charge density variation of Problem P 1.6 what is the electric charge Q contained within the spherical envelope of radius δz in terms of δz, ρ_{e0}, R, and n?

P 1.8 Derive Equation 1.39 by considering the energy stored in a spring, whose elastic constant is K, at its maximum displacement from equilibrium.

P 1.9 Suppose the atomic spacing s in Figure 1.1 is aR, where R is the atomic radius shown in Figure 1.9a and a is a dimensionless scaling factor. For what value of the scaling factor a is the constant C in Equation 1.46 equal to unity? What conclusion do you draw from this result? Can C reasonably be equal to unity in this model? [*Hint:* Assume $\lambda \sim R$ and let the surface S be that of a cubic volume of dimensions $aR \times aR \times aR$ surrounding the atomic oscillator.]

REFERENCES

1. Stefan, J., "Über die Beziehung zwischen der Wärmstrahlung und der Temperatur," *Sitzungberichte, Akademie der Wissenschaften,* Wien, Bd. 79, Pt. 2, 1879, pp. 391–428.
2. Boltzmann, L., "Ableitung des Stefan'schen Gesetzes, betreffend die Abhängigkeit der Wärmestrahlung von der Temperatur aus der electromagnetischen Lichttheorie," *Annalen der Physik und Chemie,* Bd. 22, 1884, pp. 291–294.
3. de Broglie, Prince Louis Victor Pierre Raymond, "Recherches sur la théorie des quanta," Thesis, University of Paris (Sorbonne), Paris, 1924 (published in *Annales de Physique,* 10e série, Tome III, Masson et Cie, Editeurs, Paris, 1924, pp. 22–128).

4. Gallimore, S. D., *A study of plasma ignition enhancement for aeroramp injectors in supersonic combustion applications,* PhD Dissertation, Department of Mechanical Engineering, Virginia Tech, Blacksburg, VA, May 2001.
5. Maxwell, J. C., "A dynamical theory of the electromagnetic field," *Philosophical Transactions of the Royal Society of London,* 1865, pp. 459–512.
6. Dulong, P.-L., and A.-T. Petit, "Des recherches sur la mesure des températures et sur les lois de la communication de la chaleur (Seconde partie, des lois du refroidissement)," *Annales de Chimie et de Physique,* Tome 7, 1817, p. 247.

TEXTS FOR FURTHER REFERENCE

1. *The Theory of Heat Radiation,* M. Planck (trans. Morton Masius), Dover Publications, Inc., New York, 1959.
2. *Radiative Transfer,* S. Chandrasekhar, Dover Publications, Inc., New York, 1960.
3. *Engineering Radiation Heat Transfer,* J. A. Wiebelt, Holt, Rinehart and Winston, New York, 1966.
4. *Radiative Transfer,* H. C. Hottel and A. F. Sarofim, McGraw-Hill Book Company, New York, 1967.
5. *Radiative Heat Transfer,* T. J. Love, C. E. Merrill Publishing Company, Columbus, Ohio, 1968.
6. *Radiation Heat Transfer* (augmented edition), E. M. Sparrow and R. D. Cess, Hemisphere Publishing Corporation, Washington, 1978.
7. *Transfer Processes: An Introduction to Diffusion, Convection, and Radiation* (2nd edition), D. K. Edwards, V. E. Denny, and A. F. Mills, Hemisphere Publishing Corporation, Washington, 1979.
8. *Thermal Radiation Heat Transfer,* R. Siegel and J. R. Howell, Hemisphere Publishing Corporation, Washington, 1992.
9. *Thermal Radiative Transfer & Properties,* M. Quinn Brewster, John Wiley & Sons, New York, 1992.
10. *Radiative Heat Transfer,* M. F. Modest, McGraw-Hill, New York, 1993.

2

BASIC CONCEPTS; THE BLACKBODY

This chapter begins with the introduction of three basic concepts essential to the understanding of radiation heat transfer: the solid angle, the intensity (or radiance), and the emissive power of a source. The blackbody is then defined as a perfect absorber of incident radiation and, as a consequence, is shown to be an ideal emitter. Next, the nature of blackbody radiation is explored and used to establish the existence of radiation pressure and radiation energy density. The Stefan–Boltzmann law is then derived on the basis of these concepts and classical thermodynamics. Finally, the Planck blackbody radiation distribution function is introduced, which describes the maximum amount of thermal radiation per unit wavelength that a body at a given temperature can emit in a wavelength interval $d\lambda$ about wavelength λ, and Wien's displacement law, which predates Planck's work by nearly two decades, is derived from this expression.

2.1 THE SOLID ANGLE

The *solid angle* ω (sr) is one of the most basic concepts of radiation heat transfer analysis. Its thorough understanding is essential to the understanding of several other important concepts. The unit of the solid angle is the *steradian,* abbreviated sr.

Consider a differential surface element dS located a distance r from a second differential area element dA. The line of length r connecting dA and dS intersects the normal to dA at an angle β_A and the normal to dS at an angle β_S, as shown in Figure 2.1. The dimensions of dA and dS are shown greatly exaggerated in the figure. In fact, both dA and dS are arbitrarily small so that no matter what the value of the finite length r, it is always large compared to the dimensions of dA and dS. The surface element $dS \cos \beta_S$, which is hinged to surface element dS along their common lower

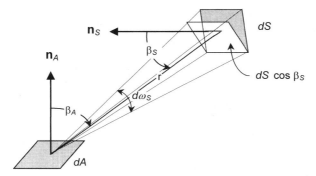

Figure 2.1 Definition of the differential solid angle

edge, is tilted toward dA so that the line r connecting dA and dS is normal to dS cos β_S. Of course, the distance between the point along r where r intersects surface element dS cos β_S and the point where r intersects surface element dS is negligibly small compared to r because dS is vanishingly small. Then the *differential solid angle* $d\omega_S$ *subtended by dS at dA* is defined as

$$d\omega_S \equiv \frac{dS \cos \beta_S}{r^2} \quad (\text{sr}) \qquad (2.1)$$

Note that the solid angle in steradians is actually a dimensionless ratio of area over length squared, just as a one-dimensional angle in radians is a dimensionless ratio of lengths.

An important special case of the solid angle, illustrated in Figure 2.2, is the differential solid angle subtended by a differential spherical sector dS at the center of a

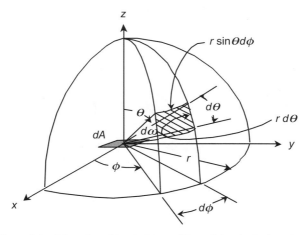

Figure 2.2 Differential solid angle subtended by a differential spherical sector at the center of a sphere

sphere of radius r. If the origin of the coordinate system is taken to be the center of the sphere, then the projected area of the differential spherical sector when viewed from the center of the sphere is

$$dS \cos\beta_S = (r \sin\theta \, d\phi)(r \, d\theta) = r^2 \sin\theta \, d\theta \, d\phi \tag{2.2}$$

since $\beta_S = 0$. Then from Equation 2.1 the differential solid angle in terms of θ and ϕ is

$$d\omega = \sin\theta \, d\theta \, d\phi \tag{2.3}$$

2.2   INTENSITY (OR RADIANCE) OF RADIATION

The term *spectral* and its synonym *monochromatic* ("mono" = one, "chrome" = color) refer to radiation confined to a vanishingly small wavelength interval $d\lambda$ centered about a specified wavelength λ. The *spectral intensity* i_λ (λ, θ, ϕ) of a plane source is the power per unit wavelength in wavelength interval $d\lambda$ centered about a specified wavelength λ, per unit projected area of the source, per unit solid angle, passing in a given direction θ, ϕ. Note that the symbol λ appears twice in the symbol for the spectral intensity. This is not redundant usage; the subscript λ reminds us that the intensity is a per-unit-wavelength quantity, and the parenthetical argument λ reminds us that the value of the spectral intensity depends on wavelength. The reader should be cautioned that in some arenas, such as the optics and atmospheric radiation communities, intensity is referred to as *radiance*.

Referring to Figure 2.3, if $d^3P(\lambda$, θ, $\phi)$ is the power (W) leaving plane surface element dA (m^2) in direction θ, ϕ and at angle β_A with respect to the normal to dA, in wavelength interval $d\lambda$ about wavelength λ, and within solid angle $d\omega_S$, then

$$i_\lambda \, (\lambda, \, \theta, \, \phi) \equiv \frac{d^3P(\lambda, \, \theta, \, \phi)}{dA \, \cos\beta_A d\omega_S d\lambda} \quad (\text{W/m}^2 \cdot \text{sr} \cdot \mu\text{m}) \tag{2.4}$$

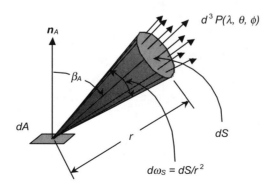

Figure 2.3   Illustration of the nomenclature for spectral intensity and directional, spectral emissive power

The superscript "3" in Equation 2.4 is required for notational consistency; for the intensity to be a finite quantity the number of differential symbols d must be the same in both the numerator and the denominator.

It is often useful to know the intensity integrated over all possible wavelengths for a given source. In such cases we define the *total intensity*

$$i(\theta, \phi) \equiv \int_{\lambda=0}^{\infty} i_{\lambda}(\lambda, \theta, \phi) \, d\lambda \quad (\text{W/m}^2 \cdot \text{sr}) \tag{2.5}$$

The word "total" is used exclusively in this context in the radiation heat transfer literature.

2.3 DIRECTIONAL, SPECTRAL EMISSIVE POWER

The *directional, spectral emissive power* $e_{\lambda}(\lambda, \theta, \phi)$ of a plane source is the power per unit wavelength in a specified wavelength interval $d\lambda$ about wavelength λ, per unit source surface area, emitted in direction θ, ϕ into the space above the source. (The reader should contemplate the difference between this definition and the definition of the spectral intensity.) Then, once again referring to Figure 2.3, if $d^3P(\lambda, \theta, \phi)$ is the power (W) emitted in direction θ, ϕ into the space above a plane differential surface element of area dA (m^2) in wavelength interval $d\lambda$ (μm) about wavelength λ, the differential directional, spectral emissive power contained in solid angle $d\omega$ (sr) is

$$dE_{\lambda} \equiv \frac{d^3P(\lambda, \theta, \phi)}{dA \, d\lambda} \quad (\text{W/m}^2 \cdot \mu\text{m}) \tag{2.6}$$

Introducing Equation 2.4 with $\beta_A = \theta$, this becomes

$$dE_{\lambda} = i_{\lambda}(\lambda, \theta, \phi) \cos\theta \, d\omega_S \quad (\text{W/m}^2 \cdot \mu\text{m}) \tag{2.7}$$

Then, in terms of the directional, spectral intensity the directional, spectral emissive power is

$$e_{\lambda}(\lambda, \theta, \phi) \equiv \frac{dE_{\lambda}}{d\omega} = i_{\lambda}(\lambda, \theta, \phi) \cos\theta \quad (\text{W/m}^2 \cdot \text{sr} \cdot \mu\text{m}) \tag{2.8}$$

2.4 HEMISPHERICAL, SPECTRAL EMISSIVE POWER

The *hemispherical, spectral emissive power* $e_{\lambda}(\lambda)$ is the power per unit source area emitted into the hemispherical ("2π") space above a plane source in a specified

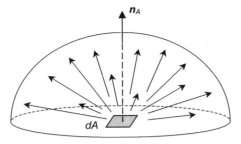

Figure 2.4   Radiation emitted by a surface element dA into the hemispherical space above dA

wavelength interval $d\lambda$ about wavelength λ, as illustrated in Figure 2.4. The hemispherical, spectral emissive power may be obtained by integrating the directional, spectral emissive power over the hemispherical space above the source,

$$e_\lambda(\lambda) = \int_{2\pi} e_\lambda(\lambda,\ \theta,\ \phi)\ d\omega \tag{2.9}$$

or

$$e_\lambda(\lambda) = \int_{2\pi} i_\lambda(\lambda,\ \theta,\ \phi)\ \cos\theta\ d\omega \tag{2.10}$$

or finally

$$e_\lambda(\lambda) = \int_{\phi=0}^{2\pi} \int_{\theta=0}^{\pi/2} i_\lambda(\lambda,\ \theta,\ \phi)\ \cos\theta\ \sin\theta\ d\theta\ d\phi \tag{2.11}$$

2.5   HEMISPHERICAL, TOTAL EMISSIVE POWER

The *hemispherical, total emissive power e* is the power per unit surface area emitted into the hemispherical space above a plane source at all wavelengths. Thus,

$$e = \int_{\lambda=0}^{\infty} e_\lambda(\lambda)\ d\lambda = \int_{\lambda=0}^{\infty} \int_{2\pi} i_\lambda(\lambda,\ \theta,\ \phi)\ \cos\theta\ d\omega\ d\lambda \tag{2.12}$$

or

$$e = \int_{\lambda=0}^{\infty} \int_{\phi=0}^{2\pi} \int_{\theta=0}^{\pi/2} i_\lambda(\lambda,\ \theta,\ \phi)\ \cos\theta\ \sin\theta\ d\theta\ d\phi\ d\lambda \quad (\text{W/m}^2) \tag{2.13}$$

2.6 SPECTRAL INTENSITY OF OUR ATOMIC OSCILLATOR

The spectral intensity of the atomic oscillator model introduced in Chapter 1 (Equation 1.46) is

$$i_\lambda(\lambda, \theta, \phi) \equiv \frac{P_{r,\lambda}(\lambda, T)}{4\pi} = 2CkTc_0\,\lambda^{-4} \quad (\text{W/m}^2\cdot\text{sr}\cdot\mu\text{m}) \tag{2.14}$$

where the factor $1/4\pi$ is introduced because the solid angle subtended by a sphere at a point source at its center is 4π. In writing Equation 2.14 we have used the fact that $\cos\beta_A = 1.0$ for all directions in the case of a spherical source when symmetry is taken into account. Note that the intensity of a point source is independent of direction and distance from the source.

2.7 THE BLACKBODY

A *blackbody* is defined as a perfect absorber of thermal radiation; that is, it absorbs all incident radiation from all directions and at all wavelengths. All other properties of a blackbody issue from this definition.

A practical laboratory blackbody is illustrated in Figure 2.5. It is fabricated from a metal, such as copper or even silver, having a high thermal conductivity and a large thermal mass. This is to make it easier to maintain a constant and uniform temperature on the walls. The interior wall temperature is monitored at several locations to assure that the temperature is uniform and to provide a means of controlling the cavity temperature by, for example, using electrical heaters imbedded in the walls. The walls are coated with a shiny black paint that absorbs most of the incident radiation and produces predominantly mirrorlike reflections for the small amount that is reflected.

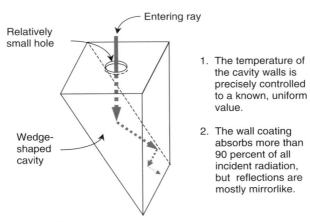

Figure 2.5 A practical laboratory blackbody

Then, combined with the wedge shape of the cavity, this assures that any entering ray will suffer several energy-absorbing reflections before escaping back through the entrance aperture. Making the aperture small compared to the surface area of the cavity minimizes the amount of entering radiation that escapes by reflection.

2.8   RADIATION WITHIN AN ISOTHERMAL ENCLOSURE

Consider an evacuated enclosure of arbitrary size and shape whose walls are maintained at a constant, uniform temperature. Now place a small blackbody, called a *test body,* somewhere within the enclosure. It is a matter of experience that after the passage of a sufficient period of time the small blackbody, which has a finite thermal mass, will be in thermal equilibrium at the temperature of the enclosure walls. This occurs as a consequence of the second law of thermodynamics, which states that a net transfer of heat will occur from a warmer body to a cooler body when the two bodies are in thermal communication with each other. In the present scenario the thermal communication is by radiation.

We can now establish that *the total radiation emitted from a blackbody depends only on its temperature.* Suppose the temperature of the enclosure walls is increased to a different uniform value and maintained at that value. As before, the test body will eventually come to a new equilibrium temperature equal to that of the enclosure walls. But for this to happen the energy stored in the body must increase. This means that at the beginning of the heating transient the test body is emitting less power than it is absorbing from its surroundings. Therefore, its rate of emission must be higher at the end of the heating transient, when it is once again in equilibrium with its surroundings, than it was at the beginning of the heating transient, when it was at a lower temperature. The inverse argument can, of course, be made if the enclosure wall temperature is decreased. Because the temperature increments in question are completely arbitrary, as are the location and orientation of the body within the enclosure, we conclude that the total power emitted from a blackbody depends only on its temperature. Further, we may conclude that the radiation field within the isothermal enclosure is itself independent of location and direction, since, if this were not the case, the equilibrium temperature of the test body would depend on its location or orientation. Consistent with these observations, we say that *the radiation field within an isothermal enclosure is uniform and isotropic,* and that the radiation field in an isothermal enclosure consists of *blackbody radiation.* It should be emphasized that it is not required for the enclosure walls to be composed of blackbodies for this to be true.

2.9   THE BLACKBODY AS AN IDEAL EMITTER

Note the distinction made here between a *perfect* absorber and an *ideal* emitter. It is important to use the two adjectives correctly. The former pertains to a value judgment ("perfection") while the latter refers to a model. We established in Section 2.7 that a blackbody is by definition a perfect absorber. Now we will prove by contradiction that *no body can emit more thermal radiation than a blackbody at the same temperature.* Our approach is to assume that a body exists that emits more heat than a blackbody at the same temperature. We then show that if such a body existed it would lead to a violation of the second law of thermodynamics.

Consider an enclosure whose walls are constructed entirely of blackbodies, i.e., bodies that absorb all radiation incident from all directions and at all wavelengths. Further, let the enclosure walls be maintained at a constant, uniform temperature. Now place a small test body—which by hypothesis is a more efficient emitter of thermal radiation than a blackbody and which is initially at the same temperature as the enclosure walls—somewhere in the interior of the enclosure. Of course, it does not matter where the body is placed because, as was established in Section 2.8, the radiation field within the isothermal enclosure is uniform and isotropic. Now, even if the test body absorbs all radiation incident upon it, it still must emit more power than it absorbs because, by hypothesis, it is a more efficient emitter than a blackbody. Then, because the body is finite in size and thus has a finite heat capacity, its stored energy and thus its temperature must decrease. But as soon as this happens a situation exists in which a spontaneous net heat transfer occurs from a cooler body (the test body) and a warmer body (the enclosure walls) without any other external effect. This is a clear violation of the second law of thermodynamics and so cannot happen. Thus, we conclude that no body can emit more thermal radiation than a blackbody at the same temperature. It is in this sense that we say that a blackbody is an ideal emitter of thermal radiation.

2.10 THE BLACKBODY AS AN IDEAL EMITTER AT ALL WAVELENGTHS

An argument similar to that given in Section 2.9 can be used to establish that no body can emit more thermal radiation in a wavelength interval $d\lambda$ about wavelength λ than a blackbody at the same temperature. In the proof of this property, the enclosure walls, now constructed of isothermal blackbodies, are covered with ideal filters that allow only radiation in the wavelength interval $d\lambda$ to pass through. It is then elementary to demonstrate that if the enclosed body emits more radiation in this wavelength interval than a blackbody at the same temperature, the second law of thermodynamics is once again violated. In constructing this proof use is made of the fact that all radiation emitted by the test body outside of the wavelength interval $d\lambda$ is reflected back to the body by the filters and reabsorbed. Therefore, net exchange between the body and the cavity walls is possible only in the wavelength interval $d\lambda$. The formal structure of this argument is left as an exercise for the student.

2.11 THE BLACKBODY AS AN IDEAL EMITTER IN ALL DIRECTIONS

No body can emit more radiation in a given direction θ, ϕ than a blackbody at the same temperature. Suppose that a test body within an isothermal enclosure is once again, at least initially, at the temperature of the walls of the enclosure, and that at some wavelength λ the radiation emitted from the test body in some direction θ, ϕ exceeds the radiation absorbed by the body at that wavelength and coming from that direction. In this situation the spectral radiation field would be locally nonuniform within the isothermal enclosure. But this possibility was excluded in Section 2.8. It might be objected that, if the test body emits *less* than a blackbody in a given direction, this too would upset the local uniformity in the radiation field. However, if this were so it would also be true that the test body would, because of reflection, absorb less than a blackbody from that direction by the same amount, so that the resulting radiation reflected from the body would exactly compensate for the less-than-ideal emission. We may also conclude from this argument that radiation emitted from a blackbody is independent of direction.

2.12 RADIATION PRESSURE

Radiation from the sun exerts a force of about 1.16×10^6 kN on the earth! In spite of this impressive demonstration of might, radiation is a rather ephemeral medium, so much so, in fact, that its ability to exert pressure was still open to question as recently as the last quarter of the 19th century.

In 1876 the Italian physicist A. Bartoli offered the following proof of the existence of radiation pressure [1]. He imagined a cylinder divided into three chambers by movable, perfectly reflecting pistons, as shown in Figure 2.6. The pistons have perfect seals but can nevertheless slide freely without friction. Apertures in the face of each piston are fitted with valves that can be slid open, also without friction, to permit adjacent chambers to communicate with each other. The left- and right-hand ends of the cylinder are closed off by blackbodies at two different temperatures, T_1 and T_2, where $T_1 < T_2$. Piston A is initially near the left-hand end of the cylinder with its aperture open and piston B is initially near the right-hand end with its aperture closed (state I). Thus, the space between the two pistons is filled with radiation whose energy density corresponds to that of a blackbody at temperature T_1. Bartoli then postulated the following cycle:

Process I–II   Close the aperture in piston A, thereby trapping radiation initially in equilibrium with the blackbody at temperature T_1 between the two pistons. Then

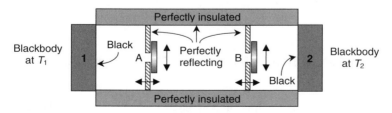

Figure 2.6   Top view of the apparatus for Bartoli's thought experiment

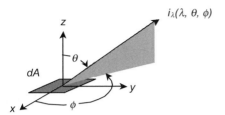

Figure 2.7 Geometry for relating radiation pressure to spectral intensity

move piston A toward piston B until the radiation energy density between the two pistons has increased to a value corresponding to that of the blackbody at temperature T_2. Then open the aperture of piston B.

Process II–III   Move piston A to the right until it makes contact with piston B, and then continue moving the two pistons together to the right until piston B makes contact with the blackbody at temperature T_2. This forces the radiation that was trapped between the two pistons into the blackbody at temperature T_2.

Process III–I   Close the aperture in piston B and open the aperture in piston A and then return the two pistons to their original positions, thus completing the cycle.

Clearly, one result of this cycle is the net transfer of heat from one reservoir to another reservoir at a higher temperature. However, the second law of thermodynamics states that it is impossible for a device to operate in a cycle and have this as its only effect. Therefore, we conclude that a net amount of work must have also been done during the cycle. But since the pistons and valves are frictionless, the work must have been done against radiation pressure.

Radiation pressure on a real or imaginary surface, like the pressure exerted by a fluid, is defined as the net rate of momentum flux through a surface. Consider the surface area element dA (real or imaginary) shown in Figure 2.7. Its projected area in the θ, ϕ direction is $dA \cos \theta$, where θ is the angle that a line passing through dA in direction θ, ϕ makes with the normal to dA. Then according to Equation 2.4, if the spectral intensity of dA in direction θ, ϕ in wavelength interval $d\lambda$ about wavelength λ is $i_\lambda(\lambda, \theta, \phi)$, the spectral power passing through dA in that direction is

$$d^3 P(\lambda, \theta, \phi) = i_\lambda(\lambda, \theta, \phi)dA \cos\theta \, d\theta \, d\lambda \qquad (2.15)$$

Now the net momentum flux $\Im$ across dA in direction θ, ϕ corresponding to the wavelength interval $d\lambda$ is the spectral power passing through dA in that direction per unit area divided by the speed of propagation c of the radiation carrying that power, or

$$d^2\Im(\lambda, \theta, \phi) = \frac{d^3 P(\lambda, \theta, \phi)}{c \, dA} \qquad (2.16)$$

and the component of this momentum normal to dA is

$$d^2\Im_n(\lambda, \theta, \phi) = \frac{d^3 P(\lambda, \theta, \phi)}{c \, dA} \cos\theta \qquad (2.17)$$

Substituting Equation 2.15 into Equation 2.17 yields

$$d^2\mathfrak{S}_n(\lambda, \, \theta, \, \phi) = \frac{1}{c} \, i_\lambda(\lambda, \, \theta, \, \phi) \cos^2\theta \, d\omega \, d\lambda \tag{2.18}$$

If we consider the beam of spectral intensity $i_\lambda(\lambda, \, \theta, \, \phi)$ to have been perfectly reflected from dA, then the momentum flux due to this beam is twice the amount given by Equation 2.18. Therefore, the pressure exerted on the perfectly reflecting surface element dA due to radiation incident from the hemispherical space above dA at all wavelengths is

$$p = \frac{2}{c} \int_{\lambda=0}^{\infty} \int_{2\pi} i_\lambda(\lambda, \, \theta, \, \phi) \cos^2\theta \, d\omega \, d\lambda \tag{2.19}$$

Finally, introducing Equation 2.5, there results

$$p = \frac{2}{c} \int_{2\pi} i(\theta, \, \phi) \cos^2\theta \, d\omega \tag{2.20}$$

In the special case of blackbody radiation we know that the intensity is independent of direction, as established in Section 2.8. In this special case Equation 2.20 becomes

$$p = \frac{2i_b}{c} \int_{\phi=0}^{2\pi} \int_{\theta=0}^{\pi/2} \cos^2\theta \, \sin\theta \, d\theta \, d\phi = \frac{4\pi}{3c} i_b \tag{2.21}$$

While we have not yet established the relationship between the total intensity of a blackbody and its temperature, we have demonstrated, in Section 2.8, that the intensity of a blackbody depends only on its temperature; that is,

$$i_b = i_b(T_b \text{ only}) \tag{2.22}$$

2.13   RADIATION ENERGY DENSITY

Another useful concept is that of the *radiation energy density,* which is defined as the radiation energy per unit volume in the limit as the volume approaches zero. It is analogous to the internal energy of a substance and has the same dimensions $[E/L^3]$ and units (J/m^3). (Recall that the concept of radiation energy density has already been evoked in Bartoli's proof of the existence of radiation pressure.)

The radiation energy density may best be understood by considering Figure 2.8, which shows a cylindrical beam of spectral intensity $i_\lambda(\lambda, \, \theta, \, \phi)$ incident to surface element dA. The differential distance $d\ell$ swept out by the beam during differential time

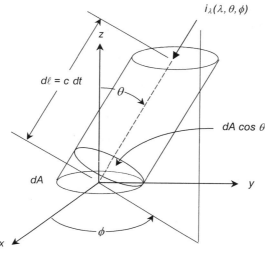

Figure 2.8 A beam of intensity $i_\lambda(\lambda,\ \theta,\ \phi)$ incident to surface element dA

increment dt is $c\ dt$, where c is the speed of light. The volume of the cylinder swept out by the beam is

$$d^2V = d\ell\ dA\ \cos\theta \tag{2.23}$$

and the amount of energy in wavelength interval $d\lambda$ contained in this volume is

$$d^4E_\lambda(\lambda,\ \theta,\ \phi) = i_\lambda(\lambda,\ \theta,\ \phi)\ dA\ \cos\theta\ d\omega\ d\lambda\ dt \tag{2.24}$$

where $d\omega$ is the solid angle subtended by the beam at dA. Thus, the radiant energy density due to radiation arriving in d^2V in wavelength interval $d\lambda$ about wavelength λ from direction $\theta,\ \phi$ is

$$d^2u_\lambda(\lambda,\ \theta,\ \phi) = \frac{d^4E_\lambda(\lambda,\ \theta,\ \phi)}{d^2V} = \frac{i_\lambda(\lambda,\ \theta,\ \phi)\ dA\ \cos\theta\ d\omega\ d\lambda\ dt}{c\ dt\ dA\ \cos\theta}$$
$$= \frac{i_\lambda\ (\lambda,\ \theta,\ \phi)d\omega\ d\lambda}{c} \tag{2.25}$$

The *total* radiation energy density for radiation arriving in this volume element at all wavelengths is

$$du(\theta,\ \phi) = \frac{1}{c}\int_{\lambda=0}^{\infty} i_\lambda(\lambda,\ \theta,\ \phi)\ d\omega\ d\lambda = \frac{1}{c}\ i(\theta,\ \phi)\ d\omega \tag{2.26}$$

where once again the total intensity $i(\theta, \phi)$ is defined by Equation 2.5. The total radiation arriving in d^2V from all directions is

$$u = \frac{1}{c} \int_{4\pi} i(\theta, \phi) \, d\omega \qquad (2.27)$$

and for blackbody radiation,

$$u_b = \frac{4\pi i_b}{c} \qquad (2.28)$$

Equations 2.21 and 2.28 can be combined to yield

$$p_b = \frac{u_b}{3} \qquad (2.29)$$

2.14   RELATIONSHIP BETWEEN RADIATION FROM A BLACKBODY AND ITS TEMPERATURE

We are now in a position to derive the form of the Stefan–Boltzmann law of radiation heat transfer, Equation 1.7. Recall that Boltzmann first provided the theoretical basis for this relationship only in 1884, even though Stefan had already established its general form five years earlier on the basis of Dulong and Petit's data (see Section 1.4).

Citing Bartoli, Boltzmann postulated that thermal radiation exerts pressure and has energy density and so could be used as the working fluid in a Carnot engine [2]. An apparatus suitable for conducting Boltzmann's thought experiment is illustrated in Figure 2.9, and the accompanying hypothetical thermodynamic cycle is shown in Figure 2.10. The engine in Figure 2.9 consists of a frictionless piston–cylinder arrangement in which the face of the piston and the walls of the cylinder are perfectly

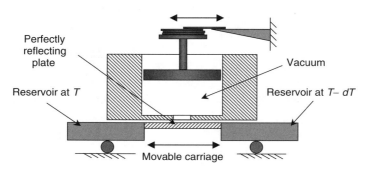

Figure 2.9   Apparatus for Boltzmann's thought experiment

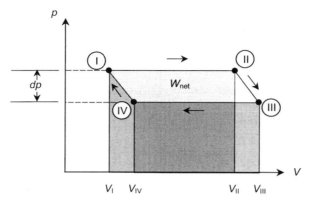

Figure 2.10 The reversible cycle described by the apparatus in Figure 2.9

reflecting. The head of the cylinder is pierced with a small hole that can be blocked with either a perfectly reflecting plate or with one of two black thermal energy reservoirs whose temperatures are T and $T - dT$. Boltzmann then imagined the following reversible cycle:

Process I–II   The process begins in state I with the thermal energy reservoir at temperature T covering the hole in the head of the cylinder. The radiation in the piston–cylinder arrangement is in equilibrium with this reservoir, a condition that sets the radiation pressure through Equations 2.22, 2.28, and 2.29. The piston is slowly pulled upward without friction by a reversible engine (not shown) external to the piston–cylinder arrangement. During this reversible isothermal expansion in which the volume increases from V_I to V_{II}, the thermal energy reservoir supplies sufficient energy to maintain the energy density, and thus the temperature and pressure, constant in the piston–cylinder arrangement. Thus, the work done by the radiation in the piston–cylinder arrangement on its surroundings is

$$_IW_{II} = p(V_{II} - V_I) \tag{2.30}$$

and the increase in stored energy in the piston–cylinder arrangement is

$$\Delta U_{I-II} = u(V_{II} - V_I) \tag{2.31}$$

Then, according to the first law of thermodynamics, the heat transferred from the thermal energy reservoir at temperature T to the contents of the piston–cylinder arrangement during process I–II is

$$_IQ_{II} = {}_IW_{II} + \Delta U_{I-II} = (p + u)(V_{II} - V_I) \tag{2.32}$$

Process II–III   The carriage on which the two thermal energy reservoirs and the perfectly reflecting plate are mounted is moved without friction so that the perfectly reflecting plate replaces the thermal energy reservoir at temperature T. Then the piston is allowed to continue upward without friction as infinitesimally small weights are removed by sliding them without friction onto adjacent shelves. The increase in volume during this reversible adiabatic expansion is dV, the decrease in temperature is dT, and the corresponding decreases in radiation pressure and radiation energy density are, respectively, dp and du. Noting that all line segments representing a differential process must be straight, the work done by the contents of the piston–cylinder arrangement on their surroundings is

$$\delta^2 W_{\text{II}-\text{III}} = -\tfrac{1}{2} dp \, dV \tag{2.33}$$

Because the process is adiabatic then, according to the first law, this also represents the decrease in the stored energy.

Process III–IV   Now the carriage is once again displaced without friction such that the thermal energy reservoir at temperature $T - dT$ replaces the perfectly reflecting plate. The piston is then slowly pushed downward by the reversible engine external to the piston–cylinder arrangement until the volume has been reduced to V_{IV} such that $V_{\text{II}} - V_{\text{I}} = V_{\text{III}} - V_{\text{IV}}$. During this reversible isothermal compression in which the volume decreases from V_{III} to V_{IV}, the thermal energy reservoir absorbs sufficient energy to maintain the energy density, and thus the temperature and pressure, constant in the piston–cylinder arrangement. The work done on the contents of the piston–cylinder arrangement during this process is

$$_{\text{III}}W_{\text{IV}} = p_{\text{III}}\,(V_{\text{III}} - V_{\text{IV}}) = (p - dp)(V_{\text{II}} - V_{\text{I}}) \tag{2.34}$$

and the decrease in stored energy of the contents of the piston–cylinder arrangement is

$$\Delta U_{\text{III}-\text{IV}} = u_{\text{III}}\,(V_{\text{III}} - V_{\text{IV}}) = (u - du)(V_{\text{II}} - V_{\text{I}}) \tag{2.35}$$

Then, from the first law of thermodynamics, the heat transferred to the thermal energy reservoir at $T - dT$ from the contents of the piston–cylinder arrangement is

$$_{\text{III}}Q_{\text{IV}} = {}_{\text{III}}W_{\text{IV}} + \Delta U_{\text{III}-\text{IV}} = (p - dp + u - du)(V_{\text{II}} - V_{\text{I}}) \tag{2.36}$$

Process IV–I   The carriage is now once again displaced without friction such that the perfectly reflecting plate replaces the reservoir at temperature $T - dT$. Then infinitesimally small weights are slid without friction from shelves onto the top of the piston until the original state is attained. The increase in the temperature of the contents of the piston–cylinder arrangement is dT and the corresponding increases in the radiation pressure and radiation energy density are dp and du. The volume decreases by dV during this adiabatic compression, and the corresponding work done on the contents of the piston–cylinder arrangement is

$$-\delta^2 W_{\text{IV}-\text{I}} = \delta^2 W_{\text{II}-\text{III}} = -\tfrac{1}{2}dp\, dV \tag{2.37}$$

Once again, because the process is adiabatic, this also represents the increase in stored energy of the contents of the piston–cylinder arrangement.

The thermal efficiency of the externally reversible cycle shown in Figure 2.10 is

$$\eta_{\text{th}} \equiv \frac{W_{\text{net}}}{Q_H} \tag{2.38}$$

where

$$
\begin{aligned}
W_{\text{net}} &= {}_\text{I}W_{\text{II}} + {}_\text{II}W_{\text{III}} + {}_\text{III}W_{\text{IV}} + {}_\text{IV}W_\text{I} \\
&= p(V_{\text{II}} - V_\text{I}) - \tfrac{1}{2}dp\, dV - (p - dp)(V_{\text{II}} - V_\text{I}) + \tfrac{1}{2}dp\, dV \\
&= dp(V_{\text{II}} - V_\text{I})
\end{aligned}
\tag{2.39}
$$

and

$$Q_H = {}_\text{I}Q_{\text{II}} = {}_\text{I}W_{\text{II}} + \Delta U_{\text{I}-\text{II}} = (p + u)(V_{\text{II}} - V_\text{I}) \tag{2.40}$$

With the introduction of Equation 2.29, Equations 2.39 and 2.40 may be written as

$$W_{\text{net}} = \tfrac{1}{3}du\,(V_{\text{II}} - V_\text{I}) \tag{2.41}$$

and

$$Q_H = \tfrac{4}{3}u\,(V_{\text{II}} - V_\text{I}) \tag{2.42}$$

Introducing these results into Equation 2.38 yields

$$\eta_{\text{th}} = \frac{du}{4u} \tag{2.43}$$

We have said that the two-temperature cycle under consideration is externally reversible. Therefore,

$$\eta_{\text{th}} = \eta_R = 1 - \frac{Q_L}{Q_H} = 1 - \frac{T_L}{T_H} = \frac{T_H - T_L}{T_H} = \frac{dT}{T} \tag{2.44}$$

Equations 2.43 and 2.44 may now be equated to yield

$$\frac{du}{4u} = \frac{dT}{T} \tag{2.45}$$

Upon integration and exponentiation we obtain

$$u = CT^4 \tag{2.46}$$

where C is a constant of integration whose value cannot be determined from the model. Recalling Equation 2.28, Equation 2.46 may be written

$$i_b = \left(\frac{cC}{4\pi}\right)T^4 \tag{2.47}$$

Finally, it can be shown that the emissive power of a blackbody is related to the blackbody intensity by (see Equation 2.78)

$$e_b = \pi i_b \tag{2.48}$$

Then introducing Equation 2.47 into Equation 2.48 yields

$$e_b = \frac{cC}{4} T^4 \tag{2.49}$$

The constant $cC/4$ is usually given the symbol σ and called the *Stefan–Boltzmann constant*. Its value is now known to be approximately 5.6696×10^{-8} W/m²·K⁴. Thus, we write the *Stefan–Boltzmann law*

$$e_b = \sigma T^4 \tag{2.50}$$

2.15 CANDIDATE BLACKBODY RADIATION DISTRIBUTION FUNCTIONS

We saw in Chapter 1 that Maxwell's equations describing electromagnetic radiation entertain solutions that are a function of time and space and that can be interpreted as a wave propagating at the speed of light. For example, for the special case of a periodic y-polarized transverse electric wave propagating in the x direction, we have from Chapter 1

$$E_{y,m} = |E_{y,m}|e^{i\omega_m(nx/c_0 - t)} \tag{1.28}$$

where the subscript m refers to the mth harmonic of the Fourier series representation of a general periodic function. The angular frequency ω (r/s) can presumably assume any positive value*; therefore, Maxwell's equations allow propagating electromagnetic waves at any frequency. We also established in Chapter 1 that while thermal radiation occupies only a miniscule portion of the entire electromagnetic spectrum, it nevertheless extends over about three orders of magnitude in wavelength.

Earlier in this chapter we introduced the concept of the blackbody, defined as a body that absorbs all incident radiation from all directions and at all wavelengths. We then argued from the standpoint of classical thermodynamics that no body at a given temperature can emit more radiation in a given direction and at a given wavelength than a blackbody at the same temperature. This brings up the intriguing question, What is the maximum amount of thermal radiation, per unit wavelength, that a blackbody at a given temperature can emit at a given wavelength?

For purposes of the present development we will take the *blackbody spectral intensity* $i_{b\lambda}(\lambda, T)$ as the basic measure of thermal radiation emitted from a blackbody at a given temperature in wavelength interval $d\lambda$ about wavelength λ. The notation implies that the blackbody intensity is independent of direction, as was established in Section 2.11. What physical constraints might reasonably be applied to the blackbody spectral intensity? Intuition (and perhaps more than a little hindsight) would suggest the following:

$$i_{b\lambda}(\lambda, T) \rightarrow 0 \text{ as } \lambda \rightarrow 0 \tag{2.51}$$

$$i_{b\lambda}(\lambda, T) \rightarrow 0 \text{ as } \lambda \rightarrow \infty \tag{2.52}$$

$$i_{b\lambda}(\lambda, T) \rightarrow 0 \text{ as } T \rightarrow 0 \tag{2.53}$$

and

$$i_{b\lambda}(\lambda, T) \rightarrow \infty \text{ as } T \rightarrow \infty \tag{2.54}$$

By the end of the 19th century some of the best minds of physics were hard at work trying to derive the correct expression for the *blackbody radiation distribution function,* as it came to be called. Experimental data were already available that seemed to verify the physical constraints represented by Equations 2.51 through 2.54; there only remained to find the mathematical expression that fit the data and, of course, for which

$$\int_{\lambda=0}^{\infty} \int_{2\pi} i_{b\lambda}(\lambda, T) \cos\theta \, d\omega \, d\lambda = \sigma T^4 \tag{2.55}$$

An early result of great historical interest is that of the German physicist Wilhelm Carl Werner Otto Fritz Franz ("Willy") Wien (1864–1928). In 1896 Wien proposed

---

* The mathematics, in fact, allows negative values of frequency. While this is a helpful notion in some fields, for example, communication theory, it has no physical meaning here.

an expression for the blackbody radiation distribution function, which, when expressed in terms of intensity, can be written [3]

$$i_{b\lambda}^W(\lambda, T) = C_1 \lambda^{-5} e^{-C_2/\lambda T} \tag{2.56}$$

where C_1 and C_2 are constants whose values depend on the system of units used. In the *Système International* (SI) system of units $C_1 = 1.191044 \times 10^8$ W·μm^4/m^2·sr and $C_2 = 14,388$ μm·K (although Wien was not able to determine the values of these constants to this accuracy*).

Wien's semiempirical expression is based on classical thermodynamics and intuition, and not a little bit on his familiarity with experimental results already in the literature. Unfortunately, it does not conform to the fourth constraint, Equation 2.54; in the limit as temperature approaches infinity, the spectral intensity given by Equation 2.56 approaches a finite value. The Wien blackbody radiation distribution function is a very good approximation for the actual blackbody radiation distribution function for sufficiently short wavelengths. However, as indicated in Figure 2.11 where it is plotted for a blackbody temperature of 6000 K, beyond the peak in the function it diverges significantly from blackbody behavior.

Lord Rayleigh (John William Strutt, the third Baron Rayleigh, 1842–1919) in 1900 [4] and Sir James Hopwood Jeans (1877–1946) in 1905 [5], working independently, derived another candidate expression for the blackbody radiation distribution function using the new Maxwell–Boltzmann statistics, a statistical view of nature that was supposed to correctly account for the ever-growing appreciation for

---

* Wien thought that these were empirical constants, but we will discover later in this chapter that they are fundamental physical constants whose values can be related to other well-established physical constants. One of these latter is the so-called *Planck's constant h*, which had not yet been discovered when Wien derived his blackbody radiation distribution function.

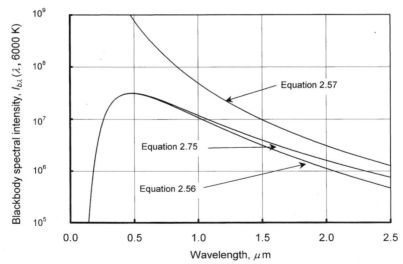

Figure 2.11   Candidate blackbody radiation distribution functions ($T = 6000$ K)

its molecular structure. Their expression, when expressed in terms of spectral intensity is

$$i_{b\lambda}^{RJ}(\lambda, T) = \frac{C_1 \lambda^{-5}}{C_2/\lambda T} \tag{2.57}$$

where C_1 and C_2 are the same constants as in Wien's expression, Equation 2.56. The Rayleigh–Jeans blackbody radiation distribution function, plotted in Figure 2.11 corresponding to a blackbody temperature of 6000 K, obviously and disastrously violates the first physical constraint, Equation 2.51, although it is in excellent agreement with the actual blackbody radiation distribution function for sufficiently long wavelengths.

Maxwell–Boltzmann statistics had already been used with impressive success to solve, in a matter of a few short months since its introduction, a wide range of problems, many of which had obstinately defied solution using Newtonian mechanics and classical thermodynamics for over a century. It was all the vogue and no one doubted its correctness. Its dramatic failure, in the hands of the two most prestigious living physicists, to correctly predict the blackbody radiation distribution function was a serious setback for physics and a disaster for classical statistics. For this reason, and because of its mathematical form, this failure was contemporaneously referred to as the "ultraviolet catastrophe."

Referring back to Section 2.6, we see that the Rayleigh–Jeans expression, Equation 2.57, is identical to Equation 2.14 if it is recognized that $2Ckc_0 = C_1/C_2$. In fact, as we shall see presently, this is true if $C = 1$ in Equation 2.14. Therefore, the description given in Chapter 1 of the emission of thermal radiation from a material substance leads to the Rayleigh–Jeans expression to within a constant. (Recall that classical statistical mechanics entered into the development in Chapter 1 with the equipartition-of-energy expression, Equation 1.40.)

In 1901 the German physicist Max Planck (1858–1947) published an article containing what turned out to be the correct blackbody radiation distribution function [6].* It may seem curious that Jeans bothered to publish his (incorrect) distribution function four years later, and this five years after Rayleigh had published the same result in the same journal! This sequence of events reflects the degree of controversy surrounding the troubled birth of Planck's *quantum statistics,* which is arguably the single most important contribution to physics in the modern age.

2.16 PLANCK'S BLACKBODY RADIATION DISTRIBUTION FUNCTION

We now undertake to outline the salient points of Planck's derivation. The English-speaking reader interested in a more detailed (and somewhat revised) treatment with-

---

* Max Planck was awarded the Nobel Prize in physics in 1918 for his discovery of quantum statistics. Planck, an outspoken opponent of Nazi policies, had the final years of his life embittered by the execution of his only son for participating in a conspiracy to assassinate Adolf Hitler.

out confronting the rigors of German syntax is referred to Planck's 1912 book, re-released in 1959 by Dover [7].

In Planck's search for the correct blackbody radiation theory he discovered a modification of Wien's relation that fitted published data for the blackbody radiation distribution function while at the same time satisfying the physical constraints represented by Equations 2.51 through 2.54. He then set out to modify classical theory to fit this empirical result. What follows is an abridged interpretation of Planck's theoretical development.

We consider oscillators, such as the atomic oscillator of Chapter 1, capable of interacting with electromagnetic radiation and that are in thermal equilibrium with their surroundings at temperature T. The total energy of such an oscillator of mass M and spring constant K is

$$e = ke + pe = \frac{p^2}{2M} + \frac{1}{2}Kx^2 \tag{2.58}$$

where p is its instantaneous momentum and x is its instantaneous displacement from equilibrium. According to classical (prequantum) statistical mechanics, the number of oscillators having values of x and p lying within the ranges of dx and dp is [8]

$$dN = NCe^{-e/kT} \, dx \, dp \tag{2.59}$$

where C is a constant of proportionality defined such that

$$N = \iint dN \tag{2.60}$$

Curves of constant energy appear as ellipses when plotted in phase space, as illustrated in Figure 2.12. Consider two energy levels, e and $e + \Delta e$, where Δe is very small. In this case the quantity $e^{-e/kT}$ may be treated as a constant over Δe so that we

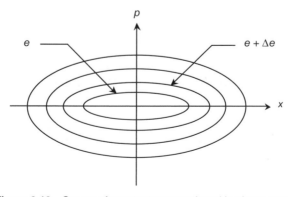

Figure 2.12 Curves of constant energy plotted in phase space

may write for the number of oscillators having energies in the interval Δe between e and $e + \Delta e$,

$$\Delta N = NCe^{-e/kT} \iint dx\, dp \tag{2.61}$$

However, the integral

$$\Delta S = \iint dx\, dp \tag{2.62}$$

is simply the area between the two ellipses representing the energies e and $e + \Delta e$. The entire area enclosed by an ellipse is

$$S = \pi x_{max} p_{max} \tag{2.63}$$

where it is easy to show that $x_{max} = \sqrt{2e/K}$ and $p_{max} = \sqrt{2Me}$. Thus,

$$S = 2\pi e \sqrt{\frac{M}{K}} \tag{2.64}$$

But the natural frequency of the atomic oscillator is

$$\nu = \frac{1}{2\pi} \sqrt{\frac{K}{M}} \tag{2.65}$$

so that

$$S = \frac{e}{\nu} \tag{2.66}$$

Therefore,

$$\Delta S = \frac{\Delta e}{\nu} \tag{2.67}$$

and the number of oscillators having energies in the interval Δe between e and $e + \Delta e$ is

$$\Delta N = \left(NC\frac{\Delta e}{\nu} \right) e^{-e/kT} \tag{2.68}$$

Now consider energy increments such that the x-p plane is divided into elliptical bands of equal area h. Then, if the rings are numbered $n = 0, 1, 2, \ldots$, the energy of an oscillator located at the inner boundary of ring n has energy

$$e = S\nu = nh\nu \tag{2.69}$$

and the number of oscillators in ring n is

$$N_n = \left(\frac{NC\,\Delta e}{\nu}\right)e^{-nh\nu/kT} = N_0 e^{-nh\nu/kT} \tag{2.70}$$

The total energy of all of the oscillators is approximately

$$
\begin{aligned}
E &= \sum_{n=0}^{\infty} N_n e_n = \sum_{n=0}^{\infty} N_n\, nh\nu \\
&= N_0 h\nu e^{-h\nu/kT}\,(1 + 2e^{-h\nu/kT} + 3e^{-2h\nu/kT} + \cdots) \\
&= N_0 h\nu e^{-h\nu/kT}\,(1 - e^{-h\nu/kT})^{-2}
\end{aligned}
\tag{2.71}
$$

The total number of oscillators is

$$N = \sum_{n=0}^{\infty} N_n = N_0\,(1 + e^{-h\nu/kT} + \cdots) = N_0\,(1 - e^{-h\nu/kT})^{-1} \tag{2.72}$$

Thus, the average energy per oscillator is

$$W = \langle e \rangle = \frac{E}{N} = \frac{h\nu}{e^{h\nu/kT} - 1} \tag{2.73}$$

Tradition has it that Planck had intended to explore the behavior of the result of his analysis in the limit as h approached zero because he, like all scientists of his time, believed that atomic oscillators could occupy a continuum of available energies. However, when he did this he recovered the well-known result from classical statistical mechanics

$$\lim_{h \to 0} \frac{h\nu}{e^{h\nu/kT} - 1} = kT \tag{2.74}$$

However, Rayleigh had already established that classical statistical mechanics cannot predict the blackbody radiation distribution at short wavelengths (the "ultraviolet catastrophe"). Therefore, Planck was led to the conclusion that h must be a finite constant.

It turns out that all of the approximations made in the foregoing analysis become exact if it is assumed that all of the oscillators in a given band have an energy corresponding to the inner ellipse bounding the band. If true, this would mean that oscillators are able to change their energies only in discrete increments of size $h\nu$. Since this is the only theory that correctly predicts observed behavior of blackbody radiation, it must be valid. This conclusion announced the birth of quantum mechanics. The quantity h was a new physical constant called, of course, Planck's constant. Its value is now known to be approximately 6.6262×10^{-27} J·s.

If Planck's expression for the average energy of an atomic oscillator in equilibrium with its surroundings at temperature T, Equation 2.73, is used to replace the classical equipartition of energy result, Equation 1.40, in Equation 1.43, there results

$$i_{b\lambda}(\lambda, T) = \frac{C_1 \lambda^{-5}}{e^{C_2/\lambda T} - 1} \tag{2.75}$$

In Equation 2.75, $C_1 = 2hc_0^2$ and $C_2 = hc_0/k$ for blackbody spectral intensity emitted into a vacuum. Introducing these more basic physical constants, Equation 2.75 can be written

$$i_{b\lambda}(\lambda, T) = \frac{2c_0^2 h \lambda^{-5}}{e^{hc_0/\lambda kT} - 1} \tag{2.76}$$

Equation 2.76 is the Planck blackbody radiation distribution function.

Equation 2.75, which is plotted in Figure 2.11, meets all of the physical constraints, Equations 2.51 through 2.54, and is in excellent agreement with experiment. Inspection of Equations 2.75 and 2.76 confirms that the blackbody spectral intensity $i_{b\lambda}(\lambda, T)$ is indeed independent of direction.

2.17   BLACKBODY DIRECTIONAL, SPECTRAL EMISSIVE POWER

From Eq. 2.8, the *directional, spectral emissive power* of a blackbody is

$$e_{b\lambda}(\lambda, T, \theta) = i_{b\lambda}(\lambda, T) \cos\theta \tag{2.77}$$

Thus, the directional, spectral emissive power of a blackbody varies as the cosine of the angle with respect to the surface normal (we refer to this angle as the *zenith* angle). The directional distributions of the blackbody directional, spectral intensity and emissive power are represented graphically in Figure 2.13.

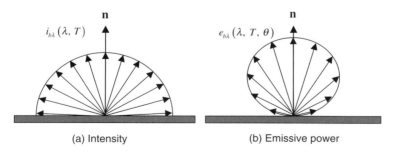

(a) Intensity   (b) Emissive power

Figure 2.13   Directional distribution of blackbody directional, spectral (a) intensity and (b) emissive power

2.18 BLACKBODY HEMISPHERICAL, SPECTRAL EMISSIVE POWER

Because the spectral intensity of a blackbody is independent of direction θ, ϕ, the *blackbody hemispherical, spectral emissive power, $e_{b\lambda}(\lambda, T)$*, is

$$e_{b\lambda}(\lambda, T) = i_{b\lambda}(\lambda, T) \bigg|_{2\pi} \cos\theta \, d\omega = \pi i_{b\lambda}(\lambda, T) \quad (\text{W/m}^2 \cdot \mu\text{m}) \qquad (2.78)$$

2.19 BLACKBODY TOTAL INTENSITY

Evaluation of the *blackbody total intensity $i_b(T)$* is complicated by the form of the integrand, Equation 2.75. We begin with a change of variables,

$$i_b(T) \equiv \int_{\lambda=0}^{\infty} i_{b\lambda}(\lambda, T) \, d\lambda = \int_{\lambda=0}^{\infty} \frac{2C_1}{\lambda^5(e^{C_2/\lambda T} - 1)} \, d\lambda = \frac{2C_1 T^4}{C_2^4} \int_0^{\infty} \frac{\eta^3}{e^\eta - 1} \, d\eta \quad (2.79)$$

where $\eta = C_2/\lambda T$. The value of the improper integral in Equation 2.79 is $\pi^4/15$, so that

$$i_b = \frac{2C_1\pi^4}{15C_2^4} T^4 = \frac{1}{\pi}\sigma T^4 \qquad (2.80)$$

where σ is the familiar *Stefan–Boltzmann constant*,

$$\sigma = \frac{2C_1\pi^5}{15C_2^4} = 5.6696 \times 10^{-8} \quad (\text{W/m}^2 \cdot \text{K}^4) \qquad (2.81)$$

2.20 BLACKBODY HEMISPHERICAL, TOTAL EMISSIVE POWER

The *blackbody hemispherical, total emissive power, e_b*, may be obtained by integrating Equation 2.78 over all wavelengths,

$$e_b(T) = \int_{\lambda=0}^{\infty} e_{b\lambda}(\lambda, T) \, d\lambda = \pi \int_{\lambda=0}^{\infty} i_{b\lambda}(\lambda, T) \, d\lambda = \pi i_b(T) \qquad (2.82)$$

and then applying Equation 2.80 to obtain

$$e_b = \pi i_b = \sigma T^4 \quad (\text{W/m}^2) \qquad (2.83)$$

As we have already seen (see Equation 2.50) this result was known before Planck derived his blackbody distribution function. Both Stefan's 1879 empirical expression and Boltzmann's 1884 analysis based on classical thermodynamics had already established what is now called the *Stefan–Boltzmann law* nearly two decades before the discovery of quantum mechanics.

2.21   THE BLACKBODY FUNCTION

When the blackbody radiation distribution function, Equation 2.75, is substituted into Equation 2.78 and the result divided by T^5, a function of only the single variable λT is produced. In other words, all Planck blackbody radiation distribution curves for all temperatures and at all wavelengths collapse onto a single curve under this change of variables; that is,

$$\frac{e_{b\lambda}(\lambda, T)}{T^5} = \frac{2\pi c_0^2 h}{(\lambda T)^5} \left(\frac{1}{e^{hc_0/k(\lambda T)} - 1} \right) = f(\lambda T \text{ only}) \quad (\text{W/m}^2 \cdot \mu\text{m} \cdot \text{K}^5) \quad (2.84)$$

Curiously, the resulting coordinates are not dimensionless as one might anticipate. The so-called "*blackbody function*" (not to be confused with the blackbody *radiation distribution* function), shown in Figure 2.14, is most useful when normalized by its total value (the area under the curve in Figure 2.14) and integrated over an interval λT of interest, yielding

$$F_{(\lambda T)_1 \rightarrow (\lambda T)_2} = F_{0 \rightarrow (\lambda T)_2} - F_{0 \rightarrow (\lambda T)_1} \quad (2.85)$$

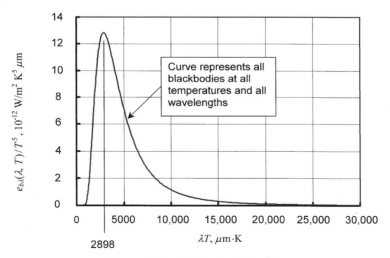

Figure 2.14   The blackbody function

where

$$F_{0\to(\lambda T)_{1,2}} \equiv \frac{\displaystyle\int_{\lambda T=0}^{(\lambda T)_{1,2}} f(\lambda T)\,d(\lambda T)}{\displaystyle\int_{\lambda T=0}^{\infty} f(\lambda T)\,d(\lambda T)} \tag{2.86}$$

The normalized blackbody function, defined by Equations 2.85 and 2.86, represents the fraction of the total power emitted from a blackbody in the interval $(\lambda T)_1 \to (\lambda T)_2$. For example, the total power emitted from a unit area of a blackbody whose temperature is T_1 in the wavelength interval $[\lambda_1, \lambda_2]$ is

$$\sigma T_1^4 \, F_{\lambda_1 T_1 \to \lambda_2 T_1} \quad (\text{W/m}^2) \tag{2.87}$$

The normalized blackbody function is very useful as a weight function for computing total surface radiation properties. In the "old days" before access to personal computers was common, tables based on Equations 2.85 and 2.86 were included as appendices in the back of radiation heat transfer texts. As a result problems based on the blackbody function had to be "worked by hand." Fortunately, this laborious approach, while instructive and perhaps even good for the soul, is no longer necessary.

It is often required when doing radiation heat transfer analysis to compute the Planck-weighted mean value of a quantity that varies with both wavelength and temperature over some specified wavelength interval. For example, suppose we have a quantity $g(\lambda, T)$ and we wish to know its Planck-weighted mean value at a given temperature T_1 over the wavelength interval $[\lambda_1, \lambda_2]$. Then we must compute

$$\langle g_{\lambda_1 \to \lambda_2}(T_1) \rangle = \frac{\displaystyle\int_{\lambda_1 T_1}^{\lambda_2 T_1} g(\lambda T_1) f(\lambda T_1)\,d(\lambda T_1)}{\displaystyle\int_0^{\infty} f(\lambda T_1)\,d(\lambda T_1)} \tag{2.88}$$

Note that normally $g(\lambda, T)$ is known rather than $g(\lambda T)$; however, once T_1 is specified it may be treated as a constant in Equation 2.88, in which case the integrals are functions of wavelength λ only.

Example Problem 2.1

What fraction of the total power emitted by the sun is emitted in the visible portion of the spectrum (0.4 to 0.7 μm)?

SOLUTION

Assume that the sun emits approximately as a blackbody at 5780 K. Then we seek the portion of its power emitted between (0.4 μm)(5780 K) $\leq \lambda T \leq$ (0.7 μm)(5780

K), or 2312 μm·K $\leq \lambda T \leq$ 4046 μm·K. Equations 2.85 and 2.86 can be combined and approximated as

$$F_{2313 \to 4046} \cong \frac{\sum\limits_{i=23}^{40} f(\lambda T)_i}{\sum\limits_{i=1}^{500} f(\lambda T)_i} \tag{2.86a}$$

The choice of the finite upper limit on the summation in the denominator of Equation 2.86a is justified by the fact that about 99.9 percent of the area under the blackbody function curve is accounted for when $\lambda T =$ 50,000 μm-K.

We use the spreadsheet shown below to evaluate Equation 2.86a and find that the fraction emitted in the visible band is 0.38, or 38 percent.

	A	B	C	D	E	F
1	100	1.1421E-64		2.15E-10	5.6533E-10	0.379909
2	200	6.4601E-35	Column A contains the product λT in μm-K			
3	300	2.2336E-25				
4	400	8.5888E-21	Cell A1: 100			
5	500	3.7597E-18	Cell A2: =A1+100 (copy into Cells A3 through A500)			
6	600	1.8327E-16				
7	700	2.6111E-15	Column B contains the blackbody function f(λT)			
8	800	1.7507E-14	Cell B1: =2*PI()*2.9979E8*2.9979E8*6.6262E-34/			
9	900	7.1732E-14	(A1*A1*A1*A1*A1*(EXP(6.6262E-34*2.9979E8/			
10	1000	2.0967E-13	(1.38E-23*A1*1.0E-6))-1))*1.0E24			
11	1100	4.8184E-13	(copy into Cells B2 through B500)			
12	1200	9.2803E-13				
13	1300	1.5649E-12	Cell D1 contains the numerator in Equation 2.86a			
14	1400	2.3827E-12				
15	1500	3.3493E-12	Cell D1: =SUM(B23:B40)			
16	1600	4.419E-12				
17	1700	5.5405E-12	Cell E1 contains the denominator in Equation 2.86a			
18	1800	6.6648E-12				
19	1900	7.7491E-12	Cell E1: =SUM(B1:B500)			
20	2000	8.7596E-12	Cell F1 contains the estimate of the value of			
21	2100	9.672E-12	Equation 2.86a			
22	2200	1.0471E-11				
23	2300	1.1148E-11	Cell F1: =D1/E1			

2.22   WIEN'S DISPLACEMENT LAW

It is interesting to know the wavelength at which the blackbody radiation distribution function peaks for a given temperature. In terms of the blackbody function introduced in the previous section, this is equivalent to asking the question, For what value of λT is the function a maximum? This in turn is equivalent to finding the root of the equation

$$\frac{\partial}{\partial(\lambda T)}\left[\frac{2\pi c_0^2 h}{(\lambda T)^5}\left(\frac{1}{e^{hc_0/\lambda kT}-1}\right)\right]=0 \tag{2.89}$$

Letting $x \equiv hc_0/\lambda kT$, Equation 2.89 leads to the transcendental equation

$$e^x = \frac{1}{1-x/5} \tag{2.90}$$

Equation 2.90 can be solved numerically* to yield $x \approx 4.965$, so that

$$(\lambda T)_{\text{peak}} = 2898 \ \mu\text{m}\cdot\text{K} \tag{2.91}$$

This result is commonly referred to as *Wien's displacement law,* although that name rightfully belongs to the more general result,

$$e_{b\lambda}(\lambda, T) = T^5 f(\lambda T) \tag{2.92}$$

Wien derived Equation 2.92 from classical thermodynamics nearly two decades before Planck's discovery of quantum mechanics and thus before the reasoning implied by Equation 2.84 produced the same result [10]. Wien postulated a simple thought experiment involving the piston–cylinder arrangement illustrated in Figure 2.15. The apparatus, which uses blackbody radiation as a working fluid, is sealed and the walls and piston face are perfect specular (mirrorlike) reflectors. The system consisting of the blackbody radiation trapped within the apparatus is imagined to undergo an adiabatic expansion in which the piston is withdrawn at velocity v. Radiation striking the piston face with wavelength λ is reflected with wavelength λ' due to the Doppler effect. It may be shown that

$$\Delta\lambda = \lambda' - \lambda = \frac{2v\lambda}{c}\cos\theta \tag{2.93}$$

where θ is the angle that the incident radiation makes with respect to the normal to the piston face.

---

* Alternatively, Siewert [9] gives the closed-form analytical expression

$$x = 4\exp\left\{-\frac{1}{\pi}\int_0^\infty\left[\tan^{-1}\left(\frac{\pi}{\ln 5 - 5 - t - \ln t}\right) - \pi\right]\frac{dt}{t+5}\right\}$$

from which $x = 4.96511\ldots.$

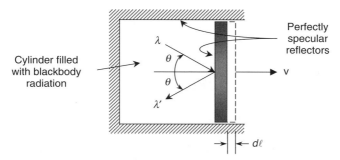

Figure 2.15 The apparatus used in Wien's thought experiment

During the adiabatic expansion

$$\delta Q = dU_b + p_b dV = 0 \qquad (2.94)$$

where the subscript b reminds us that the internal energy and pressure are due to the presence of blackbody radiation. Now multiplying Equation 2.29 by dV yields

$$p_b dV = \frac{u_b dV}{3} \qquad (2.95)$$

where, if A is the surface area of the piston face,

$$dV = A \, d\ell \qquad (2.96)$$

Then combining Equations 2.94, 2.95, and 2.96 produces

$$dU_b = \frac{-u_b A \, d\ell}{3} \qquad (2.97)$$

But we can also write

$$dU_b = d(u_b V) = A d(u_b \ell) = A(u_b d\ell + \ell \, du_b) \qquad (2.98)$$

Eliminating dU_b between Equations 2.97 and 2.98 yields

$$-\frac{u_b d\ell}{3} = u_b d\ell + \ell \, du_b \qquad (2.99)$$

which may be rearranged to obtain

$$\frac{du_b}{u_b} = -\frac{4}{3} \frac{d\ell}{\ell} \qquad (2.100)$$

Upon integration Equation 2.100 becomes

$$u_b \sim \ell^{-4/3} \tag{2.101}$$

We now seek a relationship between wavelength λ and the distance ℓ moved by the piston. Assuming that the speed of light c is much greater than the velocity v of the piston, all beams, whether approaching or receding from the piston face, have a component of velocity along the axis of the cylinder of $c \cos\theta$. Thus, a given beam will strike the piston face $c \cos\theta/2\ell$ times per second and, according to Equation 2.93, will undergo a wavelength increase $\Delta\lambda$ of $2v\lambda \cos\theta/c$ with each reflection. Therefore, the wavelength of the beam increases at a rate of

$$\frac{d\lambda}{dt} = \left(\frac{2v\lambda \cos\theta}{c}\right)\left(\frac{c \cos\theta}{2\ell}\right) = \frac{1}{\ell} v\lambda \cos^2\theta \tag{2.102}$$

If we assume that the rays are equally distributed in all directions, the average value of $\cos^2\theta$ is $\frac{1}{3}$. Therefore, on the average,

$$\frac{d\lambda}{dt} = \frac{v\lambda}{3\ell} \tag{2.103}$$

or, since $v = d\ell/dt$,

$$\frac{d\lambda}{dt} = \left(\frac{\lambda}{3\ell}\right)\frac{d\ell}{dt} \tag{2.104}$$

Then, upon separating variables and integrating, there results

$$\lambda \sim \ell^{1/3} \tag{2.105}$$

Finally, combining Equations 2.46, 2.101, and 2.105 leads to

$$\lambda T = \text{constant} \tag{2.106}$$

for each spectral component during the adiabatic expansion.

Now consider a spectral range extending from λ_1 to $\lambda_1 + d\lambda_1$ containing energy $u\lambda_1 \, d\lambda_1$ per unit volume in the piston–cylinder apparatus filled with radiation from a blackbody at temperature T_1. Let an adiabatic expansion take place that lowers the temperature to T_2 and changes the spectral range to that between λ_2 and $\lambda_2 + d\lambda_2$, as illustrated in Figure 2.16. Then at the lower and upper limits of the two spectral ranges we can invoke Equation 2.106 to write

$$\frac{\lambda_1}{\lambda_2} = \frac{T_2}{T_1} \tag{2.107}$$

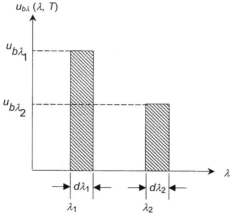

Figure 2.16 Changes in spectral energy density and wavelength provoked by an adiabatic expansion of blackbody radiation

and

$$\frac{\lambda_1 + d\lambda_1}{\lambda_2 + d\lambda_2} = \frac{T_2}{T_1} \tag{2.108}$$

Finally, Equations 2.107 and 2.108 can be combined to yield

$$\frac{d\lambda_1}{d\lambda_2} = \frac{T_2}{T_1} \tag{2.109}$$

Now, the ratio of the total energy density in the cylinder before and after the expansion is, according to Equation 2.46,

$$\frac{u_{b1}}{u_{b2}} = \frac{T_1^4}{T_2^4} \tag{2.110}$$

Then, if at the beginning of the adiabatic expansion only radiation in the wavelength interval $d\lambda_1$ were present in the cylinder, the decrease in its spectral energy would be in the same proportion as the decrease in the total energy density given by Equation 2.110; that is,

$$\frac{u_{b\lambda_1} d\lambda_1}{u_{b\lambda_2} d\lambda_2} = \frac{T_1^4}{T_2^4} \tag{2.111}$$

Combining Equations 2.109 and 2.111 there results

$$\frac{u_{b\lambda_1}}{u_{b\lambda_2}} = \frac{T_1^5}{T_2^5} \tag{2.112}$$

Finally, recalling that the blackbody spectral intensity is proportional to the blackbody spectral energy density in a given wavelength interval, Equation 2.112 can be rearranged to obtain

$$\frac{i_{b\lambda_1}}{T_1^5} = \frac{i_{b\lambda_2}}{T_2^5} = \cdots = \frac{i_{b\lambda}}{T^5} = \text{constant} \qquad (2.113)$$

But from Equation 2.106 we know that for each spectral component during an adiabatic expansion of blackbody radiation, $\lambda T = $ constant. Combining this fact with Equation 2.113 we conclude (as did Wien) that

$$i_{b\lambda}(\lambda, T) = T^5 f(\lambda T) \qquad (2.114)$$

The significance of Equation 2.114 is that, whatever the form of the function $f(\lambda T)$, the successful radiation distribution function for blackbodies at all temperatures must have the property that, when divided by T^5, the result is a function only of the product λT. Note that the Wien, Rayleigh–Jeans, and Planck distribution functions all adhere to this principle. Equation 2.114 should properly be referred to as Wien's displacement law; however, popular usage tends to give this name to Equation 2.91.

Figure 2.17 shows blackbody radiation distribution function curves corresponding to blackbodies at temperatures of 600, 700, and 800 K. The displacement of the peak in these curves toward shorter wavelength with increasing temperature, required by Equation 2.91, is evident.

Wien's displacement law explains why, as a body gets hotter and hotter, it first begins to glow dark red (at the so-called *Draper point,* 798 K) and eventually, at a sufficiently high temperature, appears more blue than red ("white hot"). This is the ba-

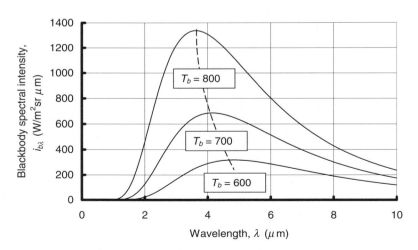

Figure 2.17 Illustration of Wien's displacement law for blackbody temperatures of 600, 700, and 800 K

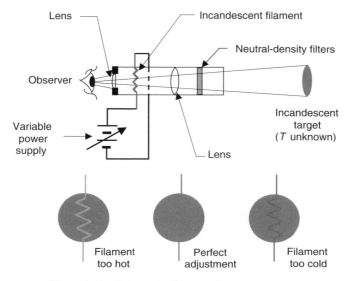

Figure 2.18   Schematic diagram of an optical pyrometer

sis of operation of one type of *optical pyrometer,* illustrated in Figure 2.18. Lenses in the optical pyrometer allow the operator to view an incandescent source of unknown temperature superimposed on an electrically heated filament. A system of changeable neutral-density filters allows the operator to attenuate the radiation from the unknown source without significantly altering its spectrum. The operator then adjusts the flow of electric current through the electrically heated filament until the filament blends into the source; that is, until the filament and the source are the same "color." The knob on the adjustable power supply that controls the filament current is calibrated in terms of source temperature.

Team Projects

TP 2.1   In this team project we will build a rudimentary radiometer suitable for demonstrating certain radiation heat transfer phenomena. It will be based on the *thermistor,* a passive electrical component having a large negative temperature coefficient of resistance. The variation of resistance with temperature for the device we will use (Radio Shack Cat. No. 271-110A) is shown in Figure 2.19. The thermistor will be incorporated in one arm of a potentiometer circuit, as shown in Figure 2.20. The potentiometer consists of a 2-V battery in series with a 10-kΩ resistor and the thermistor. The readout device can be any high-input-impedance voltage-sensitive datalogger. The datalogger monitors the voltage drop across the fixed 10-kΩ resistor, which increases as resistance of the thermistor resistance decreases.

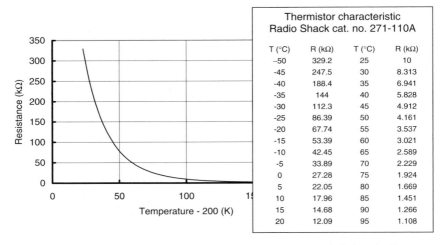

Thermistor characteristic Radio Shack cat. no. 271-110A			
T (°C)	R (kΩ)	T (°C)	R (kΩ)
–50	329.2	25	10
-45	247.5	30	8.313
-40	188.4	35	6.941
-35	144	40	5.828
-30	112.3	45	4.912
-25	86.39	50	4.161
-20	67.74	55	3.537
-15	53.39	60	3.021
-10	42.45	65	2.589
-5	33.89	70	2.229
0	27.28	75	1.924
5	22.05	80	1.669
10	17.96	85	1.451
15	14.68	90	1.266
20	12.09	95	1.108

Figure 2.19 Thermistor resistance–temperature characteristic (nominal)

The thermistor will be painted black and placed at the focus of an optical train consisting of a collimating tube and a parabolic reflector, as shown schematically in Figure 2.21. The idea here is that radiation focused on the thermistor is absorbed, thereby raising its temperature and provoking a change in its electrical resistance that is metered by the potentiometer circuit. The collimating tube limits the field-of-view of the radiometer and ensures that radiation arriving at the reflector is more or less parallel so that it may be efficiently focused on the blackened thermistor. The reflector acts like a light amplifier to increase the radiative heat flux on the thermistor.

The collimating tube should be about 10 cm long and 1 cm in diameter and its interior should be polished. One way of obtaining a polished interior is to form the collimating tube from a sheet of heavy-duty aluminum foil. The parabolic reflector could come from a flashlight. The dimensions of the cavity containing the thermistor will depend on the focal length of the reflector but should be kept small, on the order of 3 cm long.

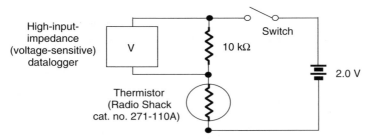

Figure 2.20 Schematic diagram for the radiometer circuit

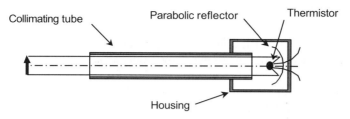

Figure 2.21 A radiometer based on reflective optics

TP 2.2 In this project we will build a simple laboratory blackbody. Our design, illustrated in Figure 2.22, is based on a standard 3-in. copper tubing elbow. The rims of the two open ends should be sanded flat. Then the fitting should be thoroughly cleaned and its interior highly polished using fine steel wool. A $\frac{1}{4}$-in.-thick square base plate 6 in. on a side should be cut from a copper plate, sanded smooth, and thoroughly cleaned. Before sanding and cleaning, the plate should be clearance drilled for a $\frac{1}{4}$-in. bolt at each of its four corners. Then one end of the copper elbow should be soldered to the center of the copper base plate.*

---

* First, a small firebrick oven must be built around the copper parts, which are held in place by a jig. Then a butane or propane torch flame is directed into the oven until the junction of the elbow and plate are sufficiently hot to melt solder (silver solder is superior but requires a higher temperature and is generally more difficult to work with). Care must be taken to thoroughly wet the surfaces to be joined with flux before applying the solder. Because only experience can guarantee a good solder joint, it is recommended that you engage a shop technician to do this step.

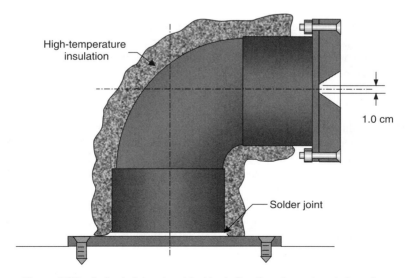

Figure 2.22 A simple laboratory blackbody (heating elements not shown)

A $4\frac{1}{2}$-in.-outside-diameter flange should now be cut from the 1/4-in.-thick copper plate and clearance drilled with four bolt holes equally spaced around its perimeter. The inside diameter should be such that it just slides over the opposite end of the elbow from the base plate. This should not be a press fit, because solder must be able to flow between the ring and the elbow. The flange should be smoothed and cleaned up and then soldered in place as shown in Figure 2.22, following the instructions given in the footnote.

After thoroughly cleaning the blackbody interior, the wall formed by the base plate should be painted with a high-temperature flat black paint such as that used on woodstove exteriors. A small brush should be used to avoid getting this paint on the interior walls of the elbow itself. Use several coats, allowing ample time for drying between coats. Next, a high-gloss, high-temperature black paint should be sprayed as evenly as possible on the interior walls of the elbow. Once again, be sure to allow time for complete drying between application of successive coats.

Next, cut a $4\frac{1}{2}$-in.-outside-diameter disk of solid high-temperature insulation about $\frac{1}{2}$ in. thick. Pierce its center with a 1.0-cm-diameter beveled hole and drill four holes to match the hole pattern in the flange. Then cut a $4\frac{1}{2}$-in.-diameter disk from a sheet of heavy-duty aluminum foil, cut four holes to match the hole pattern in the flange, and pierce its center with a 1.0-cm-diameter hole. Finally, bolt the insulating ring to the flange with the aluminum disk sandwiched in between, the shiny side facing into the blackbody cavity.

The outside surface of the elbow should now be wrapped with high-temperature-insulated nichrome heater wire, leaving ample length at the two ends. Then the entire elbow exterior surface should be covered with high-temperature insulation, either wrapped on or applied as a paste and allowed to harden. When the ends of the nichrome have been connected to a high-temperature terminal strip, the blackbody is ready for use. A Variac is used as the power source. The finished product, with the insulation removed, is shown in Figure 2.23.

TP 2.3 In this team project we will use the laboratory blackbody built in TP 2.2 to calibrate the rudimentary radiometer built in TP 2.1. This is a two-step process. First, we build the probe shown in Figure 2.24. It is based on the thermistor whose characteristics are given in Figure 2.19 (Radio Shack Cat. No. 271-110A) and whose circuit is shown in Figure 2.20. This probe will be used to accurately determine the temperature and emissive power of the blackbody as a function of the Variac power setting. The probe must be sufficiently long to extend well into the blackbody cavity but not sufficiently long to actually touch the cavity wall. The probe will be calibrated using a constant-temperature bath. With the calibrated probe held in place in the beveled hole, the Variac is set to a relatively low power setting and

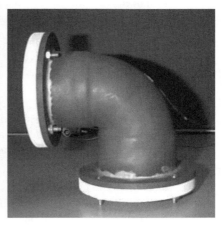

Figure 2.23 Photographs of the laboratory blackbody with the insulation removed

the value is carefully recorded. When the probe temperature attains an equilibrium value, the temperature is also recorded. This process is repeated for successively higher settings of the Variac power until a temperature range of about $25°C < T_b < 100°C$ has been obtained, where T_b is the temperature of the blackbody. The emissive power (W/m²) of the blackbody (i.e., the radiation flux issuing from the hole when the temperature probe is removed) should then be (approximately) as given by Equation 2.50. A curve is then plotted showing emissive power of the blackbody as a function of the Variac setting.

The second step is to align the radiometer built in TP 2.1 with the hole in the blackbody and to obtain the radiometer output as a function of the blackbody emissive power. Experiment with the distance of the radiometer entrance aperture from the blackbody hole; you will find a high sensitivity of the result to this distance (why?). The distance should be sufficiently large to leave room for a thin sheet of polished metal (or aluminum foil bonded to both sides of a sheet of cardboard) to be passed between the radiometer and the blackbody. This will be used to "chop" the signal so that the radiometer "zero" can be established, i.e., the value obtained when it is observing itself in the mirror. These zero values should be subtracted from the values obtained when observing the blackbody. For a repeatable posi-

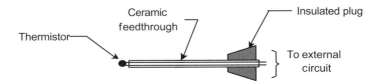

Figure 2.24 Probe for measuring blackbody temperature

tion of the radiometer with respect to the blackbody plot a curve of the radiometer output as a function of the blackbody power. Discuss the results.

Discussion Points

DP 2.1   Consider the statement, "The radiance (intensity) is preserved through an optical train." What does this mean? Is it true? Why or why not? Let your initial discussion be guided by a concrete example, such as the optical train in Figure 2.21, and then generalize from there.

DP 2.2   A practical laboratory blackbody is presented in Section 2.7. The discussion of the blackbody emphasizes its behavior as an efficient absorber. Briefly discuss its efficiency as an emitter. Why is it important that the temperature be uniform on its walls?

DP 2.3   Do you find the reasoning in Sections 2.8 and 2.9 to be in any sense circular? Please explain your answer.

DP 2.4   In Section 2.10 a proof is outlined that establishes that no body can emit more radiation in a given wavelength interval than a blackbody at the same temperature. In particular, reference is made to filters that must have pretty special properties. Elaborate on those properties of the filters not specifically mentioned in the book. Could such a set of filters really be fabricated? Are they impossible to fabricate (i.e., do they violate a law of physics) or are they simply very difficult to fabricate (i.e., their fabrication would require an inordinate amount of time, skill, and money)?

DP 2.5   Assuming that the filters discussed in Discussion Point DP 2.4 are physically possible, fill in any missing details of the formal argument in Section 2.10 establishing that no body can emit more radiation in a given wavelength interval than a blackbody at the same temperature.

DP 2.6   Why do you think it is specified in Figure 2.6 that the view of the apparatus for Bartoli's thought experiment is a top view?

DP 2.7   Spacecraft have been proposed whose propulsion system would be based on the "solar sail" concept. The idea is that radiation pressure from the sun would exert a propulsive reaction force on a large sail. Would the ideal solar sail be black, white, or silvered on the side facing the sun? What color would it be on the other side?

DP 2.8   A gimmick toy commonly available in variety shops is the "radiometer" (not to be confused with the serious scientific instrument of the same name). It consists of a partially evacuated glass bulb containing a four-vaned rotor mounted vertically on a low-friction bearing, as illustrated in Figure 2.25. One side of each vane, or paddle, is painted black and the other side is coated with a light-colored material. When illuminated by a

Figure 2.25 A "radiometer"

sufficiently bright light, the rotor begins to rotate on its axis. Which direction does it rotate (white surfaces moving forward or black surfaces moving forward)? Why?

DP 2.9 According to a footnote in Section 2.15, the concept of negative frequency is a helpful notion in communication theory. Go to the library and find a recent textbook on communication theory and see if you can find an application for which negative frequency is a helpful concept. Be prepared to share what you find with your classmates.

DP 2.10 What two equations from earlier in the chapter together imply Equation 2.55?

DP 2.11 What sort of physical quantities and mechanisms do you think might be accounted for by the constant C in Equation 2.14? In addition to the development leading to Equation 2.14, please see the next-to-the-last paragraph in Section 2.15.

DP 2.12 The denominator of Equation 2.86 is known exactly (what is its value?). Then why is its value approximated in Equation 2.86a?

DP 2.13 All three candidate blackbody radiation distribution functions (Wien, Rayleigh–Jeans, and Planck) conform to Wien's displacement law, Equation 2.114. Do all three also predict that $(\lambda T)_{\text{peak}} = 2898$ μm·K? If not, what values do the other two yield for $(\lambda T)_{\text{peak}}$? Please interpret your results.

DP 2.14 How could the optical pyrometer in Figure 2.18 be modified to work at temperatures below the Draper point?

DP 2.15 A thermistor is used in TP 2.1 and TP 2.3. In such applications the designer must avoid a phenomenon called *thermal runaway*. Self-heating within the thermistor causes its temperature to increase, thereby lowering its resistance and provoking an even higher current and thus a further increase in ohmic heating. Under certain circumstances this process may lead to a vicious circle in which the thermistor eventually overheats and fails. Using the data in Figure 2.19 and elementary principles of convection heat transfer, determine if thermal runaway is a potential problem for the circuit in

Figure 2.20. Can you formulate a stability criterion for this circuit? What could be done to increase the stability of the circuit against thermal runaway? Discuss your results in view of any simplifying assumptions you have made. Submit a brief word-processed report on your findings.

DP 2.16 A method for calibrating the rudimentary radiometer built in TP 2.1 is described in TP 2.3. What would it mean if the output of the temperature probe shown in Figure 2.24 depended on its length? Consider what such a variation would indicate about the quality of the blackbody.

DP 2.17 Calibration of the laboratory blackbody fabricated in TP 2.2 is discussed in TP 2.3. The blackbody emissive power is measured with the hole closed by the insulated plug at the base of the temperature probe, but when the blackbody is subsequently used to calibrate the radiometer built in TP 2.1 the hole is open. What effect might this have on the emissive power of the blackbody at a given Variac setting? How serious might this problem be? What design feature(s) of the blackbody address this problem?

DP 2.18 Consider Equation 2.23. Do you agree with this result? Why is it correct?

Problems

P 2.1 Demonstrate by integration of Equation 2.3 that the solid angle subtended by a hemisphere at its center is 2π, and thus that the solid angle subtended by a sphere at its center is 4π.

P 2.2 Demonstrate that the intensity of a (spherical) point source is independent of (a) direction and (b) distance from the source. [*Hint:* For part (a) we take advantage of the fact that a point source must be uniform across its (spherical) surface since by definition it is vanishingly small. For part (b) consider an observer consisting of a plane area element dS oriented normal to the line connecting the point source and dS. Find the expression for the intensity for any two large distances, say r_1 and r_2, of dS from the point source (i.e., $r^2 \gg dS$).]

P 2.3 Estimate the force exerted on the earth by radiation from the sun. Does your result agree with the value given at the beginning of Section 2.12?

P 2.4 Verify the right-hand side of Equation 2.21.

P 2.5 Show that the emissive power of a blackbody is related to the blackbody intensity by Equation 2.48.

P 2.6 To what value does Wien's blackbody radiation distribution function, Equation 2.56, converge in the limit as $T \to \infty$? Do you believe this value has any physical significance?

P 2.7 Derive Equation 2.58.

P 2.8   Verify that $x_{max} = \sqrt{2e/K}$ and $p_{max} = \sqrt{2Me}$ in Equation 2.63.

P 2.9   Where does the approximation occur in Equations 2.71?

P 2.10   Verify Equation 2.74.

P 2.11   Show that the Planck blackbody radiation distribution function, Equation 2.75, converges to the Wien blackbody radiation distribution function, Equation 2.56, for sufficiently short wavelengths.

P 2.12   Show that the Planck blackbody radiation distribution function, Equation 2.75, converges to the Rayleigh–Jeans blackbody radiation distribution function, Equation 2.57, for sufficiently long wavelengths.

P 2.13   Suppose a blackbody radiates into a medium whose index of refraction $n \equiv c_0/c$ is different from unity. In this case the entire emitted spectrum shifts according to $\lambda = c/v = c_0/nv$. Verify that the expression for the blackbody spectral intensity, Equation 2.76, becomes

$$i_{b\lambda}(\lambda, T) = \frac{2hc_0^2\lambda^{-5}}{n^2(e^{hc_0/n\lambda kT} - 1)} \tag{2.115}$$

where now λ is the wavelength *in the medium.*

P 2.14   Verify Equation 2.78.

P 2.15   Verify that

$$\int_0^\infty \frac{\eta^3}{e^\eta - 1}\, d\eta = \frac{\pi^4}{15} \tag{2.116}$$

P 2.16   Verify Equation 2.86a.

P 2.17   Verify that Equation 2.90 follows from Equation 2.89 with the suggested change of variable.

P 2.18   Solve Equation 2.90 (numerically).

P 2.19   Show that Wien's displacement law, Equation 2.91, becomes

$$n(\lambda T)_{peak} = 2898 \ \mu m \cdot K \tag{2.117}$$

where once again λ is the wavelength *in the medium.*

P 2.20   Using data from the thermistor characteristic curve of Figure 2.19, develop a theoretical calibration curve for output voltage versus blackbody emissive power for the rudimentary radiometer of TP 2.1. Take into account the offset value of voltage corresponding to the case of a mirror covering the entrance aperture when the instrument is at 25°C. Will self-heating of the thermistor be important (see DP 2.15)?

REFERENCES

1. Bartoli, A., *Exner's Rep. der Physik,* Bd. 21, 1884, p. 198; also in *Nuovo Cimento,* 15, 1883, p. 195.
2. Boltzmann, L. E., "Über eine von Hrn. Bartoli endeckte Beziehung der Wärmestrahlung zum zweiten Hauptsatze," *Annalen der Physik und Chemie,* Bd. 22, 1884, pp. 31–39.
3. Wien, W., "Über die Energievertheilung im Emissionsspektrum eines schwarzen Körpers," *Annalen der Physik,* Vierte Folge, Band 58, 1896, pp. 662–669.
4. Lord Rayleigh, "The law of complete radiation," *Philosophy Magazine,* Vol. 49, 1900, pp. 539–540.
5. Jeans, Sir James, "On the partition of energy between matter and ether," *Philosophy Magazine,* Vol. 10, 1905, pp. 91–97.
6. Planck, M., "Über das Gesetz der Energievertheilung im Normalspektrum," *Annalen der Physik,* Band 4, 1901, pp. 553–563.
7. Planck, M., *The Theory of Heat Radiation,* Dover Publications, Inc., New York, 1959.
8. Lee, J. F., F. W. Sears, and D. L. Turcotte, *Statistical Thermodynamics,* Addison–Wesley Publishing Company, Inc., Reading, Massachusetts, 1963, p. 113.
9. Siewert, C. E., "An exact expression for the Wien displacement constant," *Journal of Quantitative Spectroscopy and Radiative Transfer,* Vol. 26, 1981, p. 467.
10. Wien, W., "Temperatur und Entropie der Strahlung," *Annalen der Physik,* Zweite Folge, Band 52, 1894, pp. 132–165.

3

DESCRIPTION OF REAL SURFACES; SURFACE PROPERTIES

All surfaces of practical engineering interest deviate to some degree from blackbody behavior; that is, they do not absorb all incident radiation and they emit less than a blackbody at a given temperature. More to the point, the departure from blackbody behavior of real surfaces is more or less dependent on the wavelength interval of interest and on the direction of incidence or emission. In this chapter we define certain surface properties that describe the departure of practical surfaces from blackbody behavior. We investigate how these properties are related to each other and learn how to use them with the blackbody model to describe emission, absorption and reflection by real surfaces.

3.1 DEPARTURE OF REAL SURFACES FROM BLACKBODY BEHAVIOR

Surfaces of practical engineering interest, such as metals, paints, and plastics, generally do not absorb all incident radiation, nor do they generally emit as much radiation as a blackbody at the same temperature. Furthermore, their departure from blackbody behavior usually depends on the wavelength interval of interest as well as on the direction of emission or incidence.

Figure 3.1 compares the emission spectrum of two hypothetical nonblack surfaces at 1073 K with that of a blackbody at the same temperature. One of the hypothetical radiation distribution functions, designated in the figure as curve a, is simply a scaled-down version of the (Planck) blackbody radiation distribution function. That is, the ratio of the spectral intensity at a given wavelength to that of the Planck function at the same wavelength is the same for all wavelengths. Of course, this ratio must be less than unity. By convention, a surface whose radiation distribution function

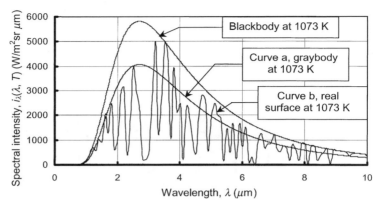

Figure 3.1 Spectral intensity of a blackbody, a graybody, and a hypothetical real surface, all at 1073 K

conforms, to within acceptable accuracy, to that of curve a, is called a *gray surface,* or a *graybody.* While no surface conforms exactly to the graybody model, it remains a very important engineering tool for radiative analysis in cases where speed and simplicity are at a premium.

In the case of the other hypothetical radiation distribution function shown in Figure 3.1, designated as curve b, the departure from blackbody behavior is dependent on wavelength. This is the more usual situation for real surfaces.

Recall that for a blackbody the intensity is independent of direction. Figure 3.2 compares the variation with zenith angle θ of the emitted intensity at a given azimuth angle ϕ_1 and wavelength λ_1 for a black surface, a diffuse surface, and a hypothetical real surface, all at the same temperature T_1. The emitted intensity from a *diffuse* surface is independent of direction and is less than the intensity of a blackbody. The figure illustrates that for all directions the emitted intensity of a real surface at a given temperature can be no more than the intensity of a black surface at the same temperature.

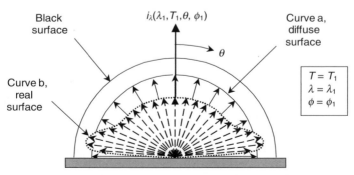

Figure 3.2 The directional, spectral intensity at a given wavelength, temperature, and azimuth angle for a black surface, a diffuse surface, and a hypothetical real surface

3.2   DIRECTIONAL, SPECTRAL EMISSIVITY

The *directional, spectral emissivity* $\varepsilon_\lambda'(\lambda,T,\theta,\phi)$ is a dimensionless quantity whose value indicates the degree to which a real surface at a given temperature emits like a blackbody in a given direction and at a given wavelength. Specifically, it is the ratio of the spectral intensity of emission from a body in direction θ,ϕ to the intensity of a blackbody at the same temperature and wavelength; that is,

$$\varepsilon_\lambda'(\lambda, T, \theta, \phi) \equiv \frac{i_{\lambda, e}(\lambda, T, \theta, \phi)}{i_{b\lambda}(\lambda, T)} \tag{3.1}$$

The prime ($'$) in the notation here identifies this surface property (and others) as being "directional." We recognize the directional, spectral emissivity to be a scaling factor relating the spectral intensity of a real surface to that of a blackbody at the same temperature. The utility of knowing the directional, spectral emissivity for a surface is that it can be multiplied by the blackbody spectral intensity corresponding to the surface temperature to obtain the intensity of spectral radiation at a given wavelength and in a given direction.

A surface is said to be *gray in a given direction* θ_1,ϕ_1 if its directional, spectral emissivity is independent of wavelength in that direction; that is, if

$$\varepsilon_\lambda' = \varepsilon_\lambda'(T, \theta_1, \phi_1) < 1.0 \qquad \text{(gray in direction } \theta_1, \phi_1) \tag{3.2}$$

Remember that the subscript λ in Equation 3.2 indicates a spectral property; the fact that, in the special case of a gray surface, the directional, spectral emissivity is independent of wavelength does not change the fact that it is a spectral quantity. Also, it should be emphasized that a given surface might exhibit gray behavior in some directions while exhibiting nongray behavior in others.

A surface is said to be *diffuse at wavelength* λ_1 if its directional, spectral emissivity is independent of direction at that wavelength; that is,

$$\varepsilon_\lambda' = \varepsilon_\lambda'(\lambda_1, T) < 1.0 \qquad \text{(diffuse at wavelength } \lambda_1) \tag{3.3}$$

Curve a in Figure 3.2 corresponds to a diffuse surface at wavelength λ_1 if the curve remains the same for all values of ϕ_1. Curve b in Figure 3.2 corresponds to a more realistic surface.

In many cases of practical engineering interest it is convenient to approximate a real surface as an equivalent diffuse surface if the corresponding loss of accuracy is acceptable. We will see in Part II of the book that radiation heat transfer analysis involves the solution of integral (rather than differential) equations. The integration process over direction tends to compensate for the departure from diffuse behavior for many real surfaces.

A surface is said to be *diffuse, gray* if its directional, spectral emissivity is independent of both direction and wavelength; that is, if

$$\varepsilon_\lambda' = \varepsilon_\lambda'(T \text{ only}) \tag{3.4}$$

This hypothesis is often acceptable in engineering practice even though in reality it is never exactly true. In a distressingly large number of actual applications, knowledge is lacking about the directional, spectral emissivity. In such situations an estimate based on the diffuse, gray assumption is often the only avenue available for heat transfer analysis. In many cases, however, a simplified analysis based on the diffuse, gray hypothesis is carried out even when estimates of the wavelength and directional dependence of surface emissivity are available. This often occurs because practitioners are not familiar with the modern statistically based methods, presented in Part III of this book, capable of exploiting available information about wavelength and directional dependence of surface properties.

3.3   HEMISPHERICAL, SPECTRAL EMISSIVITY

How much power does a plane area element dA at temperature T emit into the hemispherical space above it in wavelength interval $d\lambda$ about wavelength λ, per unit surface area and wavelength? An equivalent question is, What is the hemispherical, spectral emissive power of dA in wavelength interval $d\lambda$ about wavelength λ? The answer to both questions can be expressed in two equivalent manners. First,

$$e_\lambda(\lambda, T) = \varepsilon_\lambda(\lambda, T)\, e_{b\lambda}(\lambda, T) = \varepsilon_\lambda(\lambda, T)\pi\, i_{b\lambda}(\lambda, T) \qquad (3.5)$$

where

$$\varepsilon_\lambda(\lambda, T) \equiv \frac{e_\lambda(\lambda, T)}{\pi i_{b\lambda}(\lambda, T)} \qquad (3.6)$$

is the *hemispherical, spectral emissivity*. Second, the hemispherical, spectral emissivity can be related to the directional spectral emissivity by substituting Equation 3.1 into Equation 2.10, yielding

$$e_\lambda(\lambda, T) = \int_{2\pi} i_{\lambda,e}(\lambda,\, \theta,\, \phi)\cos\theta\, d\omega = \int_{2\pi} \varepsilon_\lambda'(\lambda,\, T,\, \theta,\, \phi)\, i_{b\lambda}(\lambda,\, T)\cos\theta\, d\omega \quad (3.7)$$

Then equating the two expressions for the hemispherical, spectral emissive power, Equations 3.5 and 3.7, and simplifying yields

$$\varepsilon_\lambda(\lambda, T) = \frac{\displaystyle\int_{2\pi} \varepsilon_\lambda'(\lambda,\, T,\, \theta,\, \phi)\cos\theta\, d\omega}{\displaystyle\int_{2\pi} \cos\theta\, d\omega} \qquad (3.8)$$

or

$$\varepsilon_\lambda(\lambda, T) = \frac{1}{\pi}\int_{2\pi} \varepsilon_\lambda'(\lambda,\, T,\, \theta,\, \phi)\cos\theta\, d\omega \qquad (3.8a)$$

The process we have used to derive Equation 3.8a is generic; we will use it again and again to relate hemispherical properties to directional properties, and to relate total properties to spectral properties. Therefore, it is worthwhile to review the procedure. The first step is to write two expressions for the relevant power quantity, one based on the definition of the property being sought (Equation 3.5 in the current example), and the other using the directional (or spectral) version of the same property (Equation 3.7 in the current example). The two expressions are then equated and the result solved for the hemispherical (or total) property in terms of the weighted integral of the directional (or spectral) property.

Note that the dimensions are the same for the two sides of Equation 3.8a because $d\omega$ and π both have the dimensions of solid angle. A good check of Equation 3.8a is to consider what would happen if the directional emissivity $\varepsilon'\lambda(\lambda, T)$ for a given hypothetical surface were independent of direction; that is, if the surface were *diffuse*. In this case we obtain from Equation 3.8a

$$\varepsilon_\lambda(\lambda, T) = \varepsilon'_\lambda(\lambda, T) \qquad \text{(diffuse surface)} \qquad (3.9)$$

as expected.

Example Problem 3.1

Hypothetical directional, spectral emissivity data for a certain surface at a given wavelength are plotted in Figure 3.3 and presented in Table 3.1. Axisymmetry is assumed; that is, it is assumed that the directional emissivity is independent of ϕ. What is the hemispherical, spectral emissivity of the surface at this wavelength?

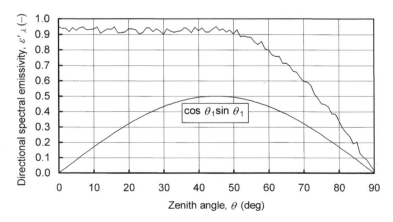

Figure 3.3 Hypothetical directional, spectral emissivity data

Table 3.1   Hypothetical directional, spectral emissivity data

θ (deg)	ε'_λ (θ)	θ (deg)	ε'_λ (θ)	θ (deg)	ε'_λ (θ)	θ (deg)	ε'_λ (θ)
0	0.903	23	0.930	46	0.907	69	0.623
1	0.925	24	0.913	47	0.905	70	0.579
2	0.921	25	0.930	48	0.902	71	0.600
3	0.909	26	0.922	49	0.897	72	0.530
4	0.908	27	0.946	50	0.915	73	0.536
5	0.945	28	0.940	51	0.906	74	0.515
6	0.949	29	0.925	52	0.886	75	0.489
7	0.910	30	0.900	53	0.902	76	0.450
8	0.936	31	0.921	54	0.890	77	0.439
9	0.939	32	0.916	55	0.871	78	0.386
10	0.919	33	0.913	56	0.852	79	0.382
11	0.920	34	0.941	57	0.823	80	0.332
12	0.937	35	0.908	58	0.841	81	0.289
13	0.934	36	0.944	59	0.819	82	0.251
14	0.932	37	0.916	60	0.808	83	0.252
15	0.912	38	0.933	61	0.772	84	0.187
16	0.930	39	0.912	62	0.757	85	0.178
17	0.930	40	0.901	63	0.749	86	0.174
18	0.922	41	0.901	64	0.726	87	0.108
19	0.916	42	0.945	65	0.703	88	0.108
20	0.922	43	0.917	66	0.682	89	0.079
21	0.912	44	0.941	67	0.653	90	0.010
22	0.934	45	0.946	68	0.639		

SOLUTION

For 91 observations we approximate the right-hand side of Equation 3.8 as the ratio of finite sums; that is,

$$\varepsilon_\lambda(\lambda, T) \cong$$

$$\frac{\frac{1}{2}\varepsilon'_{\lambda,1} \cos\theta_1 \sin\theta_1 + \sum_{i=2}^{90} \varepsilon'_{\lambda,i} \cos\theta_i \sin\theta_i + \frac{1}{2}\varepsilon'_{\lambda,91} \cos\theta_{91} \sin\theta_{91}}{\frac{1}{2}\cos\theta_1 \sin\theta_1 + \sum_{i=2}^{90} \cos\theta_i \sin\theta_i + \frac{1}{2}\cos\theta_{91} \sin\theta_{91}}$$ (3.8b)

We import the data from Table 3.1 into a spreadsheet, create columns for θ_i, $\cos\theta_i$ $\sin\theta_i$ and $\varepsilon'_\lambda(\lambda, T)$, and then carry out the operations implied by Equation 3.8b. When we do this we obtain $\varepsilon_\lambda(\lambda, T) = 0.826$. This result can be easily justified by studying Figure 3.3, which also shows the behavior of the weight factor $\cos\theta_i$, $\sin\theta_i$.

3.4   DIRECTIONAL, TOTAL EMISSIVITY

How much power does an area element dA at temperature T emit in direction θ, ϕ within solid angle $d\omega$ surrounding θ, ϕ per unit surface area at all wavelengths? The equivalent question is, What is the directional, total emissive power of dA? Once again, we offer two answers to these equivalent questions, one based on the definition of the *directional, total emissivity, $\varepsilon'(T, \theta, \phi)$*:

$$e'(T, \theta, \phi) = \varepsilon'(T, \theta, \phi)\, e'_b(T, \theta, \phi) = \varepsilon'(T, \theta, \phi) i_b(T) \cos\theta \qquad (3.10)$$

and the other in terms of the directional, spectral emissivity:

$$e'(T, \theta, \phi) = \int_{\lambda=0}^{\infty} e'_\lambda(\lambda, T, \theta, \phi)\, d\lambda = \int_{\lambda=0}^{\infty} i_\lambda(\lambda, T, \theta, \phi) \cos\theta\, d\lambda \qquad (3.11)$$

or

$$e'(T, \theta, \phi) = \int_{\lambda=0}^{\infty} \varepsilon'_\lambda(\lambda, T, \theta, \phi)\, i_{b\lambda}(\lambda, T) \cos\theta\, d\lambda \qquad (3.12)$$

Equating the two expressions for the directional, total emissive power of dA, Equations 3.10 and 3.12, and solving for the directional, total emissivity yields

$$\varepsilon'(T, \theta, \phi) = \frac{\displaystyle\int_{\lambda=0}^{\infty} \varepsilon'_\lambda(\lambda, T, \theta, \phi)\, i_{b\lambda}(\lambda, T)\, d\lambda}{\displaystyle\int_{\lambda=0}^{\infty} i_{b\lambda}(\lambda, T)\, d\lambda} \qquad (3.13)$$

From Equations 2.82 and 2.83 we know that the denominator of Equation 3.13 is $\sigma T^4/\pi$, so that we finally obtain

$$\varepsilon'(T, \theta, \phi) = \frac{\pi}{\sigma T^4} \int_{\lambda=0}^{\infty} \varepsilon'_\lambda(\lambda, T, \theta, \phi)\, i_{b\lambda}(\lambda, T)\, d\lambda \qquad (3.14)$$

Note that application of Equation 3.14 to the special case of a gray surface yields

$$\varepsilon'(T, \theta, \phi) = \varepsilon'_\lambda(T, \theta, \phi) \qquad \text{(gray surface)} \qquad (3.15)$$

as expected.

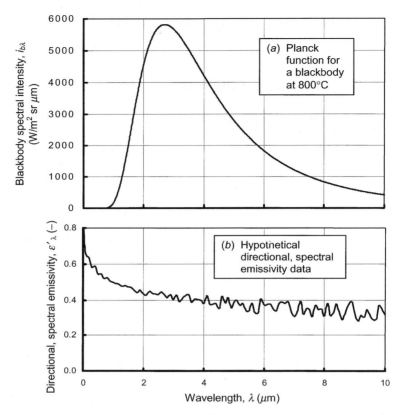

Figure 3.4 (a) Planck radiation distribution function for a blackbody at 800°C, and (b) the directional, spectral emissivity data for a hypothetical surface

Example Problem 3.2

The directional, spectral emissivity of a certain 800°C surface in a specified direction θ, ϕ in the wavelength interval $0.1 \leq \lambda \leq 10$ μm is represented in Figure 3.4, and the corresponding data appear in Table 3.2. What is the directional, total emissivity of the surface at that temperature and in the specified direction?

Solution

It is noted that we have been provided with directional, spectral emissivity data over only a small fraction of the total ($[0, \infty]$) wavelength interval. This is typically the case. Also shown in Figure 3.4 is the Planck blackbody radiation distribution function corresponding to a blackbody at 800°C (= 1073 K). Note that relatively little power is emitted by a blackbody at 1073 K outside the range of the available emissivity data. Therefore, we are justified in making the approximation

$$\varepsilon'(T, \theta, \phi) \cong \frac{\int_{\lambda=0}^{10} \varepsilon'_\lambda(\lambda, T, \theta, \phi)\, i_{b\lambda}(\lambda, T)\, d\lambda}{\int_{\lambda=0}^{10} i_{b\lambda}(\lambda, T)\, d\lambda} \tag{3.13a}$$

We approximate the right-hand side of Equation 3.13a as a ratio of finite sums,

$$\varepsilon'(T, \theta, \phi) \cong \frac{\sum_{i=1}^{101} \varepsilon'_{\lambda,i}(\lambda, T, \theta, \phi)\, i_{\lambda,i}(\lambda, T)}{\sum_{i=1}^{101} i_{\lambda,i}(\lambda, T)} \tag{3.13b}$$

Table 3.2   Directional, spectral emissivity data for the hypothetical surface of Example Problem 3.2

λ (μm)	ε'_λ (−)	λ (μm)	ε'_λ (−)	λ (μm)	ε'_λ (−)	λ (μm)	ε'_λ (−)
0.0	0.847	2.5	0.431	5.0	0.389	7.5	0.346
0.1	0.656	2.6	0.444	5.1	0.400	7.6	0.305
0.2	0.636	2.7	0.429	5.2	0.368	7.7	0.368
0.3	0.580	2.8	0.414	5.3	0.382	7.8	0.344
0.4	0.590	2.9	0.430	5.4	0.337	7.9	0.348
0.5	0.547	3.0	0.403	5.5	0.338	8.0	0.310
0.6	0.545	3.1	0.428	5.6	0.382	8.1	0.379
0.7	0.521	3.2	0.392	5.7	0.329	8.2	0.313
0.8	0.522	3.3	0.414	5.8	0.328	8.3	0.330
0.9	0.507	3.4	0.423	5.9	0.405	8.4	0.376
1.0	0.500	3.5	0.430	6.0	0.374	8.5	0.320
1.1	0.492	3.6	0.388	6.1	0.373	8.6	0.288
1.2	0.487	3.7	0.405	6.2	0.378	8.7	0.339
1.3	0.476	3.8	0.403	6.3	0.353	8.8	0.374
1.4	0.481	3.9	0.399	6.4	0.337	8.9	0.388
1.5	0.472	4.0	0.379	6.5	0.382	9.0	0.322
1.6	0.470	4.1	0.423	6.6	0.386	9.1	0.281
1.7	0.455	4.2	0.376	6.7	0.381	9.2	0.313
1.8	0.442	4.3	0.378	6.8	0.386	9.3	0.301
1.9	0.454	4.4	0.381	6.9	0.335	9.4	0.346
2.0	0.434	4.5	0.353	7.0	0.306	9.5	0.307
2.1	0.426	4.6	0.421	7.1	0.334	9.6	0.288
2.2	0.449	4.7	0.384	7.2	0.390	9.7	0.292
2.3	0.438	4.8	0.413	7.3	0.320	9.8	0.370
2.4	0.433	4.9	0.355	7.4	0.350	9.9	0.341
						10.0	0.314

We import the data from Table 3.2 into the spreadsheet, create a column for the blackbody spectral intensity i_λ, and then carry out the operations implied by Equation 3.13b. When we do this we obtain $\varepsilon'(T, \theta, \phi) \approx 0.40$.

3.5 THE HEMISPHERICALIZING AND TOTALIZING OPERATORS

It is informative to examine Equations 3.8 and 3.14 together. The first involves the so-called *hemisphericalizing* operator,

$$H[g'(\theta, \phi)] \equiv \frac{1}{\pi} \int_{2\pi} g'(\theta, \phi) \cos\theta \, d\omega \qquad (3.16)$$

where $g'(\theta, \phi)$ is a directional surface property; and the second involves the so-called *totalizing* operator,

$$T[g_\lambda(\lambda)] = \frac{\pi}{\sigma T^4} \int_{\lambda=0}^{\infty} g_\lambda(\lambda) \, i_{b\lambda}(\lambda, T) \, d\lambda \qquad (3.17)$$

where $g_\lambda(\lambda)$ is a spectral surface property. The hemisphericalizing operator H weights the directional property by the cosine of the zenith angle over 2π-space, while the totalizing operator T weights the spectral property by the Planck blackbody radiation distribution function over all wavelengths. Under certain special circumstances we will be able to use these two operators to relate hemispherical and/or total surface properties to their directional, spectral counterparts.

3.6 HEMISPHERICAL, TOTAL EMISSIVITY

What is the total power emitted by area element dA at temperature T into the hemispherical space above dA per unit surface area? The equivalent question is, What is the hemispherical, total emissive power of dA? Once again, the two answers are

$$e(T) = \varepsilon(T) \, e_b(T) \qquad (3.18)$$

and

$$e(T) = \int_{\lambda=0}^{\infty} \int_{2\pi} \varepsilon_\lambda'(\lambda, T, \theta, \phi) \, i_{b\lambda}(\lambda, T) \cos\theta \, d\omega \, d\lambda \qquad (3.19)$$

The first expression is based on the definition of the *hemispherical, total emissivity,* and the second expression is Equation 3.7 integrated over all wavelengths. Equating the two expressions for $e(T)$ and solving for the hemispherical, total emissivity yields

$$\varepsilon(T) = \frac{1}{\sigma T^4} \int_{\lambda=0}^{\infty} \int_{2\pi} \varepsilon_\lambda'(\lambda, T, \theta, \phi) \, i_{b\lambda}(\lambda, T) \cos\theta \, d\omega \, d\lambda \qquad (3.20)$$

Note that this same result could have been obtained by hemisphericalizing and total-izing the directional, spectral emissivity, that is,

$$\varepsilon(T) = T\{H[\varepsilon_\lambda' (\lambda, T, \theta, \phi)]\} = H\{T[\varepsilon_\lambda' (\lambda, T, \theta, \phi)]\} \tag{3.21}$$

3.7   THE DISPOSITION OF RADIATION INCIDENT TO A SURFACE; THE REFLECTIVITY, ABSORPTIVITY, AND TRANSMISSIVITY

When thermal radiation is incident to a surface it must be *reflected, absorbed,* or *transmitted,* as indicated schematically in Figure 3.5. Later we will see that radiation passing through a gas or other semitransparent medium will be *scattered,* which is a phenomenon similar to reflection from a surface. In fact, in some contexts the two terms are interchangeable. Figure 3.5 illustrates the first law of thermodynamics applied to thermal radiation,

$$\text{Incident} = \text{Absorbed} + \text{Reflected} + \text{Transmitted} \tag{3.22}$$

or

$$1 = \frac{\text{Absorbed}}{\text{Incident}} + \frac{\text{Reflected}}{\text{Incident}} + \frac{\text{Transmitted}}{\text{Incident}} \tag{3.23}$$

The three terms on the right-hand side of Equation 3.23 are called, respectively, the *absorptivity* α, the *reflectivity* ρ, and the *transmissivity* τ. Thus, Equation 3.23 can be rewritten

$$1 = \alpha + \rho + \tau \tag{3.24}$$

A surface or medium is said to be *opaque* if its transmissivity for radiation incident from all directions and at all wavelengths is zero. Then for opaque surfaces,

$$\alpha + \rho = 1 \qquad \text{(opaque)} \tag{3.25}$$

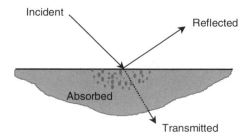

Figure 3.5   The disposition of radiation incident to a surface

or

$$\rho = 1 - \alpha \quad \text{(opaque)} \tag{3.26}$$

3.8 DIRECTIONAL, SPECTRAL ABSORPTIVITY

The *directional, spectral absorptivity* α'_λ is the fraction of the power in wavelength interval $d\lambda$ about λ incident to surface element dA in solid angle $d\omega_i$ about direction θ_i, ϕ_i which is absorbed; that is,

$$\alpha'_\lambda(\lambda, T_A, \theta_i, \phi_i) \equiv \frac{d^3 P_a(\lambda, T_A, \theta_i, \phi_i)}{i_{\lambda,i}(\lambda) \, dA \cos\theta_i \, d\omega_i \, d\lambda} \tag{3.27}$$

where $d^3 P_a(\lambda, T_A, \theta_i, \phi_i)$ is the power absorbed in wavelength interval $d\lambda$ about λ from direction θ_i, ϕ_i, and $i_{\lambda,i}$ is the incident intensity in that wavelength interval and from that direction. The symbol T_A in Equation 3.27 and elsewhere in this chapter refers to the temperature of the absorbing surface dA rather than to the temperature of the irradiating source. The geometry for this situation is shown in Figure 3.6.

Essential to the development of the Monte Carlo ray-trace method in Part III of the book is the statistical interpretation of the directional, spectral absorptivity as the probability that an energy bundle carrying energy in a specified wavelength interval $\Delta\lambda$ and incident from direction θ, ϕ will be absorbed.

3.9 KIRCHHOFF'S LAW

Consider a large enclosure whose walls are maintained at the same uniform temperature. Filters are available to place between the walls of the enclosure and any object within the enclosure. Each filter has a pass band of differential width $d\lambda$ centered about some specified wavelength λ. Now let us introduce into this enclosure a body whose directional, spectral emissivity is $\varepsilon'_\lambda(\lambda, T_A, \theta, \phi)$. We once again (see Section

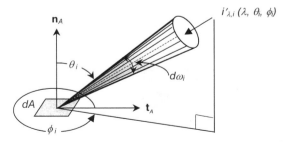

Figure 3.6 Beam radiation incident to a surface element dA

2.10) appeal to the second law of thermodynamics to argue that the body must come into thermal equilibrium with the walls after a sufficient amount of time has elapsed, regardless of the pass band of the filters in place. Now, in the presence of a given set of filters the body can emit net radiation only in the pass band of the filters, because all radiation emitted outside the pass band is reflected back to the body where it is eventually reabsorbed. For the body to remain in thermal equilibrium with the enclosure it must absorb the same amount of energy from the enclosure as it emits in the filter pass band. Also, we know that the spectral intensity is uniform and isotropic within the isothermal enclosure. We are therefore led to conclude that, for a given direction θ, ϕ,

$$d^3 P_a(\lambda, T_A, \theta, \phi) = d^3 P_e(\lambda, T_A, \theta, \phi) \tag{3.28}$$

where $d^3 P_e(\lambda, T_A, \theta, \phi)$ is the power emitted from the body in wavelength interval $d\lambda$ about λ in direction θ, ϕ. Introducing the definitions of the directional, spectral emissivity and absorptivity, Equations 3.1 (with Equation 3.4) and 3.27, respectively, into Equation 3.28 yields

$$\alpha'_\lambda(\lambda, T, \theta, \phi)\, i_{b\lambda}(\lambda, T)\, dA\, \cos\theta\, d\omega\, d\lambda \\ = \varepsilon'_\lambda(\lambda, T, \theta, \phi)\, i_{b\lambda}(\lambda, T)\, dA\, \cos\theta\, d\omega\, d\lambda \tag{3.29}$$

or

$$\alpha'_\lambda(\lambda, T, \theta, \phi) = \varepsilon'_\lambda(\lambda, T, \theta, \phi) \tag{3.30}$$

That is, the directional, spectral absorptivity is equal to the directional, spectral emissivity. This result, known as *Kirchhoff's law,* holds without constraint; it is always true.

3.10 HEMISPHERICAL, SPECTRAL ABSORPTIVITY

What is the power per unit area absorbed by a surface element dA in the wavelength interval $d\lambda$ about λ due to incident radiation from the hemispherical space above dA? The situation described is depicted in Figure 3.7.

We seek to express the *hemispherical, spectral absorptivity* in terms of the directional, spectral absorptivity just as we expressed the hemispherical, spectral emissivity in terms of the directional, spectral emissivity in Section 3.3. Our approach is the same. That is, we write two independent expressions, both of which are answers to the question posed above:

$$\frac{dP_{\lambda,a}(\lambda, T_A)}{dA} = \alpha_\lambda(\lambda, T_A) \int_{2\pi} i_{\lambda,i}(\lambda, \theta_i, \phi_i)\cos\theta_i\, d\omega_i \tag{3.31}$$

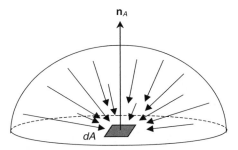

Figure 3.7   Radiation incident to a surface element dA from the hemispherical space above dA

where the subscript i refers to incident radiation and $\alpha_\lambda(\lambda, T_A)$ is the *hemispherical, spectral absorptivity;* and

$$\frac{dP_{\lambda,a}(\lambda, T_A)}{dA} = \int_{2\pi} \alpha'_\lambda(\lambda, T_A, \theta, \phi)\, i_{\lambda,i}(\lambda, \theta_i, \phi_i)\cos\theta_i\, d\omega_i \qquad (3.32)$$

Equating the two expressions for the absorbed flux and solving for the hemispherical, spectral absorptivity yields

$$\alpha_\lambda(\lambda, T_A) = \frac{\displaystyle\int_{2\pi} \alpha'_\lambda(\lambda, T_A, \theta_i, \phi_i)\, i_{\lambda,i}(\lambda, \theta_i, \phi_i)\cos\theta_i\, d\omega_i}{\displaystyle\int_{2\pi} i_{\lambda,i}(\lambda, \theta_i, \phi_i)\cos\theta_i\, d\omega_i} \qquad (3.33)$$

3.11   DIRECTIONAL, TOTAL ABSORPTIVITY

What is the total power per unit area absorbed by a surface element dA due to incident radiation from direction θ_i, ϕ_i in solid angle $d\omega_i$ about θ_i, ϕ_i? In response to this question we once again write two independent expressions, one directly invoking the definition of the *directional, total absorptivity* $\alpha'(T_A, \theta_i, \phi_i)$, and the other involving the directional, spectral absorptivity:

$$\frac{d^2P_a(T_A, \theta_i, \phi_i)}{dA} = \alpha'(T_A, \theta_i, \phi_i) \int_{\lambda=0}^{\infty} i_{\lambda,i}(\lambda, \theta_i, \phi_i)\cos\theta_i\, d\omega_i\, d\lambda \qquad (3.34)$$

and

$$\frac{d^2P_a(T_A, \theta_i, \phi_i)}{dA} = \int_{\lambda=0}^{\infty} \alpha'_\lambda(\lambda, T_A, \theta_i, \phi_i)\, i_{\lambda,i}(\lambda, \theta_i, \phi_i)\cos\theta_i\, d\omega_i\, d\lambda \qquad (3.35)$$

Equating the two expressions for the total power per unit area absorbed by a surface element dA due to incident radiation from direction θ_i, ϕ_i in solid angle $d\omega_i$ and solving for $\alpha'(T_A, \theta_i, \phi_i)$ yields

$$\alpha'(T_A, \theta_i, \phi_i) = \frac{\int_{\lambda=0}^{\infty} \alpha'_\lambda(\lambda, T_A, \theta_i, \phi_i) \, i_{\lambda,i}(\lambda, \theta_i, \phi_i) \, d\lambda}{\int_{\lambda=0}^{\infty} i_{\lambda,i}(\lambda, \theta_i \, \phi_i) \, d\lambda} \tag{3.36}$$

3.12 HEMISPHERICAL, TOTAL ABSORPTIVITY

What is the total power per unit area absorbed by a surface element dA due to radiation incident from the hemispherical space above dA? Proceeding as before, we write

$$\frac{dP_a(T_A)}{dA} = \alpha(T_A) \int_{\lambda=0}^{\infty} \int_{2\pi} i_{\lambda,i}(\lambda, \theta_i, \phi_i) \cos\theta_i \, d\omega_i \, d\lambda \tag{3.37}$$

and

$$\frac{dP_a(T_A)}{dA} = \int_{\lambda=0}^{\infty} \int_{2\pi} \alpha'_\lambda(\lambda, T_A, \theta_i, \phi_i) \, i_{\lambda,i}(\lambda, \theta_i, \phi_i) \cos\theta_i \, d\omega_i \, d\lambda \tag{3.38}$$

Equating Equations 3.37 and 3.38 and solving for the *hemispherical, total absorptivity,* there results

$$\alpha(T_A) = \frac{\int_{\lambda=0}^{\infty} \int_{2\pi} \alpha'_\lambda(\lambda, T_A, \theta_i, \phi_i) \, i_{\lambda,i}(\lambda, \theta_i, \phi_i) \cos\theta_i \, d\omega_i \, d\lambda}{\int_{\lambda=0}^{\infty} \int_{2\pi} i_{\lambda,i}(\lambda, \theta_i, \phi_i) \cos\theta_i \, d\omega_i \, d\lambda} \tag{3.39}$$

It should be pointed out here that we were not able to invoke the hemisphericalizing and totalizing operators introduced in Section 3.5 for the analysis in Sections 3.10, 3.11, and 3.12. This is because the radiation incident to surface element dA is generally not black (or diffuse, gray). However, for the special case of a surface being irradiated by a black source at temperature T_S we have

$$\alpha(T_A, T_S) = \frac{1}{\sigma T_S^4} \int_{\lambda=0}^{\infty} \int_{2\pi} \alpha'_\lambda(\lambda, T_A, \theta_i, \phi_i) \, i_{b\lambda}(\lambda, T_S) \cos\theta_i \, d\omega_i \, d\lambda$$

$$\text{(black source)} \tag{3.40}$$

or

$$\alpha(T_A, T_S) = T\{H[\alpha'_\lambda(\lambda, T_A, \theta_i, \phi_i)]\} \qquad \text{(black source)} \tag{3.41}$$

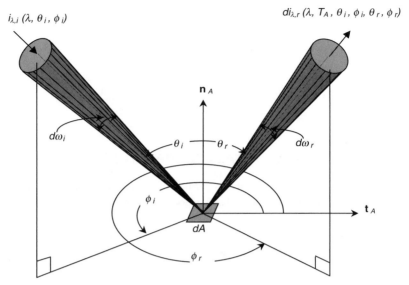

Figure 3.8 Illustration of bidirectional reflection

This result is significant because it establishes that the total absorptivity is not, strictly speaking, a surface property when the surface is irradiated by a black source because it also depends on the source temperature in this case. It can be shown that this is also true if the surface is irradiated by a diffuse, gray source.

3.13 BIDIRECTIONAL, SPECTRAL REFLECTIVITY

A spectral beam of intensity $i_{\lambda,i}(\lambda, \theta_i, \phi_i)$ is incident to area element dA from direction θ_i, ϕ_i, as shown in Figure 3.8. The flux (power per unit surface area) incident to dA due to this beam is $i_{\lambda,i}(\lambda, \theta_i, \phi_i) \cos\theta_i \, d\omega_i$. Let the intensity of the spectral beam reflected from dA in direction θ_r, ϕ_r due to this incident flux be $di_{\lambda,r}(\lambda, T_A, \theta_i, \phi_i, \theta_r, \phi_r)$. Then the *bidirectional, spectral reflectivity* is defined

$$\rho_\lambda''(\lambda, T_A, \theta_i, \phi_i, \theta_r, \phi_r) \equiv \frac{di_{\lambda,r}(\lambda, T_A, \theta_i, \phi_i, \theta_r, \phi_r)}{i_{\lambda,i}(\lambda, \theta_i, \phi_i) \cos\theta_i \, d\omega_i} \quad (3.42)$$

The bidirectional, spectral reflectivity is different from the emissivities and absorptivities we have defined to this point in that it is not dimensionless; rather, it has the dimensions of inverse solid angle (sr^{-1}). Also, note that while a finite intensity of radiation is incident to dA from direction θ_i, ϕ_i, the beam reflected in direction θ_r, ϕ_r is a differential quantity. This imbalance in size scale between the incident and reflected intensities accounts for the fact that, for most surfaces of practical engineering interest, only a very small fraction of the beam radiation incident from one given direction is reflected in some specified direction. While it is true in the case of a mirror that essentially the same intensity is reflected in the so-called specular ("mirrorlike") direction as is incident, the intensity of the reflected beam in any other direction will be negligibly small. Multiplication of the incident beam intensity by the infinitesimal solid angle in the denominator of Equation 3.42 assures that the bidirectional, spectral reflectivity is itself a finite quantity. The use of the double prime ($''$) symbol indicates that the quantity is bidirectional.

As the remaining reflectives are defined and expressed in terms of each other, it will become clear that the bidirectional, spectral reflectivity contains all information about the optical properties of a surface, including the emissivity and the absorptivity. Therefore, it is the most valuable of the surface properties.

3.14   RECIPROCITY FOR THE BIDIRECTIONAL, SPECTRAL REFLECTIVITY

Consider the isothermal enclosure shown in Figure 3.9. The black surface elements dS_1 and dS_2 are on the walls of the isothermal enclosure and subtend solid angles $d\omega_1$ and $d\omega_2$, respectively, at the nonblack area element dA.

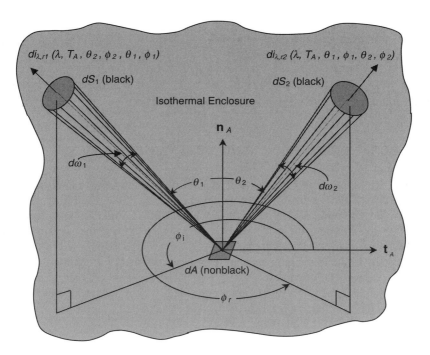

Figure 3.9   Illustration of reciprocity for bidirectional reflectivity

Now, the spectral power emitted by dS_1, reflected from dA, and absorbed by dS_2 must be equal to the spectral power in the same wavelength interval emitted by dS_2, reflected from dA, and absorbed by dS_1. If this were not so the radiation field within the isothermal enclosure would not be uniform and isotropic, in contradiction to the conclusion of Section 3.9. That is, if more radiation were passing in one direction along the path described by the two solid angles than in the other, the radiation field would not be isotropic. We then conclude that

$$di_{\lambda,r2}\, dA\, \cos\theta_2\, d\omega_2 = di_{\lambda,r1}\, dA\, \cos\theta_1\, d\omega_1 \tag{3.43}$$

But from the definition of the bidirectional, spectral reflectivity, Equation 3.42,

$$di_{\lambda,r2} = \rho_\lambda''(\lambda, T_A, \theta_1, \phi_1, \theta_2, \phi_2)\, i_{\lambda,1}(\lambda, \theta_1, \phi_1)\, \cos\theta_1\, d\omega_1 \tag{3.44}$$

and

$$di_{\lambda,r1} = \rho_\lambda''(\lambda, T_A, \theta_2, \phi_2, \theta_1, \phi_1)\, i_{\lambda,2}(\lambda, \theta_2, \phi_2)\, \cos\theta_2\, d\omega_2 \tag{3.45}$$

Furthermore, since surface elements dS_1 and dS_2 are both black,

$$i_{\lambda,1}(\lambda, \theta_1, \phi_1) = i_{\lambda,2}(\lambda, \theta_2, \phi_2) = i_{b\lambda}(\lambda, T) \tag{3.46}$$

Finally, combining Equations 3.43 through 3.46 yields

$$\rho_\lambda''(\lambda, T_A, \theta_1, \phi_1, \theta_2, \phi_2) = \rho_\lambda''(\lambda, T_A, \theta_2, \phi_2, \theta_1, \phi_1) \tag{3.47}$$

This result is referred to as *reciprocity* for the bidirectional, spectral reflectivity.

3.15   BDRF VERSUS BRDF; PRACTICAL CONSIDERATIONS

Readers familiar with the radiation heat transfer literature will have encountered the terms "BDRF" (bidirectional reflectivity function) and "BRDF" (bidirectional reflectivity distribution function). Another related term sometimes used by atmospheric scientists to describe reflection of sunlight when viewing the earth from space is "ADM" (angular distribution model). When the term BDRF is used in the literature it usually refers to the bidirectional reflectivity defined by Equation 3.42, while the similar term BRDF often refers to the ratio of the intensity reflected in a given direction to the reflected intensity of the surface suitably averaged over 2π-space. Then the *spectral bidirectional reflectivity distribution function* of a surface in wavelength interval $d\lambda$ about wavelength λ is

$$R_\lambda(\lambda, \theta_i, \phi_i, \theta_r, \phi_r) \equiv \frac{i_{\lambda,r}(\lambda, \theta_i, \phi_i, \theta_r, \phi_r)}{q_\lambda(\lambda, \theta_i, \phi_i)/\pi} \tag{3.48}$$

where

$$q_\lambda(\lambda, \theta_i, \phi_i) = \int_{\phi_r=0}^{2\pi} \int_{\theta_r=0}^{\pi/2} i_{\lambda,r}(\lambda, \theta_i, \phi_i, \theta_r, \phi_r) \cos\theta_r \sin\theta_r \, d\theta_r \, d\phi_r \quad (3.49)$$

is the flux reflected into the hemisphere above the surface in question due to a beam of radiation incident from direction θ_i, ϕ_i. Note that R_λ has a value of unity if the surface is a diffuse reflector.

Because of the fundamental importance of bidirectional reflectivity alluded to in Section 3.13 and established in Section 3.17, it is sometimes convenient to convert BRDF into bidirectional reflectivity. This brings up certain practical considerations. We begin by rewriting Equation 3.42 in a form that recognizes that, in order to be useful, the reflectivity must first be measured; that is,

$$\rho_\lambda''(\lambda, \theta_i, \phi_i, \theta_r, \phi_r) \equiv \frac{1}{i_{\lambda,i}(\lambda, \theta_i, \phi_i)\cos\theta_i} \lim_{\delta\omega \to 0} \left[\frac{\delta i_{\lambda,r}(\lambda, \theta_i, \phi_i, \theta_r, \phi_r)}{\delta\omega_i} \right] \quad (3.42a)$$

In Equation 3.42a, the limiting process is really over two solid angles, $\delta\omega_i$ and $\delta\omega_r$. In practice the incident and reflected solid angles would be defined by the geometry, as indicated in Figure 3.16 (see Problem P3.13). Thus, the accuracy of the measurement would depend on the experiment designer's ability to limit the aperture diameters, with smaller apertures generally leading to better accuracy, assuming that a sufficient level of illumination could be maintained. In view of these considerations, the practical version of Equation 3.42 would be

$$\rho_\lambda''(\lambda, \theta_i, \phi_i, \theta_r, \phi_r) \cong \frac{\bar{i}_{\lambda,r}(\lambda, \theta_i, \phi_i, \theta_r, \phi_r)}{i_{\lambda,i}(\lambda, \theta_i, \phi_i)\cos\theta_i \, \Delta\omega_i} \quad (3.42b)$$

where now $\bar{i}_{\lambda,r}$ represents the reflected intensity averaged over the finite reflected solid angle $\Delta\omega_r$.

In a similar manner Equation 3.48 can be written

$$R_\lambda(\lambda, \theta_i, \phi_i, \theta_r, \phi_r) \equiv \frac{\bar{i}_{\lambda,r}(\lambda, \theta_i, \phi_i, \theta_r, \phi_r)}{q_\lambda(\lambda, \theta_i, \phi_i)/\pi} \quad (3.48a)$$

Eliminating $\bar{i}_{\lambda,r}$ between Equations 3.42b and 3.48a and introducing Equation 3.49 yields

$$\pi \rho_\lambda'' = R_\lambda \frac{\displaystyle\int_{2\pi} i_{\lambda,r} \cos\theta_r \, d\omega_r}{i_{\lambda,i} \cos\theta_i \, \Delta\omega_i} \quad (3.50)$$

Finally, we note that the ratio

$$\frac{\displaystyle\int_{2\pi} i_{\lambda,r} \cos\theta_r \, d\omega_r}{i_{\lambda,i} \cos\theta_i \, \Delta\omega_i} = \frac{q_{\lambda,r}}{q_{\lambda,i}} \quad (3.51)$$

is the *directional-hemispherical, spectral reflectivity,* defined in Section 3.10. This quantity is somewhat easier to measure than either the BRDF or the bidirectional reflectivity, and so provides a convenient conversion factor between the two.

The reader is cautioned that the definitions of BDRF, BRDF, and ADM are not standard in the literature. Therefore, care should be exercised to verify how reflectivity is defined before using data purported to represent this quantity.

3.16 DIRECTIONAL–HEMISPHERICAL, SPECTRAL REFLECTIVITY

What fraction of the spectral power incident to dA in a beam of intensity $i_{\lambda,r}(\lambda, \theta_i, \phi_i)$ is reflected into the hemispherical space above dA? The situation implied by this question is illustrated in Figure 3.10, and the answer to the question is the *directional–hemispherical, spectral reflectivity,*

$$\rho'_\lambda(\lambda, T_A, \theta_i, \phi_i) = \frac{\int_{2\pi} i_{\lambda,r} \cos\theta_r \, d\omega_r}{i_{\lambda,i}(\lambda, \theta_i, \phi_i) \cos\theta_i \, d\omega_i} \qquad (3.52)$$

$$\rho'_\lambda(\lambda, T_A, \theta_i, \phi_i) =$$

$$\frac{\int_{2\pi_r} \rho''_\lambda(\lambda, T_A, \theta_i, \phi_i, \theta_r, \phi_r) \, i_{\lambda,i}(\lambda, \theta_i, \phi_i) \cos\theta_i \, d\omega_i \cos\theta_r \, d\omega_r}{i_{\lambda,i}(\lambda, \theta_i, \phi_i) \cos\theta_i \, d\omega_i} \qquad (3.53)$$

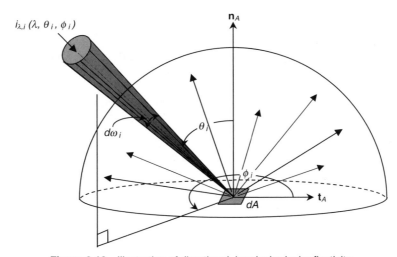

Figure 3.10 Illustration of directional–hemispherical reflectivity

or

$$\rho'_\lambda(\lambda, T_A, \theta_i, \phi_i) = \int_{2\pi_r} \rho''_\lambda(\lambda, T_A, \theta_i, \phi_i, \theta_r, \phi_r) \cos\theta_r \, d\omega_r \qquad (3.54)$$

The symbol $2\pi_r$ in the above expressions refers to integration over the reflected hemispherical space. From Equation 3.54 we now see why the bidirectional, spectral reflectivity cannot be dimensionless. If it were, then the directional–hemispherical, spectral reflectivity would have the dimensions of steradians.

3.17   RELATIONSHIP AMONG THE DIRECTIONAL, SPECTRAL EMISSIVITY; THE DIRECTIONAL, SPECTRAL ABSORPTIVITY; AND THE DIRECTIONAL–HEMISPHERICAL, SPECTRAL REFLECTIVITY

Recalling Kirchhoff's law,

$$\alpha'_\lambda(\lambda, T, \theta, \phi) = \varepsilon'_\lambda(\lambda, T, \theta, \phi) \qquad (3.30)$$

and recalling Equation 3.26 for an opaque medium, we can write

$$\rho'_\lambda(\lambda, \theta_i, \phi_i) = 1 - \alpha'_\lambda(\lambda, \theta_i, \phi_i) = 1 - \varepsilon'_\lambda(\lambda, \theta_e, \phi_e) \qquad (3.55)$$

Equation 3.55 illustrates the importance and generality of the bidirectional reflectivity. Indeed, if the bidirectional, spectral reflectivity is known for a surface, it can be processed into the directional–hemispherical, spectral reflectivity as described in Section 3.16, and then into the directional, spectral absorptivity and the directional, spectral emissivity. Other than as a tool to be used in this way, the bidirectional reflectivity has limited engineering importance. The statistical interpretation of the directional–hemispherical, spectral reflectivity is the probability that an energy bundle carrying energy in wavelength interval $\Delta\lambda$ incident from direction θ_i, ϕ_i will be reflected.

3.18   HEMISPHERICAL–DIRECTIONAL, SPECTRAL REFLECTIVITY

What fraction of the spectral power incident to dA from the hemispherical space above dA is reflected in a beam of intensity $i_{\lambda,r}(\lambda, T_A, \theta_r, \phi_r)$ in direction θ_r, ϕ_r? The situation implied by this question is illustrated in Figure 3.11. The answer to the question is the *hemispherical–directional, spectral reflectivity,* defined

$$\rho'_\lambda(\lambda, T_A, \theta_r, \phi_r) \equiv \frac{i_{\lambda,r}(\lambda, T_A, \theta_r, \phi_r)}{\dfrac{1}{\pi} \displaystyle\int_{2\pi_i} i_{\lambda,i}(\lambda, \theta_i, \phi_i) \cos\theta_i \, d\omega_i} \qquad (3.56)$$

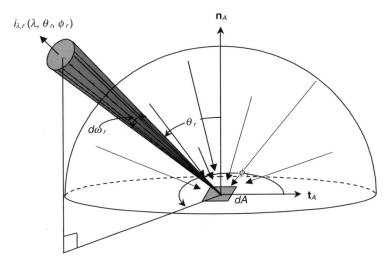

Figure 3.11   Illustration of hemispherical–directional reflectivity

The factor $1/\pi$ in the denominator of Equation 3.56 is needed for dimensional consistency since the integral is over solid angle. Thus, the denominator represents the incident intensity directionally averaged over 2π-space.

The numerator in Equation 3.56 can be replaced by noting that

$$i_{\lambda,r}(\lambda, T_A, \theta_r, \phi_r) = \int_{2\pi_i} di_{\lambda,r}(\lambda, T_A, \theta_i, \phi_i, \theta_r, \phi_r)$$

$$= \int_{2\pi_i} \rho''_{\lambda,r}(\lambda, T_A, \theta_i, \phi_i, \theta_r, \phi_r) \, i_{\lambda,i}(\lambda, T_A, \theta_i, \phi_i) \cos\theta_i \, d\omega_i \quad (3.57)$$

where the definition of the bidirectional, spectral reflectivity, Equation 3.42, has been invoked in the second equality. Then introducing Equation 3.57 into Equation 3.54, we obtain

$$\rho'_{\lambda}(\lambda, T_A, \theta_r, \phi_r) \equiv \frac{\int_{2\pi_i} \rho''_{\lambda}(\lambda, T_A, \theta_i, \phi_i, \theta_r, \phi_r) i_{\lambda,i}(\lambda, T_A, \theta_i, \phi_i) \cos\theta_i \, d\omega_i}{} \quad (3.58)$$

In the special case where the radiation incident to dA is isotropic, the incident intensity can be removed from the integrals in Equation 3.58 and divided out of the equation. The denominator then becomes unity and Equation 3.58 becomes

$$\rho'_{\lambda}(\lambda, T_A, \theta_r, \phi_r) \equiv \int_{2\pi_i} \rho''_{\lambda}(\lambda, T_A, \theta_i, \phi_i, \theta_r, \phi_r)$$

$$\times \cos\theta_i \, d\omega_i \quad \text{(isotropic radiation)} \quad (3.59)$$

PAGE 90
Equation 3.58 should read

$$\rho'_\lambda(\lambda, T_A, \theta_r, \phi_r) \equiv \frac{\int_{2\pi_i} \rho''_\lambda(\lambda, T_A, \theta_i, \phi_i, \theta_r, \phi_r) i_{\lambda,i}(\lambda, T_A, \theta_i, \phi_i) \cos\theta_i \, d\omega_i}{\frac{1}{\pi} \int_{2\pi_i} i_{\lambda,i}(\lambda, \theta_i, \phi_i) \cos\theta_i \, d\omega_i}$$

PAGE 91
Equation 3.61 should read

$$\rho_\lambda(\lambda, T_A) \equiv \frac{\int_{2\pi} i_{\lambda,r}(\lambda, T_A, \theta_r, \phi_r) \cos\theta_r \, d\omega_r}{\int_{2\pi} i_{\lambda,i}(\lambda, \theta_i, \phi_i) \cos\theta_i \, d\omega_i}$$

PAGE 92
Equation 3.62 should read
$$\rho_\lambda(\lambda, T_A) \equiv$$

$$\frac{\int_{2\pi}\int_{2\pi} \rho''_\lambda(\lambda, T_A, \theta_i, \phi_i, \theta_r, \phi_r) \, i_{\lambda,i}(\lambda, T_A, \theta_i, \phi_i) \cos\theta_i \, d\omega_i \cos\theta_r \, d\omega_r}{\int_{2\pi} i_{\lambda,i}(\lambda, \theta_i, \phi_i) \cos\theta_i \, d\omega_i}$$

Equation 3.63 should read

$$\rho_\lambda(\lambda, T_A) \equiv \frac{\int_{2\pi} \rho'_\lambda(\lambda, T_A, \theta_i, \phi_i) \, i_{\lambda,i}(\lambda, T_A, \theta_i, \phi_i) \cos\theta_i \, d\omega_i}{\int_{2\pi} i_{\lambda,i}(\lambda, \theta_i, \phi_i) \cos\theta_i \, d\omega_i}$$

Equation 3.66 should read

$$\rho'(T_A, \theta_i, \phi_i) = \frac{\int_{\lambda=0}^{\infty} \rho'_\lambda(\lambda, T_A, \theta_i, \phi_i) \, i_{\lambda,i}(\lambda, \theta_i, \phi_i) \, d\lambda}{\int_{\lambda=0}^{\infty} i_{\lambda,i}(\lambda, \theta_i, \phi_i) \, d\lambda}$$

PAGE 94
Equation 3.70 should read

$$\tau(\ell) = \frac{\int_0^{\infty} i_{\lambda,i}(\lambda, \ell) \, \tau_\lambda(\lambda, \ell) \, d\lambda}{\int_0^{\infty} i_{\lambda,i}(\lambda, \ell) \, d\lambda}$$

Equation 3.71 should read

$$\tau(\ell) = \frac{\pi \int_0^{\infty} i_{b\lambda}(\lambda, T) \, \tau_\lambda(\lambda, \ell) \, d\lambda}{\sigma T^4} \qquad \text{(gray radiation)}$$

3.19   RECIPROCITY BETWEEN THE DIRECTIONAL–HEMISPHERICAL, SPECTRAL REFLECTIVITY AND THE HEMISPHERICAL–DIRECTIONAL, SPECTRAL REFLECTIVITY

For the special case of isotropic irradiation of dA, the right-hand sides of Equations 3.54 and 3.59 are equal, in which case

$$\rho_\lambda'(\lambda, T_A, \theta_r, \phi_r) \equiv \rho_\lambda'(\lambda, T_A, \theta_i, \phi_i) \qquad \text{(isotropic radiation)} \qquad (3.60)$$

Equation 3.60 is a statement of reciprocity between the directional–hemispherical, spectral reflectivity and the hemispherical–directional, spectral reflectivity for the case of isotropic irradiation.

3.20   (BI)HEMISPHERICAL, SPECTRAL REFLECTIVITY

What fraction of the spectral power incident to dA from the hemispherical space above dA is reflected back into the hemispherical space above dA? The situation described is illustrated in Figure 3.12. The answer to the question posed is the *(bi)hemispherical, spectral reflectivity,*

$$\rho_\lambda(\lambda, T_A) \equiv \frac{\displaystyle\int_{2\pi} i_{\lambda,r}(\lambda, T_A, \theta_r, \phi_r) \cos\theta_r \, d\omega_r}{} \qquad (3.61)$$

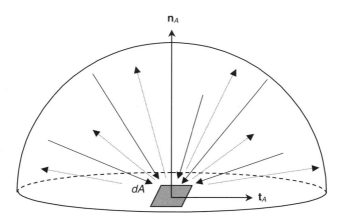

Figure 3.12   Illustration of (bi)hemispherical reflectivity

Following a process similar to that leading to Equation 3.58, Equation 3.61 may be written

$$\rho_\lambda(\lambda, T_A) \equiv$$

$$\frac{\int_{2\pi}\int_{2\pi} \rho_\lambda''(\lambda, T_A, \theta_i, \phi_i, \theta_r, \phi_r)\, i_{\lambda,i}(\lambda, T_A, \theta_i, \phi_i)\cos\theta_i\, d\omega_i\, \cos\theta_r\, d\omega_r}{} \qquad (3.62)$$

or

$$\rho_\lambda(\lambda, T_A) \equiv \frac{\int_{2\pi} \rho_\lambda'(\lambda, T_A, \theta_i, \phi_i)\, i_{\lambda,i}(\lambda, T_A, \theta_i, \phi_i)\cos\theta_i\, d\omega_i}{} \qquad (3.63)$$

3.21 TOTAL REFLECTIVITY

Any spectral reflectivity can be "totalized" in the sense illustrated by Equation 3.36 for the directional, total absorptivity. Note once again that this totalizing process is different from that defined by the more restrictive totalizing operator of Section 3.3. Recall that this latter applies, strictly speaking, only in the case of irradiation of dA by blackbody radiation. The totalizing process for reflectivities is illustrated here by considering the *directional–hemispherical, total reflectivity*. We begin by asking, what is the total power reflected into the hemispherical space above area element dA due to the incident radiation contained in the beam of total intensity $i_i(\theta_i, \phi_i)$? The two independent but equivalent responses are

$$dP(T_A, \theta_i, \phi_i) = \rho'(T_A, \theta_i, \phi_i)\cos\theta_i\, d\omega_i \int_{\lambda=0}^{\infty} i_{\lambda,i}(\lambda, \theta_i, \phi_i)\, d\lambda \qquad (3.64)$$

and

$$dP(T_A, \theta_i, \phi_i) = \cos\theta_i\, d\omega_i \int_{\lambda=0}^{\infty} \rho_\lambda'(\lambda, T_A, \theta_i, \phi_i)\, i_{\lambda,i}(\lambda, \theta_i, \phi_i)\, d\lambda \qquad (3.65)$$

Equating the two expressions for $P(T_A, \theta_i, \phi_i)$ and solving for the directional–hemispherical, total reflectivity yields

$$\rho'(T_A, \theta_i, \phi_i) = \frac{\int_{\lambda=0}^{\infty} \rho_\lambda'(\lambda, T_A, \theta_i, \phi_i)\, i_{\lambda,i}(\lambda, \theta_i, \phi_i)\, d\lambda}{} \qquad (3.66)$$

Of course, the remaining spectral reflectivities can be totalized by substituting them into Equation 3.66 in place of the directional–hemispherical, spectral reflectivity.

3.22 PARTICIPATING MEDIA AND TRANSMISSIVITY

If a medium is not opaque, at least some of the radiation incident to its bounding surfaces will penetrate into the interior. Such a medium is referred to as a *participating medium* if the radiation passing through it is absorbed or scattered within the medium, and the medium itself emits radiation from its volume. Familiar examples are the atmosphere, jet engine exhaust plumes, clouds, lenses, filters, and bodies of water. Discussion of radiation through gases ("gaseous radiation") is deferred to Chapter 6.

In this book use of the term *transmissivity* is limited to plane-parallel, solid participating media, such as window glazing and filters. The term "plane-parallel" means that the bounding surfaces of the medium are both plane and parallel, and that the lateral extent of the medium is large compared to its thickness. Snell's law of refraction, treated elsewhere in this book, predicts that a beam intersecting a bounding surface of a participating medium will undergo a predictable change of direction. However, a beam passing through a plane-parallel medium suffers no net change of direction. Treatment of the special case of refractive optical elements, such as lenses and prisms, whose geometry is conceived to change the direction of an incident beam, is also deferred to Chapter 6 and, more particularly, to Part III of the book. Under the limitation of plane-parallel, solid participating media the term "directional transmissivity" is redundant.

3.23 SPECTRAL TRANSMISSIVITY

The ability of a solid participating medium, such as window glass or a filter, to transmit radiation is often a strong function of wavelength. Indeed, this is the essential property of a filter. The *spectral transmissivity* is defined as the ratio of the transmitted spectral intensity to the incident spectral intensity in the same wavelength interval $d\lambda$ about wavelength λ; that is,

$$\tau_\lambda\,(\lambda,\,\ell) \equiv \frac{i_{\lambda,t}(\lambda)}{i_{\lambda,i}(\lambda)} \tag{3.67}$$

In Equation 3.67 the subscripts t and i refer to the transmitted and incident intensities, and the symbol ℓ is the physical thickness of the plane layer of participating medium.

3.24 TOTAL TRANSMISSIVITY

It is often of interest to know the total intensity that would be transmitted by a solid, plane-parallel participating medium, such as window glazing or a filter, due to radiation incident with a specified wavelength distribution. This quantity can be computed two different ways. First, Equation 3.67 could be used directly, yielding

$$i_t(\ell) = \int_0^\infty i_{\lambda,t}(\lambda, \ell) \, d\lambda = \int_0^\infty i_{\lambda,i}(\lambda, \ell)\tau_\lambda(\lambda, \ell) \, d\lambda \tag{3.68}$$

Alternatively, we could write

$$i_t(\ell) = \tau(\ell) \int_0^\infty i_{\lambda,i}(\lambda, \ell) \, d\lambda \tag{3.69}$$

where the quantity $\tau(\ell)$ is the *total transmissivity*. Then, equating Equations 3.68 and 3.69 and solving for the total transmissivity, we have

$$\tau(\ell) = \frac{\int_0^\infty i_{\lambda,i}(\lambda, \ell) \, \tau_\lambda(\lambda, \ell) \, d\lambda}{} \tag{3.70}$$

It is important to realize here that the value of the total transmissivity depends on the wavelength distribution of the incident radiation. For the important special case of radiation incident from a graybody at temperature T, Equation 3.70 becomes

$$\tau(\ell) = \frac{\pi \int_0^\infty i_{b\lambda}(\lambda, T) \, \tau_\lambda(\lambda, \ell) \, d\lambda}{} \qquad \text{(gray radiation)} \tag{3.71}$$

Team Project

TP 3.1   In this project we will use the rudimentary radiometer we built in TP 2.1 (or any available radiometer) to observe a variety of sources through filters and reflected from various surfaces. Our goal is to gain understanding of the thermal radiation behavior of real surfaces and of various common filters. Obtain a rectangular one-gallon metal can such as those commonly containing commercial cleaning fluids or solvents (be sure to properly dispose of any remaining contents). Select a can having a highly polished exterior surface. Using masking tape and spray paint, paint two horizontal bands around the can, each taking up about one-third of the can's vertical surface, as shown in Figure 3.13. One of the bands should be painted black and the other white. Leave the third band unpainted. Modify the lid as shown in the figure so that tap water can be circulated through the can (be sure that the inlet tube extends to within an inch or two of the bottom). A photograph of the finished product appears in Figure 3.14.

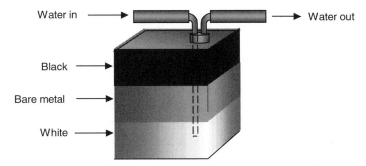

Figure 3.13 A one-gallon can modified as a radiant heat source

With hot water flowing through the can, use the radiometer to observe each of the three bands successively (this should be done in a cool, dimly lit room). Arrange the three readings in descending order. Now take the can out into full sunlight and let cold water flow through it. Use the radiometer to observe sunlight reflected from the three surfaces, and once again arrange the three readings in descending order. Based on these observations discuss paint schemes you have observed on commercial airplanes and elsewhere. If you have ever visited the Caribbean you may have noticed that most automobiles are either painted white or a light pastel color. Comment on this observation. Finally, if you laid three metal plates out in the sun; one polished, one painted black, and one painted white; which would get the hottest? Which would be the coolest?

Now repeat the above set of experiments except this time hold a piece of common window glass over the entrance aperture of the radiometer (being careful not to block the direct sunlight incident to the can with the glass during the experiment conducted outside). How did the results change? What does this say about the spectral transmissivity of window glass?

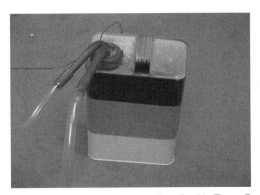

Figure 3.14 Photograph of the apparatus described in Team Project TP 3.1

Discussion Points

DP 3.1   How would you design an experiment to measure the directional, total emissivity of sheet aluminum at temperatures in the range from 200 to 400 K? What components would you need? Do you think there might be an interaction between the size of the test specimen and the maximum angle with respect to the normal for which you can obtain data? Discuss this point. Provide a labeled sketch of your conceptual design.

DP 3.2   A surface may be considered "gray" in one or more limited wavelength intervals, such as in the visible, but nongray in other wavelength intervals. Also, a surface may be considered "diffuse" in some limited solid angle range, such as within a few degrees of normal, but directional in others. Similarly, a surface may be considered "gray" in some limited solid angle range and nongray at other angles, while a surface may be considered "diffuse" in some limited wavelength interval and directional at other wavelengths. Make sketches (similar, respectively, to Figures 3.1 and 3.2) of these first two cases.

DP 3.3   Climatologists have proposed various disaster scenarios involving positive feedback. For example, it has been speculated that if the arctic oceans cool below a certain critical temperature for some reason, a thermal runaway will occur resulting in a complete freeze-over of the oceans. Can you reproduce their reasoning?

DP 3.4   It has been speculated that the Greenland ice sheet could be melted by spreading a thin layer of coal dust over some critical fraction of its surface. Presumably more solar radiation would be absorbed, resulting in surface heating and local melting of the ice. The idea is that the regional air temperature would increase, thereby provoking even more melting, with the effect that the region would grow in extent until all of the ice melted. Critically discuss the idea.

DP 3.5   Both Discussion Points DP 3.3 and DP 3.4 supposedly involve instabilities provoked by positive feedback. However, in one case the feedback cycle may be easily broken, whereas in the other it may not be. Which is which?

DP 3.6   In Example Problem 3.1 why do you think the denominator of Equation 3.8b is approximated rather than simply replacing it with its known exact value π?

DP 3.7   In Example Problem 3.2 why do you think the denominator of Equation 3.13b is approximated as a finite series rather than just replaced with its known exact value (see Equation 3.13) of $\sigma T^4/\pi$?

DP 3.8   Equation 3.21 seems to imply that the totalizing and hemisphericalizing operators are associative, at least for the case of the hemisperical, total emissivity. Work out an algebraic proof of this result. [*Hint:* Do you remember Fubini's theorem from your study of mathematics?]

DP 3.9   Kirchhoff's law, Equation 3.30, represents a special case (directional, spectral) where the emissivity is equal to the absorptivity. Can you think of other cases where an emissivity is equal to an absorptivity?

DP 3.10   Many of the data available for surface properties are for the special case of normal viewing and/or normal illumination with beam radiation. Why do you think this is so?

Problems

P 3.1   Derive Equation 3.8a starting with Equation 3.8 [*Hint:* see Discussion Point DP 3.6].

P 3.2   Show that the hemispherical, spectral emissivity is equal to the directional, spectral emissivity for the special case of a diffuse surface; i.e., verify Equation 3.9.

P 3.3   Derive Equation 3.13b.

P 3.4   Show that the directional, total emissivity is equal to the directional, spectral emissivity for the special case of a gray surface; i.e., verify Equation 3.13.

P 3.5   Show that the hemispherical, total emissivity is equal to the directional, spectral emissivity for the special case of a diffuse, gray surface.

P 3.6   Estimate the hemispherical, total emissivity corresponding to a surface temperature of 325 K for a surface whose components of the directional, spectral emissivity at that temperature are represented in Figure 3.15. Assume that the components of the directional, spectral emissivity shown in Figure

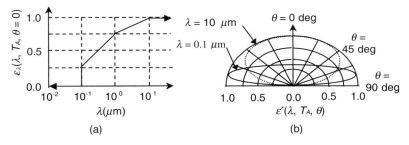

Figure 3.15   Components of the directional, spectral emissivity of a hypothetical surface representing (a) the variation with wavelength when measured normal to the surface and (b) the variation with zenith angle at 0.1 and 10 μm

3.15 are multiplied together to obtain the directional, spectral emissivity; that is, $\varepsilon'_\lambda = \varepsilon_\lambda(\lambda,T_A,\theta = 0)\, \varepsilon'(\lambda,T_A,\theta)$. Clearly state all other assumptions. (*Note:* This problem is suitable for solution by mature students only.)

P 3.7   The directional, spectral emissivity of a certain hypothetical surface is given by

$$\varepsilon'_\lambda (\lambda, T_A, \theta) = \Lambda(\lambda)\, T(T_A)\, \Theta(\theta) \tag{3.72}$$

where

$$\Lambda(\lambda) = \begin{cases} 0.0, & \lambda \le 0.05 \ \mu m \\ 0.5\lambda, & 0.05 < \lambda \le 0.05 \ \mu m \\ 0.25 + (\lambda - 0.5)\, 0.01\lambda, & 0.5 \ < \lambda \le 5.5 \ \mu m \\ 0.525, & \lambda > 5.5 \ \mu m \end{cases} \tag{3.73}$$

$$T(T_A) = \frac{300 + 0.04T_A}{340}, \qquad 300 < T_A < 1000 \ K \tag{3.74}$$

and

$$\Theta(\theta) = \begin{cases} 0.9 \cos \theta, & 0 \le \theta < 45 \ deg \\ 0.6364e^{-\theta/9.0}, & 45 \ deg \le \theta \le 90 \ deg \end{cases} \tag{3.75}$$

Enter the functions Λ, T, and Θ in a spreadsheet and then plot each of them over their range of definition. Also, plot the directional, spectral emissivity as a function of wavelength over the range $0 \le \lambda \le 6.0 \ \mu m$ for several values of zenith angle and surface temperature. Then compute the hemispherical, total emissivity of this surface at a temperature of 800 K. Would the hemispherical, total emissivity of this surface increase or decrease if the surface temperature was increased in the vicinity of 800 K? Defend your answer.

P 3.8   Show that the hemispherical, spectral absorptivity is equal to the hemispherical, spectral emissivity for the special case of a diffuse surface.

P 3.9   For what special circumstances can we write

$$\alpha_\lambda(\lambda,T_A) = \frac{1}{\pi} \int_{2\pi} \alpha'_\lambda(\lambda, T_A, \theta_i, \phi_i) \cos\theta_i \, d\omega_i \tag{3.76}$$

P 3.10   Show that the directional, total absorptivity is equal to the directional, total emissivity for the special case of a gray surface.

P 3.11 Show that for a diffuse, gray surface the hemispherical, total absorptivity is equal to the hemispherical, total emissivity.

P 3.12 The directional, spectral emissivity of a certain diffuse surface at $T_A = 300$ K may be represented by Equation 3.72 with $\Theta(\theta) = T(T_A) = 1.0$, and with $\Lambda(\lambda)$ as given in Problem P 3.7. Consider the case where the surface is irradiated by graybody radiation. For what range of source temperatures T_S will the values of the hemispherical, total absorptivity and the hemispherical, total emissivity be within 10 percent of each other? Plot $\alpha(T_A, T_S)$ and $\varepsilon(T_A)$ versus T_S for a range of T_S that includes these limits. Clearly indicate these limits on the graph.

P 3.13 Consider the schematic representation of a bidirectional reflectometer shown in Figure 3.16. A collimated infrared ($\lambda = 3.28 \ \mu$m) light source and the detector are constrained to move within the same hemispherical envelope centered above the sample surface whose bidirectional reflectivity is to be measured. The apertures of the light source and the detector are both 10 mm in diameter, and both are $R = 300$ mm from the sample. The power of the light source is maintained constant in a small wavelength interval about 3.28 μm during the measurement process. The entire apparatus is contained within a cold, blackened chamber to minimize the effects of stray radiation. The surface sample is sufficiently large to ensure that the spot of light on the surface is not larger than the surface for illumination zenith angles up to 80 deg. Also, the field-of-view of the detector completely encompasses the spot of light on the sample at this angle. The apparatus is conceived so that the detector and the light source can be positioned and swiveled to allow the detector to view the light source directly on-axis at a distance of $R = 300$ mm. Assuming that the detector output voltage is directly proportional to the radiative power incident to its aperture, show that the ratio of the detector output voltage V when viewing the sample to the detector output voltage V_0 when directly viewing the source is

$$\frac{V}{V_0} = \frac{\rho_\lambda'' \cos\theta_r A_S}{R^2} \tag{3.77}$$

In Equation 3.77, θ_r is the reflected zenith angle and A_S is the source aperture area.

P 3.14 Bidirectional reflectivity data for gray duct tape at a wavelength of 3.28 μm are provided on the compact disc packaged with the book. Assuming these data were obtained from a bidirectional reflectometer similar to the one shown in Figure 3.16 and described in Problem P 3.13, use them to compute $\rho_\lambda''(\lambda = 3.28 \ \mu$m, $\theta_i = 20$ deg, $\phi_i = 0$ deg, $\theta_r = 30$ deg, $\phi_r = 180$ deg) and $\rho_\lambda'(\lambda = 3.28 \ \mu$m, $\theta_i = 20$ deg, $\phi_i = 0$ deg).

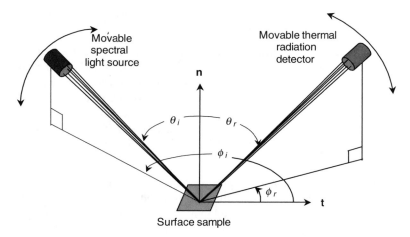

Figure 3.16   Apparatus for measuring the bidirectional, spectral reflectivity

P 3.15   Show that for a diffuse reflector

$$\rho'_\lambda(\lambda, T_A, \theta_i, \phi_i) = \pi \, \rho''_\lambda(\lambda, T_A, \theta_i, \phi_i) \qquad (3.78)$$

P 3.16   Show that for diffuse irradiation of a surface element dA

$$\rho'_\lambda(\lambda, T_A, \theta_r, \phi_r) = \frac{i'_{\lambda,r}(\lambda, T_A, \theta_r, \phi_r)}{i'_{\lambda,i}(\lambda, T_A)} \qquad (3.79)$$

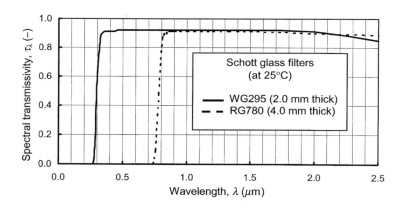

Figure 3.17   Spectral transmissivity of two Schott glass filters at 25°C

Table 3.3 Spectral transmissivity of several Schott glass filters at 25°C (all are 2.0 mm thick except RG780, which is 4.0 mm thick)

λ(μm)	WG295	GG395	GG495	RG610	RG630	RG695	RG715	RG780
0.27	0							
0.28	0.05							
0.29	0.29							
0.3	0.61							
0.32	0.86							
0.34	0.9	0						
0.36	0.91	0.01						
0.38	0.91	0.31						
0.4	0.91	0.68						
0.42	0.91	0.82						
0.44	0.91	0.87	0					
0.46	0.92	0.88	0.01					
0.48	0.92	0.88	0.27					
0.5	0.92	0.89	0.8					
0.52	0.92	0.9	0.88					
0.54	0.92	0.9	0.9					
0.56	0.92	0.9	0.9					
0.58	0.92	0.9	0.91	0				
0.6	0.92	0.9	0.91	0.33	0			
0.62	0.92	0.9	0.91	0.83	0.26			
0.64	0.92	0.9	0.91	0.89	0.85			
0.66	0.92	0.9	0.91	0.9	0.91	0	0	
0.68	0.92	0.9	0.91	0.91	0.91	0.27	0.01	
0.7	0.92	0.9	0.91	0.91	0.91	0.84	0.25	
0.72	0.92	0.9	0.91	0.91	0.91	0.91	0.74	
0.74	0.92	0.9	0.91	0.91	0.92	0.92	0.89	0
0.76	0.92	0.9	0.91	0.91	0.92	0.92	0.91	0.1
0.78	0.92	0.9	0.91	0.91	0.92	0.92	0.92	0.48
0.8	0.92	0.9	0.91	0.91	0.92	0.92	0.92	0.81
0.82	0.92	0.9	0.91	0.91	0.92	0.92	0.92	0.89
0.84	0.92	0.9	0.91	0.91	0.92	0.92	0.92	0.9
0.86	0.92	0.9	0.91	0.91	0.92	0.92	0.92	0.91
0.88	0.92	0.9	0.91	0.91	0.92	0.92	0.92	0.91
0.9	0.92	0.9	0.91	0.91	0.92	0.92	0.92	0.91
1	0.92	0.9	0.91	0.91	0.92	0.92	0.92	0.91
1.2	0.92	0.9	0.91	0.91	0.92	0.92	0.92	0.91
1.5	0.92	0.9	0.91	0.91	0.92	0.92	0.92	0.91

Source: Used by permission of Schott Glass Technologies, Inc.

P 3.17 Under what special set of circumstances can we write $\rho = 1 - \varepsilon$?

P 3.18 Consider the two Schott glass filters whose spectral transmissivities are plotted in Figure 3.17 and tabulated in Table 3.3. Use a spreadsheet approach to estimate the fraction of incident solar radiation that passes through each filter. What is the total transmissivity in each case? Assume that the sun is a blackbody at 5780 K and neglect solar radiation whose wavelength is beyond the range of the transmissivity data.

4

RADIATION BEHAVIOR
OF SURFACES

In Chapter 3 we defined the four "surface" radiation properties: emissivity, absorptivity, reflectivity, and transmissivity. We then established relationships among them and among their directional, hemispherical, spectral, and total forms. In this chapter we explore the behavior of these properties on the basis of available theory and on the basis of measurements. The goals of this chapter are to provide a solid basis for beginning the study of surface radiation properties and for making informed radiation heat transfer modeling decisions.

4.1 INTRODUCTION TO THE RADIATION BEHAVIOR OF SURFACES

The magnitude, angular distribution, and wavelength dependence of the surface properties defined in Chapter 3 are very sensitive to the physical and chemical state of the surface.* The primary surface conditions that influence the behavior of a solid surface are its bulk electrical properties (electrical conductor or electrical nonconductor), its roughness (smooth, polished, sanded, sand blasted, turned, lapped, honed, ground, peened, etc.), its chemical condition (reduced, oxidized, anodized, galvanized, etc.), its degree of contamination (clean or dirty, dry or oily, etc.), and the surface grain structure (annealed, cold rolled, hot rolled, etc.). Further, in applications for which predictions of future radiation behavior are required, it may be necessary to account for aging. Unfortunately, the subjective adjectives normally used to de-

---

* The concept of a surface is useful but somewhat artificial. In fact, radiation is emitted, absorbed, and scattered within a relatively thin layer of atoms, molecules, fibers, or particles ranging from several wavelengths to several hundred wavelengths thick. In this regard, please see the first footnote in Chapter 6.

103

scribe the mechanical and chemical state of a surface are open to a broad range of interpretation. For this reason property measurements of the particular surface or surfaces of immediate interest are recommended when a highly precise radiation heat transfer analysis is contemplated.

Modern statistically based radiation heat transfer analytical tools, which are the subject of Part III of this book, are fully capable of treating problems dealing with (bi)directional and spectral surface properties. Unfortunately, the detailed knowledge of the directional and spectral behavior of surfaces of engineering interest currently lags behind our ability to use such information. The formulation of improved general models for the directional and spectral behavior of engineered surfaces, and the experimental determination of the parameters within these models, remain at the forefront of radiation heat transfer research. The goal of this chapter is to provide an intelligent basis for beginning the study of surface radiation properties.

4.2   SOLUTION TO MAXWELL'S EQUATIONS FOR AN ELECTRICALLY CONDUCTING MEDIUM (r_e FINITE)

Maxwell's wave equations, introduced in Chapter 1, are the point of departure for understanding surface radiation properties. The wave description of thermal radiation was introduced in Section 1.7, and a free-space solution to Maxwell's wave equations for electrical nonconductors was considered in Section 1.8. The reader is encouraged to review those two sections before continuing.

We now consider the case where the propagating medium is a homogeneous, isotropic electrical conductor. In this case the term $\mathbf{E}/r_e$ in Maxwell's first equation, Equation 1.16a, representing a *current density*,

$$\mathbf{J} = \frac{\mathbf{E}}{r_e} \quad (\text{A/m}^2) \tag{4.1}$$

is not negligible. When electrical current flows through a resistance, electrical energy is converted irreversibly into sensible heat according to

$$\dot{q} = \mathbf{J} \cdot \mathbf{E} = \frac{\mathbf{E}}{r_e} \cdot \mathbf{E} = \frac{|E|^2}{r_e} \quad (\text{W/m}^3) \tag{4.2}$$

Thus, as an electromagnetic wave propagates through a conducting medium it loses power; that is, some of its power is left behind as sensible heat in the electrically conducting medium through which it passes. This is the macroscopic view of the mechanism of absorption of radiation within a body.

In the example of a right-propagating, *transverse-magnetic-polarized** electromagnetic wave, absorption of radiation within a material substance may be

---

* The magnetic field vector for a transverse-magnetic-polarized electromagnetic wave is oriented in a plane normal to the page and containing the direction of propagation.

modeled by multiplying Equation 1.28 by an exponential attenuation factor; that is,

$$E_y = [|E_y|e^{i\omega(nx/c_0 - t)}]e^{-\omega\kappa x/c_0} \tag{4.3}$$

where the subscript "m" in Equation 1.28 has been suppressed for notational simplicity. In Equation 4.3, κ is the *extinction coefficient* in the medium. The extinction of an electromagnetic wave is represented in Figure 4.1, which clearly demonstrates that absorption is not a "surface" phenomenon. Equation 4.3 can be rearranged as

$$E_y = |E_y|e^{i\omega[(n + i\kappa)x/c_0 - t]} \tag{4.4}$$

where $m = n + i\kappa$ is the *complex index of refraction.* Note that if the order of the terms $x\sqrt{\mu\varepsilon}$ and t had been reversed in Equation 1.20, nothing would have changed in the subsequent analysis except that now the complex index of refraction would be defined with a minus $(-)$ sign rather than a plus $(+)$ sign before the imaginary part. The reader is cautioned that some authorities, in fact, define the complex index of refraction this way.

It can be demonstrated by direct substitution that Equation 4.4 satisfies Maxwell's equations with r_e finite, Equations 1.16a and 1.16b, if

$$c_0^2\mu\varepsilon = (n + i\kappa)^2 - \frac{i\mu\lambda_0 c_0}{2\pi r_e} \tag{4.5}$$

When the real and imaginary parts of Equation 4.5 are equated there results

$$n^2 - \kappa^2 = \mu\varepsilon c_0^2 \quad \text{and} \quad n\kappa = \frac{\mu\lambda_0 c_0}{4\pi r_e} \tag{4.6}$$

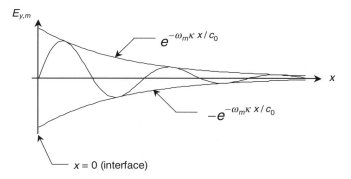

Figure 4.1 Extinction of a *y*-polarized electromagnetic wave propagating in the positive-*x* direction through an electrical conductor

Finally, Equations 4.6 can be solved for n^2 and κ^2 to obtain

$$n^2 = \frac{\mu\varepsilon c_0^2}{2}\left\{1 + \left[1 + \left(\frac{\lambda_0}{2\pi c_0 r_e \varepsilon}\right)^2\right]^{1/2}\right\} \tag{4.7}$$

and

$$\kappa^2 = \frac{\mu\varepsilon c_0^2}{2}\left\{-1 + \left[1 + \left(\frac{\lambda_0}{2\pi c_0 r_e \varepsilon}\right)^2\right]^{1/2}\right\} \tag{4.8}$$

where $\lambda_0 = c_0/\nu$ is the wavelength in a vacuum of an arbitrary harmonic of the wave described by Equation 1.25. In deriving Equations 4.7 and 4.8 the permeability μ and the permittivity ε have been treated as real quantities. In fact, both are generally complex, although both are real for ideal dielectrics.

4.3 REFLECTION FROM AN IDEAL DIELECTRIC SURFACE

To this point we have applied Maxwell's equations to the propagation of electromagnetic radiation through a homogeneous medium. Of equal interest is what occurs at an interface between two media having different indices of refraction. Consider the ideal surface, shown in Figure 4.2, consisting of a plane interface lying in the y, z plane between two electrically nonconducting (i.e., dielectric) media having different real indices of refraction, n. That is, $r_e \rightarrow \infty$ so that $\kappa = 0$. In Figure 4.2, $E_{p,i}$ is the electric component of a transverse–magnetic–(TM-)polarized, periodic electromagnetic wave incident to the interface. Following convention, the subscript p is

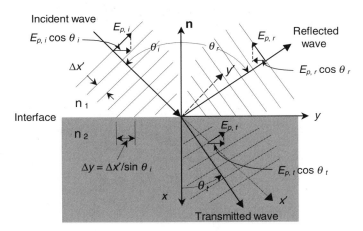

Figure 4.2 A transverse-magnetic (TM) p-polarized wave incident to an ideal surface formed by the interface between two nonconducting media having different indices of refraction

used to remind us that the electric field vector for a TM-polarized wave is in a plane parallel to the *x, y* plane, called the *plane of incidence.* Without loss of generality we consider only the real part of the incident electric field,

$$\text{Re}[E_{p,i}] = |E_{p,i}| \cos\left[\omega\left(\frac{n_1 x'}{c_0} - t\right)\right] \tag{4.9}$$

Since $\phi_i = \tan^{-1}[\text{Im}(E_{p,i})/\text{Re}(E_{p,i})]$, this is equivalent to saying that the arbitrary phase angle ϕ_i of the incident electric field is zero.

Now, the *y* component (parallel to the interface) of the electric field must sum to zero at the interface [cf. Reference 1, p. 318], or

$$|E_{p,i}| \cos\theta_i \cos\left[\omega\left(\frac{n_1\, y\, \sin\theta_i}{c_0} - t\right)\right]$$
$$-|E_{p,r}| \cos\theta_r \cos\left[\omega\left(\frac{n_1\, y\, \sin\theta_r}{c_0} - t\right)\right] \tag{4.10}$$
$$= |E_{p,t}| \cos\theta_t \cos\left[\omega\left(\frac{n_2\, y\, \sin\theta_t}{c_0} - t\right)\right] \qquad (x = 0)$$

In writing Equation 4.10 we have recognized that the reflected and transmitted components of the electric field are also real at the interface since the index of refraction is real. That is, neither is shifted in phase by virtue of being reflected or transmitted. Now *t* and *y* are independent variables and both are independent of θ_i, θ_r, and θ_t. Therefore, for Equation 4.10 to be generally true the factors of the form cos[·] in the equation must all be equal. But this can be true only if their arguments are equal; that is, if

$$n_1 \sin\theta_i = n_i \sin\theta_r = n_2 \sin\theta_t \tag{4.11}$$

Equation 4.11 leads directly to the *law of specular reflection,*

$$\theta_r = \theta_i \qquad \text{(specular reflection)} \tag{4.12}$$

and *Snell's law,**

$$\frac{\sin\theta_i}{\sin\theta_t} = \frac{n_2}{n_1} \tag{4.13}$$

---

* Some sources refer to this result as the Snell–Descartes law. The Dutch mathematician Willebrord Snell (1591–1626) discovered the relationship experimentally in 1621 but did not publish his discovery. The French philosopher René Descartes (1596–1650) first published the relationship in his *Dioptrique* in 1635 without citing Snell as the source. It is not clear whether Descartes discovered the relationship independently or, as some believe, possessed a copy of Snell's unpublished manuscript [2].

Also, since the factors of the form cos[·] in Equation 4.10 are all equal, Equation 4.10 can be rewritten

$$|E_{p,i}| \cos\theta - |E_{p,r}| \cos\theta = |E_{p,t}| \cos\theta_t \tag{4.14}$$

where now $\theta = \theta_i = \theta_r$.

The magnetic field strength vector **H** is perpendicular to the electric field strength vector **E,** as indicated in Figure 1.7. Therefore, the magnetic field strength vector associated with the incident electric field $E_{p,i}$ is perpendicular to the plane of incidence. This perpendicular component of the magnetic field strength is also conserved at the interface [cf. Reference 1, p. 318]; that is,

$$|H_{p,i}| + |H_{p,r}| = |H_{p,t}| \qquad (x = 0) \tag{4.15}$$

In writing Equation 4.15 it has been recognized that, as in Equation 4.10, the cos[·] factors that might have been shown are once again equal and so have divided out. Of course, there are no $\cos\theta_i$, $\cos\theta_r$ or $\cos\theta_t$ factors to consider because the transverse–magnetic field is already parallel to the interface. Like E_p, the quantity H_p is in general a complex quantity. However, H_p is in phase with E_p at incidence and so can also be considered real without loss of generality.

Now for the special case of the *p*-polarized electromagnetic wave propagating in the positive-x' direction, Equation 1.16b becomes

$$\frac{\partial E_p}{\partial x'} = -\mu \frac{\partial H_p}{\partial t} \tag{4.16}$$

where

$$E_p = |E_p| e^{i\omega(x'/c - t)} \tag{4.17}$$

and

$$H_p = |H_p| e^{i\omega(x'/c - t)} \tag{4.18}$$

With the introduction of Equations 4.17 and 4.18, Equation 4.16 becomes

$$|H_p| = \frac{\text{n}}{\mu c_0} |E_p| \tag{4.19}$$

Then, assuming that $\mu = \mu_0$, a constant, at the interface,* substitution of Equation 4.19 into 4.15 yields

$$\text{n}_1 |E_{p,i}| + \text{n}_1 |E_{p,r}| = \text{n}_2 |E_{p,t}| \qquad (x = 0) \tag{4.20}$$

---

* The variation of the magnetic permeability from its value in a vacuum is negligible except for ferromagnetic materials, which are generally electrical conductors.

Now, eliminating $|E_{p,t}|$ between Equations 4.14 and 4.20 yields

$$\frac{|E_{p,r}|}{|E_{p,i}|} = \frac{\cos\theta/\cos\theta_t - n_1/n_2}{\cos\theta/\cos\theta_t + n_1/n_2} \tag{4.21}$$

which becomes, after introducing Snell's law, Equation 4.13,

$$\frac{|E_{p,r}|}{|E_{p,i}|} = \frac{\tan(\theta - \theta_t)}{\tan(\theta + \theta_t)} \tag{4.22}$$

Recall that natural (unpolarized) radiation can be resolved into two orthogonal electromagnetic waves. Then an analysis similar to that leading to Equation 4.22 can be carried out for an incident *transverse-electric*-(TE-) polarized electromagnetic wave whose electric field strength vector is perpendicular to the plane of incidence. This leads to

$$\frac{|E_{s,r}|}{|E_{s,i}|} = -\frac{\sin(\theta - \theta_t)}{\sin(\theta + \theta_t)} \tag{4.23}$$

where now, following convention, the subscript "s" indicates a TE-polarized wave. Now the ratio of the reflected power to the incident power for these two components of polarization incident to the ideal interface of two nonconducting (dielectric) media is in each case simply the square of the ratio given respectively in Equations 4.22 and 4.23; that is,

$$\rho'_{p,\lambda} = \left(\frac{|E_{p,\lambda,r}|}{|E_{p,\lambda,i}|}\right)^2 \tag{4.24}$$

and

$$\rho'_{s,\lambda} = \left(\frac{|E_{s,\lambda,r}|}{|E_{s,\lambda,i}|}\right)^2 \tag{4.25}$$

The reflectivity components given by Equations 4.24 and 4.25 can be considered directional–hemispherical because they predict the fraction of power incident from a given angle that is reflected into the hemispherical space above the interface. These reflectivity components are spectral because the index of refraction n and thus the angle θ_t are generally functions of wavelength.

Finally, for naturally (or randomly) polarized radiation we can write

$$\begin{aligned}
\rho'_\lambda &= \frac{1}{2}(\rho'_{p,\lambda} + \rho'_{s,\lambda}) \\
&= \frac{1}{2}\left[\frac{\tan^2(\theta - \theta_t)}{\tan^2(\theta + \theta_t)} + \frac{\sin^2(\theta - \theta_t)}{\sin^2(\theta + \theta_t)}\right] \\
&= \frac{1}{2}\frac{\sin^2(\theta - \theta_t)}{\sin^2(\theta + \theta_t)}\left[1 + \frac{\cos^2(\theta + \theta_t)}{\cos^2(\theta - \theta_t)}\right]
\end{aligned} \tag{4.26}$$

The result represented by Equation 4.26 is known as *Fresnel's equation*. Of course, θ_t can be eliminated from Equation 4.26 in favor of n_1/n_2 using Snell's law, Equation 4.13.

4.4 EMISSIVITY FOR AN OPAQUE DIELECTRIC

The term *opaque dielectric* might at first seem to be an oxymoron; however, it is reasonable to consider the limiting case of a very thick ($t \gg \lambda$), weakly conducting material for which the theory underlying Fresnel's equation is locally valid but for which absorption does nonetheless occur over a sufficiently thick layer. (This is equivalent to the common practice of assuming incompressible flow in the analysis of each stage of an axial-flow compressor.) The interface of such a material might be expected to obey the relation for an opaque surface,

$$\varepsilon'_\lambda = \alpha'_\lambda = 1 - \rho'_\lambda \tag{4.27}$$

where Kirchhoff's law has been invoked.

Figure 4.3 shows the directional, spectral emissivity (= absorptivity) for a thick dielectric predicted using Fresnel's equation, Equation 4.26, with Equation 4.27. The notation supposes that radiation passes from a dielectric whose index of refraction is n_1 into a dielectric whose index of refraction is n_2. The results in Figure 4.3 can be considered spectral because in general the index of refraction depends on wavelength. Data for air ($n_1 = 1.0$) over black glass ($n_2 = 1.517$) are shown to be in excellent agreement with the theory for $n_2/n_1 = 1.5$. The salient point of Figure 4.3 is that electromagnetic theory predicts that, for sufficiently small values of the ratio of index of refraction, electrical nonconductors (i.e., dielectrics) emit and absorb more efficiently for directions near the normal than they do for grazing angles.

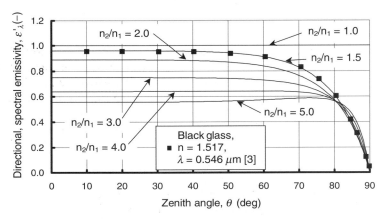

Figure 4.3 Directional, spectral emissivity for a thick dielectric according to Fresnel's equation, Equation 4.26, with Equation 4.27

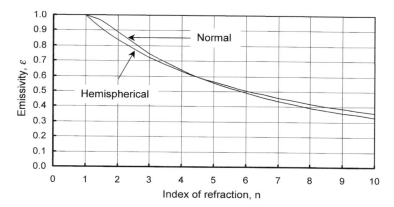

Figure 4.4 Theoretical normal and hemispherical emissivity as a function of the index of refraction for a "thick" dielectric

Figure 4.4 illustrates the variation of both the normal ($\theta = 0$) and hemispherical emissivity of a thick dielectric as a function of its index of refraction n assuming that the medium into which it is emitting is a vacuum or a material such as air for which n ≈ 1.0. These results were also obtained using Fresnel's equation with Equation 4.27. In the case of the hemispherical emissivity, integration over 2π-space was carried out using Equation 3.8b in a spreadsheet environment. These theoretical results show that the emissivity decreases with an increase in the index of refraction. Also, for sufficiently small values of the index of refraction—less than about 4.4—the normal emissivity exceeds the hemispherical, while for values beyond this threshold value the inverse is true. In fact, most dielectric materials of normal engineering interest have refractive indices of less than 4.4, and so it can be deduced that for most dielectric materials the normal emissivity exceeds the hemispherical emissivity. In this context note that the directional emissivity curve in Figure 4.3 representing $n_2/n_1 = 4$ is concave down throughout the range of emission angles, while the curve representing $n_2/n_1 = 5$ is concave up out to about 70 deg. Finally, Figure 4.4 establishes that the difference between the normal and hemispherical emissivity for the range of index of refraction shown is relatively small, and may be considered negligible in many engineering applications.

Figure 4.5 shows the measured directional–hemispherical, spectral reflectivity for a white paint. Incidence is from near normal ($\theta_i = 12.5$ deg). Most white paints are electrically nonconducting. Dielectric materials such as ceramics and other electrical insulators have low emissivity and absorptivity in the visible and near infrared, but high emissivity and absorptivity in the far infrared (not shown in the figure). Most white-appearing dielectrics are more or less black for infrared radiation (freshly fallen snow, for example!). The behavior illustrated in Figure 4.5 is typical of a commercial white paint, and explains why most automobiles in the Caribbean and the upper surfaces of many commercial airliners are painted white or pastel. Comparison of Figure 4.5 with Figure 4.4 suggests that the index of refraction of this white paint in-

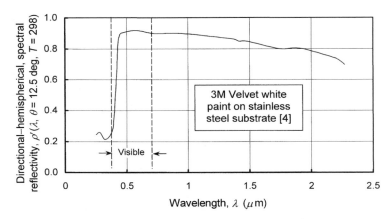

Note: Data obtained from Ref. 4; original source is P.R. Cheever, J.K. Miles, and John Romanko, "In Situ Measurements of Spectral Reflectance of Thermal Control Coatings Irradiated in Vacuo" in *Progress in Astronautics and Aeronautics*, Vol. 20, *Thermophysics of Spacecraft and Planetary Bodies—Radiation Properties of Solids and the Electromagnetic Radiation Environment in Space*, edited by Gerhard B. Heller, Copyright 1967 by Academic Press Inc. Used with permission.

Figure 4.5   Typical directional–hemispherical, spectral reflectivity of commercial white paint [4]

creases precipitously from a value of about 2.7 in the ultraviolet to a value exceeding 10 in the visible.

4.5   BEHAVIOR OF ELECTRICAL CONDUCTORS (METALS)

We now consider the case of radiation passing through an ideal plane interface from a dielectric, where $m_1 = n_1$ and the direction of propagation is x', into an electrical conductor, where $m_2 = n_2 + i\kappa_2$ and the direction of propagation is x'', as illustrated in Figure 4.6. In this case Equation 4.3 must be modified to take into account the fact that the wave passing into the conducting medium is *inhomogeneous*; that is, that its phase angle varies in direction x'' and its amplitude varies in direction x normal to the interface. Then the version of Equation 4.3 that describes the situation illustrated in Figure 4.6 is

$$E_{p,t} = [|E_{p,t}|e^{i\omega(\alpha x''/c_0 - t)}]e^{-\omega\beta x/c_0} \tag{4.28}$$

where α is a surrogate for the index of refraction in the direction of propagation in the electrical conductor, and β is a surrogate for the extinction coefficient in the direction normal to the interface, with the values of α and β related to n_2 and κ_2 in a way to be determined. The quantity c_0/α is called the *phase velocity*. Note that by comparison with Equation 4.3, $\alpha = n_2$ and $\beta = \kappa_2$ for normal incidence, i.e., when $x'' = x' = x$.

For other than normal angles of incidence the coefficients α and β can be related to n_2 and κ_2 through an analysis analogous to the one leading to Equations 4.5 and 4.6, except that now the transmitted electric field is a function of two space variables, x and x''. The usual procedure is to express x in terms of x'' and y''; that is, $x = x'' \cos\gamma - y'' \sin\gamma$. Under this change of variables Equation 4.28 then describes a function of time t and two orthogonal space variables, x'' and y''; that is,

$$E_{p,t} = [|E_{p,t}|e^{i\omega(t - \alpha x''/c_0)}]e^{-\omega\beta(x'' \cos\gamma - y'' \sin\gamma)/c_0} \tag{4.29}$$

or

$$E_{p,t} = |E_{p,t}|e^{i\omega t}e^{-\omega(i\alpha + \beta \cos\gamma)x''/c_0}e^{\omega\beta \sin\gamma y''/c_0} \tag{4.30}$$

For the electric wave represented by Equation 4.30, the first three Maxwell's equations, Equations 1.16a, 1.16b, and 1.16c (with the charge density ρ_e assumed spatially uniform) can be combined to yield

$$\nabla^2 E_{p,t} = \mu\varepsilon \frac{\partial^2 E_{p,t}}{\partial t^2} + \frac{\mu}{r_e} \frac{\partial E_{p,t}}{\partial t} \tag{4.31}$$

Introduction of Equation 4.30 into Equation 4.31 then yields

$$\frac{\omega^2 (i\alpha + \beta \cos\gamma)^2}{c_0^2} + \omega^2 (\beta \sin\gamma)^2 = -\mu\varepsilon\omega^2 + i\omega \frac{\mu}{r_e} \tag{4.32}$$

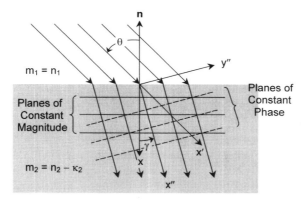

Figure 4.6 An inhomogeneous wave propagating through an electrical conductor after incidence from a dielectric

Finally, when the real and imaginary parts of Equation 4.32 are equated and the resulting pair of equations solved simultaneously with Equations 4.6, there results

$$\alpha^2 - \beta^2 = n_2^2 - \kappa_2^2 \quad \text{and} \quad \alpha\beta\cos\gamma = n_2\kappa_2 \tag{4.33}$$

where

$$\frac{\sin\gamma}{\sin\theta} = \frac{n_1}{\alpha} \tag{4.34}$$

or

$$\cos\gamma = \sqrt{1 - \left(\frac{n_1}{\alpha}\right)^2 \sin^2\theta} \tag{4.35}$$

In principle, Equations 4.33 and 4.35 can now be solved for α, β, and γ, although in practice the algebra is rather daunting.* However, it is clear from these equations that for the special case of normal incidence ($\theta = 0$) $\alpha = n_2$ and $\beta = \kappa_2$, as has already been pointed out. In fact, even for incidence at grazing angles to a metal surface from air ($n_1 \cong 1$) the value of $\cos\gamma$ may be sufficiently near unity to assume $\alpha = n_2$ and $\beta = \kappa_2$ with negligible error. This is because, as a surrogate for n_2, α is large compared to n_1 for many metals, and of course $\sin^2\theta$ is bounded by unity. For example, for aluminum at wavelengths exceeding 5.0 μm, $\cos\gamma$ is greater than 0.99.

By analogy with the formulation leading to Equation 4.13 it is hypothesized that we may also write a form of Snell's law for the situation depicted in Figure 4.6, i.e.,

$$\frac{\sin\theta}{\sin\theta_t} = \frac{m_2}{m_1} = \frac{n_2 + i\kappa_2}{n_1} \tag{4.36}$$

However, because θ and n_1 are both real variables, $\sin\theta_t$ must be complex. This means that θ_t is an artificial angle and so cannot be the same as the real angle γ. Bearing this qualification of θ_t in mind, a development parallel to that leading to Equation 4.21 produces the analogous result

$$\frac{E_{p,r}}{E_{p,i}} = \frac{\cos\theta/\cos\theta_t - n_2/m_2}{\cos\theta/\cos\theta_t + n_1/m_2} \tag{4.37}$$

except that now $E_{p,r}$ and $E_{p,i}$ are complex quantities. For the important special case of incidence from a vacuum or air, where the value of n_1 can be considered unity, the hypothetical form of Snell's law, Equation 4.36, can be rearranged to obtain

$$\cos\theta_t = \sqrt{1 - \sin^2\theta_t} = \sqrt{1 - \frac{\sin^2\theta}{m^2}} \tag{4.38}$$

---

* For example, 85 pages are required to print out the Mathematica solution.

In writing Equation 4.38 the subscript "2" on the complex index of refraction has been suppressed without danger of ambiguity since $m_1 = n_1 = 1$. With the introduction of $m = n + i\kappa$ into Equation 4.38 we have

$$\cos\theta_t = \sqrt{1 - \frac{n\sin^2\theta}{n^2 + \kappa^2} - i\frac{\kappa\sin^2\theta}{n^2 + \kappa^2}} \qquad (4.39)$$

It is now recognized that, since $\sin^2\theta \le 1$ and for most metals $\kappa \gg 1$,

$$1 \gg \frac{n\sin^2\theta}{n^2 + \kappa^2} \quad \text{and} \quad 1 \gg \left| i\frac{\kappa\sin^2\theta}{n^2 + \kappa^2} \right| \qquad (4.40)$$

so that $\cos\theta_t \cong 1.0$. Then to the extent that this *homogeneous wave approximation* is valid we may write

$$\frac{E_{p,r}}{E_{p,i}} = \frac{(n - i\kappa)\cos\theta - 1}{(n - i\kappa)\cos\theta + 1} = \frac{(n\cos\theta - 1) - i\kappa\cos\theta}{(n\cos\theta + 1) - i\kappa\cos\theta} \qquad (4.41)$$

Finally, noting that the flux carried by an electromagnetic wave is proportional to the square of the electric field strength E, and that $E^2 = E \cdot E^*$, we have

$$\rho'_{p,\lambda} = \frac{(n\cos\theta - 1)^2 + (\kappa\cos\theta)^2}{(n\cos\theta + 1)^2 + (\kappa\cos\theta)^2} \qquad (4.42)$$

A similar procedure can be applied to an incident transverse electric wave to obtain

$$\rho'_{s,\lambda} = \frac{(n - \cos\theta_i)^2 + \kappa^2}{(n + \cos\theta_i)^2 + \kappa^2} \qquad (4.43)$$

Then for randomly polarized (= unpolarized) radiation incident to an opaque surface we have as before

$$\varepsilon'_\lambda = \alpha'_\lambda = 1 - \rho'_\lambda \qquad (4.27)$$

with

$$\rho'_\lambda = \frac{1}{2}(\rho'_{p,\lambda} + \rho'_{s,\lambda}) \qquad (4.44)$$

Once again the wavelength dependence implied by the notation comes in through $n(\lambda)$ and $\kappa(\lambda)$.

Figure 4.7 compares measurement of the directional, spectral emissivity for polished platinum at 2.0 μm with the behavior predicted using Equations 4.42 through 4.44 with Equation 4.27. The excellent agreement between measurement and theory tends to validate the assumptions leading to Equations 4.42 and 4.43.

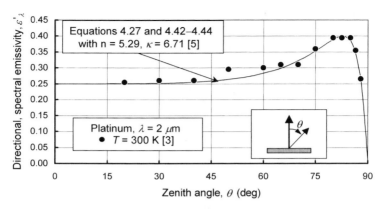

Figure 4.7 Comparison of the predicted (curves) and measured (symbols) directional, spectral emissivity (= absorptivity) of platinum at 2.0 μm

Comparison of Figures 4.3 and 4.7 makes it clear that electrical conductors are generally less efficient emitters and absorbers than dielectrics. It is also evident from comparison of Figures 4.3 and 4.7 that, in contrast to dielectrics, electrical conductors emit and absorb more efficiently at grazing angles than near the normal. Figure 4.8 effectively summarizes the theoretical (and observed) behavior of electrical conductors and nonconductors. Figure 4.8a shows the brightness variation as observed by the human eye for an incandescent dielectric sphere, and Figure 4.8b shows the same thing for an incandescent metal sphere. For comparison, Figure 4.8c shows the apparent brightness of an incandescent spherical blackbody. The metal sphere in Figure 4.8b appears brighter to the eye near the edge, whereas the dielectric sphere in Figure 4.8a appears brighter at the center. Of course, the black sphere is uniformly bright throughout.

In principle, a theoretical expression for the hemispherical, spectral emissivity could be obtained by substituting Equations 4.42 through 4.44 with Equation 4.27 into Equation 3.8a and carrying out the indicated integration. We have avoided the mathematical difficulties that would entail by carrying out the indicated integration numerically in a spreadsheet environment using Equation 3.8b, and the results are

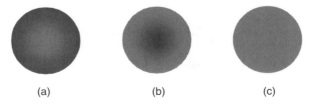

Figure 4.8 Relative brightness variation for incandescent (a) dielectric, (b) metal, and (c) black spheres

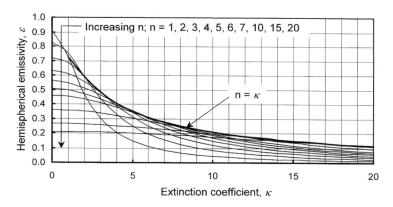

Figure 4.9 Theoretical variation of hemispherical emissivity with extinction coefficient and index of refraction for electrical conductors

shown in Figure 4.9. When presented in this form it is clear that for a given value of the index of refraction, the hemispherical emissivity decreases with an increase in the extinction coefficient; that is, as the behavior of the emitting material becomes more metallike. It is interesting that the dark line corresponding to $n = \kappa$ defines the upper limit to the hemispherical emissivity for a given value of n and κ; i.e., for a given wavelength.

For the special case of normal incidence, Equations 4.42 through 4.44 with Equation 4.27 become

$$\alpha'_{n,\lambda} = \varepsilon'_{n,\lambda} = 1 - \frac{(n-1)^2 + \kappa^2}{(n+1)^2 + \kappa^2} = \frac{4n}{(n+1)^2 + \kappa^2} \qquad (4.45)$$

or considering that $n \gg 1$ for most metals,

$$\alpha'_{n,\lambda} = \varepsilon'_{n,\lambda} \cong \frac{4n}{n^2 + \kappa^2} \qquad (4.46)$$

Figure 4.10 shows the theoretical variation of the normal emissivity with index of refraction for electrical conductors, based on Equation 4.45. As in the case of the hemispherical emissivity, the normal emissivity of electrical conductors is also shown to decrease with an increase in the extinction coefficient for a given value of the index of refraction. Note in both Figures 4.9 and 4.10 that values of emissivity associated with metals ($\varepsilon < {\sim}0.2$) are not reached until the extinction coefficient reaches a value of around 10. Indeed, for most metals in the infrared the extinction coefficient exceeds this value. As in the case of the hemispherical emissivity for metals, for a given value of the extinction coefficient the peak in the normal emissivity occurs when $n = \kappa$.

Figure 4.11 shows the theoretical ratio of hemispherical to normal emissivity of electrical conductors as a function of the extinction coefficient for a range of val-

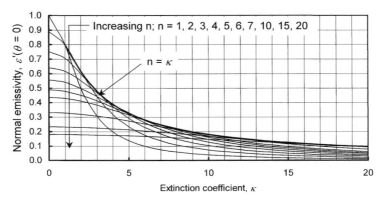

Figure 4.10 Theoretical variation of normal emissivity with extinction coefficient and index of refraction for electrical conductors

ues of the index of refraction. We first remark that for metals in the infrared ($\kappa > \sim10$, $\lambda > \sim5$) the hemispherical emissivity generally exceeds the normal emissivity by about 20 percent. This is of course due to the shape of the directional emissivity curve for metals, illustrated in Figure 4.7, which exhibits a pronounced peak at around 80 deg. Finally, we note that only at relatively small values of κ ($< \sim3$, i.e., relatively poor electrical conductors) does the normal emissivity exceed the hemispherical emissivity, and then only for sufficiently small values of the real part of the index of refraction. Finally, the role of the condition $n = \kappa$ is once again interesting, this time establishing the lower bound on the hemispherical-to-normal emissivity ratio.

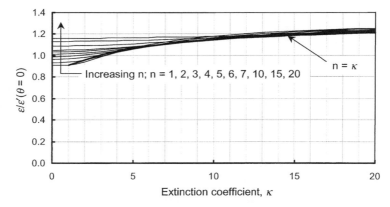

Figure 4.11 Theoretical variation of the ratio of hemispherical emissivity to normal emissivity with extinction coefficient and index of refraction for electrical conductors

4.6   THE DRUDE FREE-ELECTRON MODEL FOR METALS; DISPERSION THEORY

In Chapter 1 we considered a model for emission from the regular crystal matrix represented in Figure 1.1. In that discussion it was convenient to consider each atom to be surrounded by a spherical cloud of electrons bound more or less tightly to the atomic nucleus. A more realistic model for electrical conductors (metals), however, would divide the electron cloud into two regions, with the electrons in the inner region tightly bound to the nucleus while those in the outer region remain free to migrate from atom to atom under the influence of an applied electric field. This is the essential hypothesis of the *Drude free-electron theory* for metals. Then, consistent with the Drude free-electron view, the range of application of the model in Chapter 1 is limited to dielectrics.

The Drude free-electron theory assumes that the electronic contribution to the emissivity of a metal dominates. Consistent with this view, the complex index of refraction of interest is that given by *dispersion theory,* which relates the harmonic motion of the free electrons to the periodic electric field in the metal [6],

$$n + i\kappa = \sqrt{1 - \frac{\omega_p^2}{\omega^2 + i\omega/\tau}} \tag{4.47}$$

In Equation 4.47, $\omega = 2\pi\nu$ is the angular frequency of the incident radiation, ω_p is the so-called *plasma frequency* for the metal, and τ is the *relaxation time* required for a transient electric current to decay to $1/e$ of its initial value after the electric field is removed. The plasma frequency is given by [7]

$$\omega_p = \sqrt{\frac{N_e e^2}{m_e \varepsilon_0}} \quad \text{(r/s)} \tag{4.48}$$

where N_e is the effective number density of free electrons, e is the fundamental electronic charge, and m_e is the optical effective mass of the electron; and the relaxation time is given by

$$\tau = \frac{m_e}{r_e N_e e^2} \quad \text{(s)} \tag{4.49}$$

where r_e is the direct-current electrical resistivity of the host metal. Equating the real and imaginary parts of Equation 4.47 yields two equations in n and κ,

$$n^2 - \kappa^2 = 1 - \frac{\omega_p^2 \tau^2}{\omega^2 \tau^2 + 1} \tag{4.50}$$

and

$$2n\kappa = \frac{\omega_p^2 \tau^2}{\omega^2 \tau^2 + 1} \left(\frac{1}{\omega\tau} \right) \tag{4.51}$$

Subject to the assumptions that n and κ are real and positive, Equations 4.50 and 4.51 can be solved to yield [6]

$$n = \frac{1}{\sqrt{2}} \left\{ \left[(1 - Q)^2 + \left(\frac{Q}{\omega \tau} \right)^2 \right]^{1/2} - Q + 1 \right\}^{1/2} \qquad (4.52)$$

and

$$\kappa = \frac{1}{\sqrt{2}} \left\{ \left[(1 - Q)^2 + \left(\frac{Q}{\omega \tau} \right)^2 \right]^{1/2} + Q - 1 \right\}^{1/2} \qquad (4.53)$$

where

$$Q = \frac{\omega_p^2 \tau^2}{\omega^2 \tau^2 + 1} \qquad (4.54)$$

Equations 4.52 through 4.54, with Equations 4.48 and 4.49, can now be introduced into Equation 4.45 (or 4.46) to obtain an estimate for the normal, spectral emissivity of an ideal metal surface in terms of fundamental physical properties.

Figure 4.12 shows the spectral, normal absorptivity of gold as calculated using the Drude theory described above. The values of ω_p and τ have been obtained from Ordal et al. [8]. Also shown in the figure are the data of Hass and Hadley [9], Padalka and Shklyarevskii [10], Toscano and Cravalho [11], and Bennett and Ashley [12]. As shown for the case of gold in Figure 4.12, the Drude free-electron theory generally tends to underestimate measured values of normal, spectral absorptivity.

The *anomalous skin effect* theory [13,14] is an attempt to account for the fact that, near the surface of a metal, free electrons may be accelerated by steep gradi-

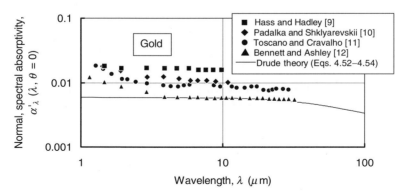

Figure 4.12 Comparison of measured normal spectral absorptivity of polished gold with Drude theory prediction

ents in the electric field between collisions, thereby leading to a local departure from Ohm's law, and so compromising the equilibrium assumption inherent to Equation 4.47. Toscano and Cravalho [15] show that when the anomalous skin effect is accounted for, somewhat better agreement is obtained with the data in Figure 4.12.

4.7 HAGEN–RUBENS APPROXIMATION FOR METALS

For metals (electrical conductors, r_e small) and at sufficiently long wavelengths ($\lambda \geq$ 5 μm, i.e., in the infrared), Equations 4.7 and 4.8 lead to the *Hagen–Rubens approximation* [16],

$$\text{n} \cong \kappa \cong \sqrt{\frac{\lambda_0 \mu_0 c_0}{4 \pi r_e}} \tag{4.55}$$

where now λ_0 is the wavelength in a vacuum in meters, μ_0 is the permeability of free space ($= 4\pi \times 10^{-7}$ N·s^2/C^2), c_0 is the speed of light in a vacuum ($= 2.9979 \times 10^8$ m/s), and r_e is the direct-current electrical resistivity of the metal in ohm-meters. With the approximation n $\cong \kappa$, Equation 4.45 reduces to

$$\varepsilon'_\lambda(\lambda, \, \theta = 0) = \frac{4\text{n}}{2\text{n}^2 + 2\text{n} + 1} \tag{4.56}$$

Figure 4.13 gives a comparison between the normal, spectral emissivity of several metals measured in the infrared with the Hagen–Rubens approximation, Equations 4.55 and 4.56. It is immediately clear that the slopes of the theoretical lines are roughly the same as those of the data trend, and that theory and measurements produce the same relative ranking of emissivity among the metals shown. The theoretical curve for platinum agrees well with the data where there is overlap, while an obvious discontinuity in level occurs between the theory and the data trend for copper. Agreement between theory and data for the other three metals lies between these two extremes, but is acceptable considering the simplicity of the theory. We may conclude that the Hagen–Rubens approximation offers a fairly reliable tool for extending measurements in the near infrared, where they are relatively easy to obtain, into the far infrared, where they are more difficult to obtain.

Recall that the heavy curves in Figures 4.9, 4.10, and 4.11 represent the special case addressed by the Hagen–Rubens approximation, i.e., n $= \kappa$. It has been noted in Figures 4.9 and 4.10 that the heavy curves form an envelope that establishes the theoretical upper limit on the emissivity for a given value of the complex part of the index of refraction, while in Figure 4.11 the heavy curve establishes a lower limit of the ratio of hemispherical to normal emissivity. Thus, we conclude that the Hagen–Rubens approximation provides a useful estimate of the maximum emissivity obtainable for a smooth metal having a known extinction coefficient.

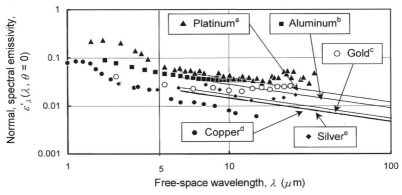

Notes:

[a] $r_e = 10.8 \times 10^{-8}$ $\Omega \cdot$m at 300 K [5]; Data at 311 K from Ref. 17, Curve 9, p. 546 (converted from reflectivity data from: R.J. Hembach, L. Hemmer-Dinger, and A.J. Katz, NASA-SP-31, 1963.)

[b] $r_e = 6.13 \times 10^{-8}$ $\Omega \cdot$m at 600 K [5]; Data at 599 K from Ref. 17, Curve 4, p. 14 (original source: P.M. Reynolds, "Spectral Emissivity of 99.7% Aluminium between 200 and 540°C" in *British Journal of Applied Physics*, Vol. 12, No. 3, March 1961.)

[c] $r_e = 2.255 \times 10^{-8}$ $\Omega \cdot$m at 298 K [5]; Data at 298 K from Ref. 17, Curve 31, p. 263 (converted from reflectivity data from: K. Schocken and J.A. Fountain, Proceedings of Conference on Spacecraft Coatings Development, NASA-TM-X-56167, 20, 1964.)

[d] $r_e = 1.678 \times 10^{-8}$ $\Omega \cdot$m at 293 K [5]; Data at 294 K from Ref. 17, Curve 41, p. 157 (original source: Elwood Allen Anderson, *The Spectral Emittance of Metals in the Infrared*, M.S. Thesis, University of California, 1962.)

[e] $r_e = 1.617 \times 10^{-8}$ $\Omega \cdot$m at 298 K [5]; Data at 298 K from Ref. 17, Curve 9, p. 634 (converted from reflectivity data from: C.H. Leigh, RAD-TR-62-33, NASA-CR-53235, N64-17590, 1962.)

Figure 4.13 Normal, spectral emissivity of several polished metals predicted using the Hagen–Rubens approximation and compared with data

Finally, the Hagen–Rubens approximation can be used to estimate the temperature dependence of the normal ($\theta = 0$), total emissivity,

$$\varepsilon(T_A, \theta = 0) = \frac{\pi}{\sigma T_A^4} \int_{\lambda=0}^{\infty} \varepsilon_\lambda'(T_A, \lambda, \theta = 0) \, i_{b\lambda}(\lambda, T) \, d\lambda \qquad (4.57)$$

Figure 4.14 shows the normal, spectral emissivity for aluminum at 600 and 800 K based on Equations 4.55 and 4.56. Also shown is the Planck function at those two temperatures. The emissivity curves in Figure 4.14 are based on the Hagen–Rubens approximation and so should be considered highly suspect for wavelengths shorter than about 5 μm. However, the Planck function contains very little energy at short wavelengths at the temperatures considered (and at lower temperatures) and, since the Planck function acts as a weighting factor in the calculation implied by Equation

4.57, any error associated with inaccuracy of the Hagen–Rubens approximation at short wavelengths is minimized at these temperatures.

Inspection of Figure 4.14 and Equation 4.57 reveals that two effects combine to produce an increase in normal, total emissivity with temperature: (1) the increase in the normal, spectral emissivity with temperature at a given wavelength, and (2) the shift of the peak in the Planck blackbody radiation distribution function toward shorter wavelengths with increasing temperature. Figure 4.15 shows the measured variation of the normal, total emissivity as a function of temperature for platinum, aluminum, and silver. The trend of increasing total emissivity with temperature in each case is as expected.

Example Problem 4.1

Estimate the normal, total emissivity of aluminum at 600 and 800 K using the Hagen–Rubens approximation, Equations 4.55 and 4.56 with Equation 4.57.

SOLUTION

We begin by creating a spreadsheet with wavelength in column A. The wavelength interval used will depend on the slope of the Planck function and the spectral emissivity as well as on the desired accuracy. The curves shown in Figure 4.14 and the results reported here were computed using a 0.01-μm interval, although less wavelength resolution, say 0.1 μm, would probably have been adequate. In columns B, C, and D, respectively, we enter Equations 4.55 and 4.56 and the Planck function, Equation 2.76. For the direct-current electrical resistivity at 600 and 800 K we use [5] 6.13×10^{-8} $\Omega \cdot$m and 8.70×10^{-8} $\Omega \cdot$m, respectively. Then in column E we form the in-

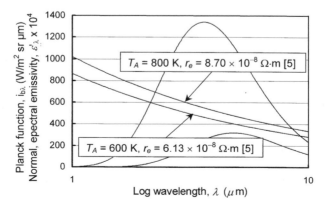

Figure 4.14 Normal, spectral emissivity of aluminum (Hagen–Rubens approximation) and the Planck function at 600 and 800 K

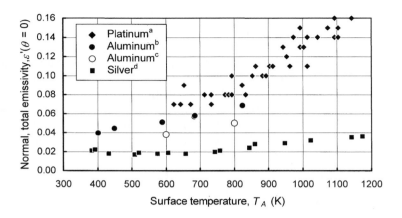

Notes:

[a] Ref. 17, Curve 1, p. 531 (original source: A.H. Sully, E.A. Brandes, and R.B. Waterhouse, "Some Measurements of the Total Emissivity of Metals and Pure Refractory Oxides and the Variation of Emissivity with Temperature," *British Journal of Applied Physics,* Vol 3, No. 3, March 1952.)

[b] Ref. 17, Curve 21, p. 11 (original source: E. Schmidt, V.A.W. Hauzeitschr, and A.G. Erftwerk, *Aluminum,* Vol. 3, 1930.)

[c] From Example Problem 4.1

[d] Ref. 17, Curve 3, p. 624 (original source: G. A. Zhorov, "Emissivity of Metals During Heating in Air," *High Temperature,* (a translation of *Teplofizika Vysokikh Temperatur,* Vol. 5, No. 3, May-June, 1967), A. E. Sheindlin, Ed.)

Figure 4.15 Measured temperature dependence of the normal, total emissivity of platinum, aluminum, and silver

tegrand of Equation 4.57, i.e., the product of columns C and D. Finally, at the bottom of column E we form the sum of column E, which will be our estimate of the total (normal) emissivity. Rounding the results to two significant figures we obtain 0.038 and 0.050 at 600 and 800 K, respectively. These values are somewhat lower than measurements reported near these temperatures, as shown in Figure 4.15.

4.8 INTRODUCTION TO THE OPTICAL BEHAVIOR OF REAL SURFACES

To this point we have treated surfaces as plane interfaces between regions having different real or complex indices of refraction. This approach has led to results that, in the case of very carefully prepared surfaces, are more or less in agreement with measurements. Even when measurements are seen to differ from the values predicted by theory, the predicted trends with wavelength and temperature have been shown to be in good agreement with measurements.

"Surface" radiation properties, especially in the case of metals, are normally determined in the first few wavelengths below the surface. For visible and near infrared radiation, which generally dominate radiation heat transfer, the peak wavelength is on the order of 1.0 to 10 μm. It follows that for wavelengths of this order of magnitude the radiation surface properties for metals are determined within the first 0.01 mm of depth! Therefore, surface preparation often dominates the bulk optical properties n and κ in determining the radiation properties. This fact severely limits the usefulness of theoretical results based on the definition of a surface as a plane interface between two semi-infinite materials having specified complex indices of refraction. While such models may be quite valuable for understanding observed trends, they are generally unreliable for obtaining values to be used in accurate heat transfer analysis.

Figure 4.16 illustrates the high level of variability in the spectral emissivity that can be expected for a typical metal depending on its surface preparation. The figure shows measured variations with wavelength of (a) hard–coat anodized aluminum, (b) "as delivered" commercially finished aluminum, and (c) vapor-deposited aluminum on polished glass. Commercially finished metals may have a thin coating of oil picked up during the final cold rolling and shearing processes, and of course layers of dust and oxidation will have built up during storage and transit. Each of these would have the effect of laying up a thin layer of dielectric material on top of the pure

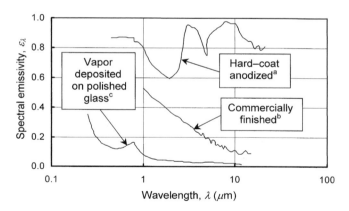

Notes:

[a] Hemispherical, Data from Ref. 18, Fig. 18-11, p. 179.

[b] Near normal, Ref. 17, Curve 8, p. 29 (Converted from reflectivity data from R.V. Dunkle and J. T. Gier, *Progress Report for the Year Ending June 27, 1953,* Institute of Engineering Research, University of California, Berkeley, 1953.).

[c] Hemispherical, Data from Ref. 18, Fig. 18-8, p. 176 (converted from reflectivity).

Figure 4.16 Typical observed spectral emissivity of hard-coat anodized, commercially finished and vapor deposited aluminum

metal that would tend to increase the emissivity at a given wavelength. Of course, a commercially finished sample would also be much less smooth than the vapor-deposited sample. Thus, the curve labeled "commercially finished" in Figure 4.16 may be thought of as the combined influence of several mechanisms, any one of which would have the effect of increasing the emissivity of the surface.

A metal surface is described as "anodized" when a controlled oxidation process has been employed to build up a relatively thick layer of aluminum oxide. The surface acts as the anode for the flow of electrical current through a chemical bath in which it is immersed. A dye is often added to the bath to produce a specific color. For the curve labeled "hard-coat anodized" in Figure 4.16 no dye has been added and the anodized specimen would have a dull matte appearance. In general, the emissivity and absorptivity increase with the thickness of the oxide layer, while the reflectivity decreases. The band structure in the spectral behavior of the anodized specimen may be attributed both to the wavelength-dependent optical properties of the dielectric layer and to its thickness. Radiation incident at a given wavelength may penetrate the dielectric layer, be reflected from the pure metal substrate, and pass back through the substrate. This would lead to constructive and destructive interference between the incident and reflected waves to an extent and in a way depending on wavelength. The result would be a wavelength-dependent interference filtering effect.

Polished metals have low absorptivity and emissivity and high reflectivity in the visible and infrared. In general, the emissivity and absorptivity decrease with increasing wavelength, as predicted by the Hagen–Rubens approximation. This effect is enhanced by the fact that at longer wavelengths the relative smoothness of a polished metal surface increases. This interplay between surface roughness and wavelength is illustrated in Figure 4.17, which shows the ratio of bidirectional, spectral reflectivity for a roughened nickel sample to that of a polished sample as a function of wavelength for an incident and reflected angle of 10 deg [19]. As the surface me-

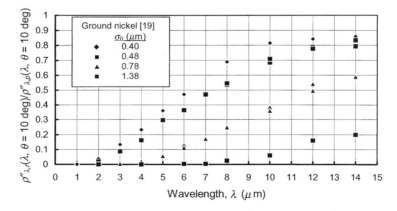

Figure 4.17 Effect of surface roughness on the bidirectional reflectivity of ground nickel (subscripts on reflectivity: r = roughened, p = polished) [19]

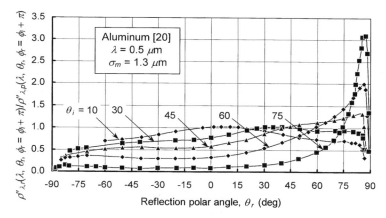

Figure 4.18 Illustration of off-specular peaking of reflection from a roughened surface ($\sigma_0/\lambda \gg$ 1.0) [20]

chanical roughness increases at a given wavelength, the ability of the surface irregularities to scatter radiation in other than the specular direction increases so that the bidirectional reflectivity in the specular direction decreases. As the wavelength of the radiation increases for a given degree of roughness, represented in the figure by the root-mean-square variation of the surface topography, σ_0, the surface irregularities become less important.

Another effect of surface roughness is illustrated in Figure 4.18, which shows the ratio of measured bidirectional reflectivity for a rough surface to the bidirectional reflectivity for a polished surface of the same metal [20]. Each curve represents a different angle of incidence of beam radiation. The horizontal axis represents the angle at which reflected intensity is measured. For a polished surface the peak in the measured bidirectional reflectivity is expected to occur at an angle of reflection equal to the angle of incidence. What is observed for roughened surfaces, however, is that the peak in bidirectional reflectivity occurs at an angle of reflection slightly greater than the angle of incidence if the latter is sufficiently large. Theory [21] supports the idea that radiation incident from a sufficiently steep angle with respect to the normal evidently fails to illuminate the deeper valleys, which are shadowed by neighboring peaks in the relief. This leads to preferential reflections in the direction of larger zenith angles and a peak in the bidirectional reflectivity at an angle larger than the specular angle.

It has already been pointed out that commercially finished metals are generally somewhat roughened, contaminated, and oxidized. While theoretical treatments are available to account for surface topography, surface contamination is simply too broad a category to lend itself to a general theoretical formalism. Oxidation, especially when viewed as a coating on a pure-metal substrate, is amenable to theoretical treatment; however, a meaningful discussion of theoretical approaches for surface coatings must await the development of radiative transport theory in Chapter 6.

4.9 SURFACE TOPOGRAPHY EFFECTS

The topography of a surface is characterized by its height and slope distributions, with the standard deviation of the height distribution normally represented by the symbol σ, and the standard deviation of the slope distribution represented by the symbol m. The nomenclature is illustrated in Figure 4.19a. The symbol σ_0 represents the root-mean-square *optical roughness,*

$$\sigma_0 = \sqrt{\frac{1}{L} \int_0^L [h(x)]^2 \, dx} \qquad (4.58)$$

where $h(x)$ is the local height above a datum plane and L is the *correlation length.* For the case of periodic roughness, such as shown in Figure 4.19b, the correlation length is simply the period of one cycle of the height function. For random height distribu-

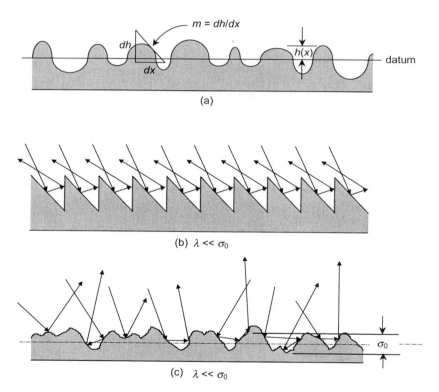

Figure 4.19 Cross sections of (a) a hypothetical surface showing the nomenclature for characterizing surface topography, and rays scattered by hypothetical (b) regularly and (c) irregularly roughened surfaces whose optical roughness scales are large compared to the wavelength of the incident radiation

tions a correlation function is typically based on the Gaussian distribution function [22],

$$g_G(x) = e^{-x^2/L^2} \tag{4.59}$$

In this case the correlation length is the length L for which

$$\int_0^L h(x)h(x + L)dx = \frac{1}{e} \int_0^L [h(x)]^2 dx \tag{4.60}$$

It is sometimes convenient to distinguish between the optical roughness σ_0 and the *mechanical roughness* σ_m, with the latter often being determined using a stylus similar to a phonograph pick-up and the former being a theoretical ideal.

Equations 4.58 through 4.60 imply that the surface topography varies only in one direction, as might be the case on the surface of, say, a turned work piece. In general, however, the surface topography will be two-dimensional, in which case Equation 4.58 would be written

$$\sigma_0 = \sqrt{\frac{1}{L_x L_y} \int_0^{L_y} \int_0^{L_x} [h(x,y)]^2 \, dx \, dy} \tag{4.61}$$

Similar modifications would be applied to Equations 4.59 and 4.60 in the case of a two-dimensional surface topography. Expressions similar to Equations 4.58 and 4.61 can be written for the slope distribution to define its root-mean-square value m_0.

The topography of a surface is said to be *stationary* if the values of σ_0 and m_0 are independent of location x,y on the surface. Furthermore, a surface is said to be *randomly rough* if its height and slope distributions are Gaussian. The idealization of a stationary, randomly rough surface often forms the basis for theoretical study, thereby playing a role in surface roughness theory similar to that played by the diffuse surface in surface-to-surface radiation heat transfer theory. That is, it is a useful model from which much can be learned—and which occasionally even approximates reality.

Finally, a dimensionless optical roughness parameter σ_0/λ is used to establish an optical roughness spectrum extending from *optically smooth* surfaces for which $\sigma_0/\lambda \ll 1$ at one extreme to *optically rough* surfaces for which $\sigma_0/\lambda \gg 1$ at the other. Optically smooth surfaces are also referred to as "slightly rough," and optically rough surfaces as "very rough." We now consider analytical and numerical models available to predict the surface optical properties across this spectrum.

4.9.1 Radiative Modeling of Optically Rough Surfaces ($\sigma_0/\lambda \gg 1$)

Examples of optically rough surfaces are shown in Figures 4.19b and 4.19c. Figure 4.19b shows a regular pattern of grooves such as might result from a machining process, while Figure 4.19c shows a more random roughness. Note that in Figure 4.19a the slope distribution is continuous, while in Figure 4.19c the topography features discontinuities. Figure 4.19c illustrates that a surface topography may exhibit more than one scale of roughness; i.e., the surface elements whose mean slopes define the primary roughness scale may themselves be rough. A fourth surface topography, similar to that in Figure 4.19b, can be imagined in which the surface consists of randomly arranged straight-line segments, or planes. This latter topography has been widely used in surface roughness modeling.

A clear distinction needs to be made between the "macroscopic surface," defined by an imaginary plane at the local mean height, and the "microscopic surface," defined by the interface that separates gray from white in Figure 4.19. Engineers are normally referring to the macroscopic surface, indicated by the "datum" plane in Figure 4.19a, when they speak of "a rough (smooth) surface." In the discussion that follows care will be taken to always use the appropriate modifier when the possibility for misunderstanding exists.

Figures 4.19b and 4.19c show arrows, representing "rays," which appear to suffer reflections whose direction is determined by the angle of incidence with, and the local slope of, the microscopic surface. In other words, we can imagine that the microscopic surface is at least partially specular. Figure 4.19b is especially intriguing because it suggests the possibility of machining surfaces to produce a preferred bidirectionality. In fact, regularly distressed surfaces, such as the one depicted in Figure 4.19b, were among the first to be modeled.

The most primitive rough-surface radiation models feature one-dimensional arrays of regularly spaced grooves in otherwise plane surfaces. An early contributor to this approach was the seminal article by Sparrow and Jonsson [23] in which the authors report a model for the directional absorptivity of a long, rectangular-cross-section channel whose walls and floor are either specular or diffuse. Shortly thereafter, Perlmutter and Howell [24] used a similar approach to investigate the directional emissivity of a surface consisting of parallel, truncated V-grooves having perfectly specular reflecting side walls and a black floor. They showed that such surfaces emit in a limited range of zenith angles determined by the depth of the groove and the groove opening angle. In an early effort to describe two-dimensional surface roughness, Torrance and Sparrow [21] modeled surfaces as arrays of randomly oriented V-grooves whose opening angles had a Gaussian distribution. The model was able to predict the off-specular peak phenomena illustrated in Figure 4.18. Smith and Hering [25] modeled rough surfaces as symmetrical V-shaped grooves having specularly reflecting walls with randomly distributed slopes. The model, which considered both shadowing and multiple reflections, was also able to reproduce experimental trends including the off-specular peak phenomenon. Christie and DeVriendt [26] considered the case of mutually orthogonal V-grooves meeting in such a way as to produce a regular two-dimensional array of square-based pyramids. Their model produced moderate agreement with measurements. More recently, Hesketh et al. [27] used a variation of the geometric-optics model of Sparrow and Jonsson to obtain acceptable agreement with their measurements of the directional, spectral emissivity of a silicon surface striated with parallel, rectangular-cross-section microgrooves.

The walls and floors of the grooves in these primitive models were gray and either specular or diffuse. With the exception of the model by Hesketh et al., in all cases radiation was assumed to be randomly polarized and incoherent, permitting use of the principles of geometric optics. That is, radiation interaction with the surfaces was completely described by the geometry and by the surface optical properties defined in Chapter 3. Although the Monte Carlo ray-trace method could have been put to effective use in implementing this generation of models, the more traditional net exchange method and its variations, presented in Part II of this book, were most often used.

The more sophisticated radiation models for optically rough surfaces are based on application of the principles of electromagnetic theory developed earlier in this chapter. In a hybrid effort Treat and Wildin [28] used the surface topography of Torrance and Sparrow [21] with a semi-empirical expression for the reflectivity based on Fresnel's equation, Equation 4.26. The semi-empirical relation, which has three adjustable constants determined by fitting the relation to measured values of bidirectional reflectivity, gives results that conform well to the fitted data.

The most general approach to the modeling of optically rough surface radiation is variously referred to as the Kirchhoff approximation, the tangent-plane method, or the physical optics method. This approach recognizes that, for sufficiently large values of σ_0/λ and sufficiently small values of λ/L, it is the slope distribution rather than the height distribution that determines the emission and scattering by a rough surface, an idea implicit in the Treat–Wildin model. The Kirchhoff approximation, or Kirchhoff method [29], consists of approximating the electric field at a point on the surface as the value it would have on a plane tangent to the surface at that point. The tangent plane in question is shown as the hypotenuse of the triangle in Figure 4.19a. Subject to this approximation one need only know the statistical distribution of the surface slopes to arrive at a scattering or emission model for the surface. That is, the electric field above the surface is simply the sum of the incident and reflected fields, with the later being related to the former through Equations 4.21 and 4.22. Sánchez-Gil and Nieto-Vesperinas [30] used the Kirchhoff approach in conjunction with one-dimensional surface profiles generated using a Monte Carlo method. The advantage here is that theoretical results obtained can be compared with measured data on the basis of the surface statistics, σ_0, m_0, and L, and the bulk optical properties of the substrate. While Sánchez-Gil and Nieto-Vesperinas did not compare their theoretical results with actual measurements, the results they obtained for angular distributions of reflected and transmitted intensity for dielectrics are consistent with observed trends and expectations.

4.9.2 Radiative Modeling of Optically Smooth Surfaces ($\sigma_0/\lambda \ll 1$)

When surfaces are "slightly rough" ($\sigma_0/\lambda \ll 1$) the modeling method of choice is the Rayleigh–Rice, or perturbed boundary condition, method. This approach was first proposed by Lord Rayleigh [31] in the context of acoustic scattering (although Rayleigh does recognize its possible applicability to electromagnetic radiation). Later Rice [32] applied the same principle to the scattering of electromagnetic radiation. The development presented here is based on the very readable and well-documented article by Schiffer [33].

The Rayleigh–Rice perturbation method assumes not only that the optical roughness σ_0/λ is small, but also that the slopes are small, i.e., that σ_0/L is small. Subject to these constraints the complex scattered electric field can be represented as an expansion of the form

$$E_s = E_s^{(0)} + E_s^{(1)} + E_s^{(2)} + \cdots \tag{4.62}$$

where the superscript indicates the order of $2\pi\sigma_0/\lambda$ in the expression for $E_s^{(n)}$, i.e., $(2\pi\sigma_0/\lambda)^0$, $(2\pi\sigma_0/\lambda)^1$, $(2\pi\sigma_0/\lambda)^2$, and so forth. Then it might be expected that the scattered electric field could be well approximated by the first few terms of the series represented by Equation 4.62 since $\sigma_0/\lambda \ll 1$. Considering that E_s is complex, the mean-square magnitude of the scattered electric field, which is to say the scattered power, is [33]

$$\langle |E_s|^2 \rangle = |E_s^{(0)}|^2 + \langle |E_s^{(1)}|^2 \rangle + 2 \operatorname{Re} [E_s^{(0)} \cdot \langle E_s^{(2)} \rangle *] + \cdots \tag{4.63}$$

the mean-square first-order terms involving products of $E_s^{(0)}$ and $E_s^{(1)}$ are zero. The zeroth-order term on the right-hand side of Equation 4.63 represents the power specularly reflected by a smooth surface, and the first of the two second-order terms represents the diffusely scattered incoherent power. The second second-order term on the right-hand side, which is usually negative, represents the reduction in the specularly reflected power due to surface roughness. Higher order terms are neglected. Armed with these interpretations of the perturbation terms it is possible, by computing their distributions with scattering angle for ranges of values of the roughness parameters and material optical properties, to gain insight into the effect of slight roughness on loss of specularity.

Finally, when roughness is very slight indeed, it is still possible to discern and predict departures from specular behavior due to a phenomenon called *surface polaritons* [34, 35]. Surface polaritons can be described as highly localized resonant electric fields set up on a metallic surface and organized around surface irregularities. When present they can interact with the incident electric field and, under certain circumstances, produce an unexpectedly strong backscattering peak.

4.9.3 Radiative Modeling When $\sigma_0/\lambda \sim 1$

It is often the case in physics and engineering that an otherwise intractable problem can be solved for the limiting values of certain key parameters. We have seen that the interaction of thermal radiation with a rough surface can be treated in the optically rough ($\sigma_0/\lambda \gg 1$) and optically smooth ($\sigma_0/\lambda \ll 1$) limits, but what can be done when $\sigma_0/\lambda \sim 1$? At least one team of researchers chose to jump through the horns of the dilemma. In a remarkable *tour de force* Dimenna and Buckius [36] have rigorously solved Maxwell's equations for the case of an interface between a vacuum on one side and either a metal or a dielectric on the other. The form of the interface, a one-dimensional ramp function, is similar to that shown in Figure 4.19b, except that now the peak in the ramp function can be moved horizontally from the arrangement shown in the figure to its mirror image. This introduces an additional parameter, that being the horizontal position of the peak relative to the correlation length. The surface is represented as a Fourier series truncated at 25 terms. Numerical solutions for the magnetic (in the case of transverse-magnetic polarization) and electric (in the case of transverse-electric polarization) field are obtained in the volume above and below the interface and matched at the interface. The fields for the two polarizations are expressed in terms of bidirectional reflectivity and bidirectional transmissivity. Finally, the two polarizations of reflectivity and the two polarizations of transmissivity are averaged to obtain the unpolarized quantities. Also, integration of the bidirectional reflectivity over the scattering hemisphere followed by subtraction of the result from unity yields the directional emissivity. The results allow direct comparison of the behavior of a metal (aluminum) and dielectric (silicon) surface having identical surface roughness and illumination.

Figure 4.20 summarizes the discussion of radiative modeling of rough surfaces.

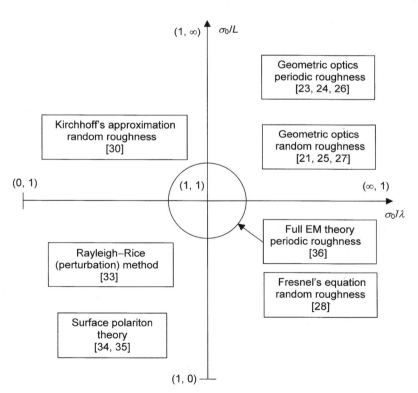

Figure 4.20 Summary of theoretical approaches to modeling surface roughness effects on incident thermal radiation

Team Project

TP 4.1   Go to the library and/or visit the Internet to find experimental data similar to those shown in the figures listed. Set up a notebook organized with a copy of the figure from this book followed by the article or articles containing suitable experimental data. Recent data are more valuable than older data, but both are useful. Concentrate on Figures 4.3, 4.5, 4.7, 4.12, 4.13, 4.15, 4.16, and 4.17.

Discussion Points

DP 4.1   Describe how surface roughness, chemical condition, degree of contamination, and surface grain structure might be expected to influence surface radiation behavior. For example, would you expect increasing mechanical surface roughness σ_0 at a given wavelength λ to produce more diffuse or less diffuse emissivity? Why?

DP 4.2   Reconcile the microscopic view of the absorption of electromagnetic radiation, presented in Chapter 1, with the macroscopic view presented in Section 4.2.

DP 4.3   The electric component of an electromagnetic wave propagating through an electrically conducting medium provokes a current flow and an associated conversion of electromagnetic power into heat, as described in Section 4.2. This is the absorption mechanism for electromagnetic radiation incident to a material substance. Explain this mechanism on the basis of the model in Chapter 1.

DP 4.4   What is the basis of Equation 4.10?

DP 4.5   What is the basis of Equation 4.15?

DP 4.6   Modern jet fighter aircraft use deflector plates mounted external to the engine exhaust nozzle to deflect the jet exhaust, thereby providing a strong moment for executing rapid, low-speed maneuvers. Unfortunately, these deflector plates are heated by the jet exhaust and become excellent targets for infrared ("heat seeking") search-and-track (IRST) weapons systems. Consider whether the exposed surfaces of the deflector plates should be metal or ceramic, assuming that either would maintain its required mechanical strength when heated?

DP 4.7   Both Equation 4.34 and Equation 4.36 are versions of Snell's law. Discuss how they are different and how they are alike.

DP 4.8   Scientific satellites are usually fabricated from aluminum alloys but bare aluminum surfaces are rarely left exposed to space. They are usually wrapped with gold-plated Mylar or some similar material. Can you explain this in terms of the material in this chapter?

DP 4.9 A scaling factor can often be applied to estimate the hemispherical, spectral emissivity from the normal ($\theta = 0$), spectral emissivity for a given material. How would this work? For example, what would the scaling factor be for a diffuse surface? For metals in the infrared (Refer to Figure 4.11)?

DP 4.10 The term "interference filtering effect" is used at one point in Section 4.8. What is an interference filter and how does it work?

DP 4.11 In Figure 4.15 the emissivity of the three metals is ranked in order of decreasing direct-current electrical resistivity, with the metal having the highest emissivity at a given temperature also having the highest resistivity. Justify this in terms of the Hagen–Rubens approximation.

Problems

P 4.1 Demonstrate by direct substitution that Equation 4.4 satisfies Maxwell's equations, Equations 1.16, if Equation 4.5 is valid.

P 4.2 Verify Equations 4.6.

P 4.3 Derive Equations 4.7 and 4.8 from Equations 4.6.

P 4.4 Show that $\kappa \rightarrow 0$ as $r_e \rightarrow \infty$.

P 4.5 Verify Equation 4.19.

P 4.6 Derive Equation 4.22 from Equation 4.21 and Snell's law, Equation 4.13.

P 4.7 Derive Equation 4.23.

P 4.8 Derive Equation 4.26 from Equations 4.22 through 4.25.

P 4.9 Eliminate θ_t from Equation 4.26 in favor of n_1/n_2 using Snell's law, Equation 4.13.

P 4.10 Use the result from Problem P 4.9 with Equation 4.27 to reproduce Figure 4.3.

P4.11 Use Equation 4.27 and Equations 4.42 through 4.44 to reproduce the theoretical result in Figure 4.7.

P 4.12 Use the material in Section 4.4 to develop a plot of the ratio of hemispherical emissivity to normal emissivity for dielectrics for a range of values of n_2/n_1.

P 4.13 Use the material in Section 4.5 to estimate the ratio of hemispherical emissivity to normal emissivity for several metals.

P 4.14 Plot the values of the ratios of n and κ from Equations 4.7 and 4.8 to their values obtained from Equation 4.55 for a few realistic values of r_e for a range of wavelengths. In your presentation include results from the literature for a few common metals.

P 4.15 Assume the hemispherical, spectral emissivity of the commercial white paint depicted in Figure 4.5 is insensitive to temperature over the temperature range 250 K $< T <$ 450 K. Prepare a graph that provides an estimate of the variation of the hemispherical, total emissivity between these two temperatures.

P 4.16 A thin plate is insulated on its backside and coated on its front side with the commercial white paint whose hemispherical, spectral emissivity is depicted in Figure 4.5. Estimate the equilibrium temperature of the plate if it were to be exposed to solar radiation.

P 4.17 A thick layer of the commercial white paint whose hemispherical, spectral emissivity is depicted in Figure 4.5 is irradiated by a graybody at a temperature of 1200 K. Estimate the hemispherical, total absorptivity of the layer.

P 4.18 A polished aluminum sheet at a temperature of 273 K is irradiated by a graybody whose temperature is 700 K. Compare the normal, total emissivity with the normal, total absorptivity of the sheet. Comment on the result.

P 4.19 A sheet of polished copper is left out in the sun at noon on a hot day. Estimate the equilibrium temperature attained by the sheet if the back of the sheet is insulated. [Is it reasonable to assume that $\alpha = \varepsilon$ in this situation? Use 0.20 for the hemispherical, total absorptivity for the sunlight, but you are left to your own devices where the emissivity is concerned. Remember, however, that the emissivity would be expected to be temperature dependent.]

P 4.20 The index of refraction of a certain dielectric at a wavelength of 2.0 μm is known to be 1.5. Estimate its hemispherical emissivity into air (n $\cong$ 1.0) at this wavelength.

P 4.21 Estimate the hemispherical emissivity of polished platinum at 300 K for a wavelength of 2.0 μm. [*Hint:* See Figure 4.7.]

REFERENCES

1. Plonsey, R., and R. E. Collin, *Principles and Applications of Electromagnetic Fields,* McGraw-Hill Book Company, Inc., New York, 1961.
2. Born, M., and E. Wolf, *Principles of Optics: Electromagnetic Theory of Propagation, Interference and Diffraction of Light,* 6th corr. ed., Pergamon Press, New York, 1989.
3. Brandenberg, W. M., "The reflectivity of solids at grazing angles," *Measurement of Thermal Radiation Properties of Solids,* J. C. Richmond (ed.), NASA SP-31, 1963.
4. Touloukian, Y. S., and D. P. DeWitt, and R. S. Hernicz (eds.), *Thermal Properties of Matter,* Vol. 9, *Thermal Radiative Properties: Coatings,* IFI/Plenum, New York, 1973 (Curve 3, p. 539).

5. Lide, D. R. (ed.), *CRC Handbook of Chemistry and Physics,* 79th ed., 1998, CRC Press LLC, Cleveland, 1998: Weaver, J. H., and H. P. R. Frederikse, "Optical properties of metals and semiconductors" (Figure 4.7), and "Electrical resistivity of pure metals" (Figures 4.13 and 4.14), ©CRC Press, with permission.

6. Fowles, G. R., *Introduction to Modern Optics,* 2nd ed., Holt, Rinehart and Winston, Inc., New York, 1975, p. 162.

7. Arnold, G. S., "Absorptivity of several metals at 10.6 μm: empirical expressions for the temperature dependence computed from Drude's theory," *Applied Optics,* Vol. 23, No. 9, 1 May 1984, 1434–1436.

8. Ordal, M. A., L. L. Long, R. J. Bell, S. E. Bell, R. R. Bell, R. W. Alexander, and C. A. Ward, "Optical properties of the metals Al, Co, Cu, Au, Fe, Pb, Ni, Pd, Pt, Ag, Ti, and W in the infrared and far infrared," *Applied Optics,* Vol. 22, 1983, pp. 1099–1120.

9. Hass, G., and L. Hadley, "Optical properties of metals," *American Institute of Physics Handbook,* 2nd ed., D. E. Gray (ed.), McGraw-Hill Book Company, New York, 1963, p. 6–119.

10. Padalka, V. G., and I. N. Shklyarevskii, "Determination of the microcharacteristics of silver and gold from the infrared optical constants and the conductivity at 82 and 295°K," *Optics and Spectroscopy,* Vol. X, No. 1, January 1961, pp. 285–288.

11. Toscano, W. M., and E. G. Cravalho, "Experimental measurements of the monochromatic near normal reflectance of gold at cryogenic temperatures," *Heat Transfer 1974,* Vol. 1, *Proceedings of the Fifth International Heat Transfer Conference,* Vol. 1, Tokyo, Japan, 1974, pp. 16–20.

12. Bennett, J. M., and E. J. Ashley, "Infrared reflectance and emittance of silver and gold evaporated in ultrahigh vacuum," *Applied Optics,* Vol. 4, No. 2, February 1965, pp. 221–224.

13. Reuter, G. E. H., and E. H. Sondheimer, "The anomalous skin effect in metals," *Proceedings of the Royal Society, London, Series A,* Vol. 195, 1948, pp. 336–364.

14. Dingle, R. B., "The anomalous skin effect and the reflectivity of metals," *Physica,* Vol. 19, 1953, pp. 311–347.

15. Toscano, W. M., and E. G. Cravalho, "Thermal radiation properties of the noble metals at cryogenic temperatures," *ASME Transactions, Series C, Journal of Heat Transfer,* Vol. 98, No. 3, August 1976, pp. 438–445.

16. Hagen, E., and H. Rubens, "Metallic reflection," *Annalen der Physik,* Vol. 1, No. 2, 1900, pp. 352–375.

17. Touloukian, Y. S., and D. P. DeWitt (eds.), *Thermophysical Properties of Matter,* Vol. 7, *Thermal Radiative Properties: Metallic Elements and Alloys,* IFI/Plenum, New York, 1970.

18. Janssen, J. E., and R. H. Torborg, "Measurement of spectral reflectance using an integrating hemisphere," *Measurement of Thermal Radiation Properties of Solids,* J. C. Richmond (ed.), NASA SP-31, 1963.

19. Birkebac, R. C., and E. R. G. Eckert, "Effects of roughness of metal surfaces on angular distribution of monochromatic reflected radiation," *ASME Transactions, Series C, Journal of Heat Transfer,* Vol. 87, No. 1, February, 1965, pp. 85–94.

20. Torrance, K. E., and E. M. Sparrow, "Off-specular peaks in the directional distribution of reflected thermal radiation," *ASME Transactions, Series C, Journal of Heat Transfer,* Vol. 88, No. 2, May 1966, pp. 223–230.

21. Torrance, K. E., and E. M. Sparrow, "Theory of off-specular reflection from roughened surfaces," *Journal of the Optical Society of America,* Vol. 57, No. 9, 1967, pp. 1105–1114.

22. Schiffer, R., "Reflectivity of a slightly rough surface," *Applied Optics,* Vol. 26, No. 4, 15 February 1987, pp. 704–712.

23. Sparrow, E. M., and V. K. Jonsson, "Thermal radiation absorption in rectangular-groove cavities," *ASME Transactions, Series E, Journal of Applied Mechanics,* Vol. 30, No. 2, June 1963, pp. 237–244.

24. Perlmutter, M., and J. R. Howell, "A strongly directional emitting and absorbing surface," *ASME Transactions, Series C, Journal of Heat Transfer,* Vol. 3, No. 3, August 1963, pp. 282–283.

25. Smith, T. F., and R. G. Hering, "Bidirectional reflectance of a randomly rough surface," Paper No. 71-465, AIAA 6th Thermophysics Conference, Tullahoma, TN, 26–28 April 1971.

26. Christie, F. A., and A. B. DeVriendt, "Bidirectional reflectance from surfaces formed by the ruling of orthogonal parallel V-grooves," Paper No. 72-55, AIAA 10th Aerospace Sciences Meeting, San Diego, CA, 17–19 January 1972.

27. Hesketh, P. J., B. Gebhart, and J. N. Zemel, "Measurements of the spectral and directional emission from microgrooved silicon surfaces," *Transactions of ASME, Journal of Heat Transfer,* Vol. 110, No. 3, August 1988, pp. 680–686.

28. Treat, C. H., and M. W. Wildin, "Investigation of a model for bidirectional reflectance of rough surfaces," Paper No. 69-67, *AIAA 7th Aerospace Sciences Meeting,* 20–22 January 1969, New York.

29. Beckmann, P., and A. Spizzichino, *The Scattering of Electromagnetic Waves from Rough Surfaces,* Pergamon Press Ltd., Oxford, UK, 1963, p. 20.

30. Sánchez-Gil, J. A., and M. Nieto-Vesperinas, "Light scattering from random dielectric surfaces," *Journal of the Optical Society of America A,* Vol. 8, No. 8, August 1991, pp. 1270–1286.

31. Rayleigh, J. W. S., *The Theory of Sound,* Vol. 2, Dover Publications, New York, 1945, pp. 89–96.

32. Rice, S. O., "Reflection of electromagnetic waves from slightly rough surfaces," *Communications in Pure and Applied Mathematics,* Vol. 4, 1951, pp. 4808–4816.

33. Schiffer, R., "Reflectivity of a slightly rough surface," *Applied Optics,* Vol. 26, 15 February 1987, pp. 704–712.

34. Gu, Z.-H., and A. A. Maradudin, eds., *Proceedings of the SPIE: Scattering and Surface Roughness,* Vol. 3141, 1997.

35. Celli, V., A. A. Maradudin, A. M. Marvin, and A. R. McGurn, "Some aspects of light scattering from a randomly rough metal surface," *Journal of the Optical Society of America A,* Vol. 2, No. 12, December 1985, pp. 2225–2239.

36. Dimenna, R. A., and R. O. Buckius, "Electromagnetic theory predictions of the directional scattering from triangular surfaces," *ASME Transactions, Journal of Heat Transfer,* Vol. 116, No. 3, August 1994, pp. 639–645.

5

WAVE PHENOMENA IN THERMAL RADIATION

Most of the tools available for modeling radiation heat transfer are based, either directly or indirectly, on geometric optics. Well-known examples are the net-exchange method and the classical Monte Carlo ray-trace method, both of which are strictly applicable only when the wavelength of the radiation involved is short compared to the physical dimensions of the enclosures being analyzed. However, some cases of engineering interest, especially in the field of engineering optics and instrument design, involve phenomena that the "raylike" description of thermal radiation is unable to account for. This chapter introduces the basic ideas behind these phenomena and sets the stage for their inclusion in the statistical treatment that follows in Part III of this book.

5.1 LIMITATIONS TO THE GEOMETRICAL VIEW OF THERMAL RADIATION

No less of an authority than Max Planck, writing in his 1912* classic, *The Theory of Heat Radiation,* states that in radiation heat transfer analysis

> [I]t will be assumed that the linear dimensions of all parts of space considered, as well as the radii of curvature of all surfaces under consideration, are large compared with the wave lengths of the rays considered. With this assumption we may, without appreciable error, entirely neglect the influence of diffraction caused by the bounding surfaces, and everywhere apply the ordinary laws of reflection and refraction of light.

* Re-released by Dover in 1959 [1].

In Chapter 1 we introduced Maxwell's equations and considered their solution for the special case of the propagation of electromagnetic radiation in free space. We concluded Section 1.10 by stating that the wave nature of thermal radiation may be ignored when modeling most radiation heat transfer problems. (Review of Sections 1.7 through 1.10 is recommended before continuing.) This is because electromagnetic waves often may be assumed to arrive at a given location within an enclosure with broad and more or less random distributions of direction, wavelength, phase, and polarization. In other words the radiation within the enclosure is considered to be spatially and spectrally well mixed so that, for any randomly selected pair of waves arriving at the same point, constructive interference is just as likely to occur as destructive interference. A further underlying assumption, alluded to in the quotation by Planck cited above, is that the wavelengths associated with spectral intervals containing significant power are short compared to the dimensions and radii of curvature of the surfaces of the enclosure.

The cumulative effect of the assumptions outlined in the preceding paragraph is to eliminate the need for all information unique to the wave description of thermal radiation (electric field strength, magnetic field strength, coherence, polarization). The result is that the principles of geometric optics, which treat thermal radiation as a collection of rays, are completely adequate for describing the actual thermal radiative behavior. The popular and widely used "net-exchange" method, which is treated in Part II of this book, and the classical Monte Carlo ray-trace method, which is the subject of Part III of this book, are examples of approaches to radiation heat transfer analysis that are based, either implicitly or explicitly, on geometric optics. Unfortunately, the geometric optics model may not always adequately represent reality. In particular, one or more of the assumptions of the model might be violated if a disproportionate amount of the total radiation energy density within the enclosure is

1. At a wavelength comparable to the dimensions or radii of curvature of the enclosure
2. Monochromatic
3. Collimated
4. Coherent (as in the case of a laser source), or
5. Polarized

While none of these conditions is likely to apply to the typical radiation heat transfer problems encountered by most engineers (e.g., spacecraft thermal control, furnace wall heat flux, and atmospheric reentry), any or all of them might well apply to the design or analysis of an optical system, such as a radiometer, intended to measure thermal radiation in the far infrared that may be monochromatic, collimated, coherent, and/or polarized.

In Chapter 4 we exploited the solution of Maxwell's equations at the plane interface between two media having different indices of refraction in order to derive the law of specular reflection (Equation 4.12) and Snell's law of refraction (Equation 4.13), and to predict certain trends for the surface radiation properties of electrical

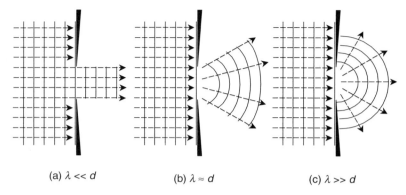

(a) $\lambda \ll d$ (b) $\lambda \approx d$ (c) $\lambda \gg d$

Figure 5.1 The disposition of electromagnetic radiation passing through a slit of width d (a) when $\lambda \ll d$, (b) when $\lambda \approx d$, and (c) when $\lambda \gg d$

conductors and nonconductors. Thus, we are already aware of situations where the raylike description of thermal radiation is inadequate. However, also in Chapter 4 we appealed to a raylike description of thermal radiation in an attempt to explain certain phenomena, such as off-specular peaking, associated with surface roughness. In these cases we imagined rays ricocheting about from peak to peak in a surface topography characterized by a mechanical roughness whose scale was large compared to the wavelength of the incident radiation. This dichotomy of interpretation naturally invokes the question, When may we use the geometric model and when must we consider wavelike behavior?

It is instructive at this point to consider three cases of electromagnetic radiation passing through a slit formed by two parallel knife-edges, as illustrated in Figure 5.1. In Figure 5.1a the wavelength of the light is short compared to the width of the slit, and we observe that a block of rays passes through the slit without significant change of direction. However, in Figure 5.1b, where the wavelength is on the same order of size as the width of the slit, we observe that the rays suffer a change of direction that is more pronounced the closer they pass to the knife edge. Finally, in Figure 5.1c, where the wavelength is long compared to width of the slit, the rays diverge to form a cylindrical fan as they pass through the slit. In this case, one can imagine the slit to be a periodically pulsating line source. From this figure we conclude that an important relationship exists between the wavelength of the radiation and the dimensions of the aperture through which it passes. As we shall see, geometric optics will take us only so far in understanding this phenomenon.

5.2 DIFFRACTION AND INTERFERENCE

In this section we consider two related phenomena, *diffraction* and *interference,* that become important when Planck's short-wavelength assumption is invalid. This may occur either because the enclosure in question is very small or because a significant amount of the energy in the enclosure is in the far infrared.

The terms "diffraction" and "interference" both refer to the algebraic summation of the electric and magnetic components at a point in space occupied by two or more electromagnetic waves having the same frequency and polarization. Interference is said to be *constructive* if the waves are in phase and *destructive* if they are out of phase. Destructive and constructive interference are illustrated in Figure 5.2. In Figure 5.2a two separate *incoherent* TM-polarized electromagnetic waves arrive at the same point and, because they have the same amplitude but are 180 deg out of phase, sum up to zero. In Figure 5.2b two separate *coherent* TM-polarized waves arrive at the same point and, because they have the same amplitude and are exactly in phase, combine to form a wave of amplitude twice that of the two original waves.

The distinction between the terms "diffraction" and "interference" is small. The term "interference" is typically used when the number of waves involved is small, as in the discussion surrounding Figure 5.2, while the term "diffraction" is usually reserved for large-scale wave interactions involving large numbers of waves. Thus, for example, a *diffraction pattern* results when a wave front is broken into an array of cylindrical line sources by a *diffraction grating* and the resulting individual waves interfere behind the grating, as illustrated in Figure 5.3.

Figure 5.3 depicts a series of monochromatic electromagnetic waves, represented by parallel lines, approaching a diffraction grating from the left. The grating may be idealized as a large number (only two of which are shown in the figure) of closely spaced parallel slits etched in a thin metal plate. Here the width of each slit is small compared to the wavelength of the approaching light, as in Figure 5.1c. When the waves reach the grating they are "diffracted" by the slits and leave as expanding cylindrical waves whose axes are parallel and lie within the slits. In other words, each slit acts as a line source of electromagnetic radiation at the wavelength of the original waves incident to the grating. When the expanding cylindrical waves reach a screen that is oriented parallel to the diffraction grating, they interfere either constructively or destructively, depending on the path lengths of the rays arriving at a given vertical position on the screen. For the correct slit spacing and grating-to-

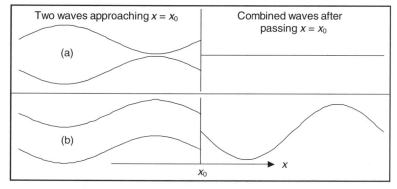

Figure 5.2 Illustration of (a) destructive and (b) constructive interference between two y-polarized electromagnetic waves propagating in the positive-x direction

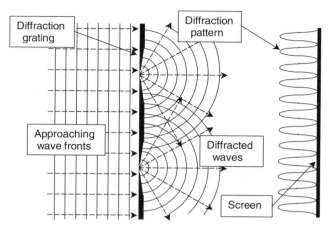

Figure 5.3 Diffraction of a monochromatic electromagnetic wave by a diffraction grating (only two of many slits shown)

screen distance a pattern of parallel light and dark lines, whose spacing is a measure of the wavelength of the incident light, will appear on the screen.

Augustin Jean Fresnel (1788–1827) extended Christian Huygens' (1629–1695) principle to explain the results of an 1801 experiment by Thomas Young (1773–1829). Young had perplexed the adherents to the corpuscular theory of light by demonstrating that a diffraction pattern results if a beam of light from a single source is split into two beams that then follow paths of different length to the same point on a screen. He created the two beams by passing the light from a narrow slit through two closely etched slits in a thin plate. The results of Young's experiment, illustrated schematically in Figure 5.4, are consistent only with the wave view of light

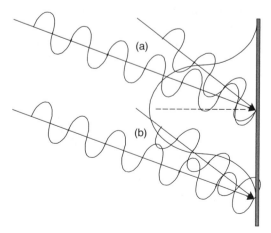

Figure 5.4 Illustration of the principle behind Young's diffraction experiment: (a) constructive and (b) destructive interference

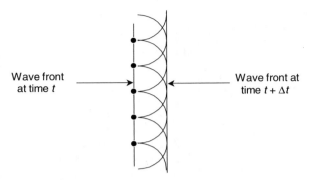

Figure 5.5 Illustration of Huygens' principle

and so were quite controversial.* In Figure 5.4a the difference in path length is an integer multiple of the wavelength, whereas in Figure 5.4b the difference is an integer multiple of wavelength plus one-half wavelength.

Huygens' principle for a propagating electromagnetic wave, illustrated in Figure 5.5, states that every point on the wave front acts as the source of a spherically spreading wave, and that the envelope of these waves forms the new wave front at a later moment in time. In the figure only a few widely spaced point sources (the dots) have been shown, but in reality there would be an unlimited number of such points whose spacing would be arbitrarily small. Fresnel pointed out that the presence of a diffraction grating in the path of a propagating plane wave would have the effect of creating lines of point sources that collectively behave as line sources of cylindrically spreading wave fronts.

Diffraction can be classified into two regimes: the *Fresnel,* or *near-field,* regime and the *Fraunhofer,* or *far-field,* regime. When the screen is in the near field of the slit, rays passing through the slit can arrive at points on the screen directly behind the slit along paths that differ in length by multiples of one-half wavelength. However, when the screen is in the far field of the slit, rays passing through the slit and arriving at the screen in the region directly behind the slit have essentially the same length, measured in terms of half-wavelengths. Therefore, in the far-field, or Fraunhofer, regime diffraction as described by Fresnel cannot occur over screen distances of less than the slit width.

The distribution of relative intensity of monochromatic radiation at wavelength λ arriving at a location x on a screen located a distance z behind a slit of width a is given by [2]

$$\frac{i_\lambda(\lambda, z, x)}{i_{\lambda,0}} = \frac{1}{\lambda z} \left| \int_{-a/2}^{a/2} e^{j(\pi/\lambda z)(x'-x)^2} dx' \right|^2 \tag{5.1}$$

* Recall that Maxwell did not tie electricity and magnetism together with his celebrated wave equations until 1864.

where $i_{\lambda,0}$ is the undiffracted intensity. The dimensionless quantity $\Delta\xi = a(2/\lambda z)^{1/2}$ may be used to classify diffraction into its two regimes, with a value of unity taken as the boundary between the Fresnel and Fraunhofer regimes. Then, if $\Delta\xi$ is greater than 1.0, diffraction is said to be in the Fresnel regime, and if $\Delta\xi$ is less than 1.0, diffraction is said to be in the Fraunhofer regime.

Figure 5.6 illustrates diffraction of 0.6328-μm (red) light by a 100-μm slit for slit-to-screen distances of 316, 645, 2580, and 31,600 μm. These patterns are symmetrical about the center of the slit, so results are shown only for negative values of the distance x measured along the screen from the center of the slit. These results, computed using Equation 5.1, correspond respectively to values of $\Delta\xi$ of 10.0, 5.0, 3.5, and 1.0. The rectangular box in the figure represents the relative light pattern that would appear on the screen in the absence of diffraction ($\Delta\xi \to \infty$). Note that the area under each of the curves, which represents the radiant power passing through the slit, must be equal to the area of the box. It is clear that, as the value of the quantity $\Delta\xi$ increases, the actual light pattern on the screen more nearly approximates the ideal "box" distribution. From this we deduce that for sufficiently short wavelengths and slit-to-screen distances, and for sufficiently wide slits, it might be possible to ignore diffraction and to consider radiation through the slit to be "raylike." However, as we approach the Fraunhofer regime the geometrical optics assumption clearly becomes less tenable.

From Figure 5.6 and its discussion we may conclude that a significant fraction of the monochromatic radiation of wavelength λ passing through a slit of width a will deviate from its geometric path by the time it reaches a plane a distance z beyond the slit if $\Delta\xi < 1.0$. While this conclusion is based on a long slit, a similar conclusion can be drawn for other aperture shapes such as circular, rectangular, and so forth, when the appropriate characteristic dimension is substituted for the slit width a. In summary, it is good practice in the design of optical systems to maintain $\Delta\xi = a(2/\lambda z)^{1/2} \gg 1$, if for no other reason than to facilitate the design process itself.

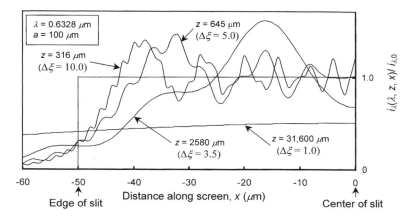

Figure 5.6 Diffraction of red light by a 100-μm slit for several slit-to-screen distances, z

Example Problem 5.1

Figure 5.7 depicts the essential features of a scanning broadband radiometer for use in monitoring the earth's radiative energy balance from low earth orbit. It consists of a collimating baffle, a 1.0-cm-diameter entrance aperture, a Cassegrain telescope, and a thermistor–bolometer thermal radiation detector. Two such instruments would be employed in the intended application: a "total" channel responsive over essentially all of the thermal spectrum (0.1 to 100 μm), and a filtered channel that responds only to the solar spectrum (0.1 to 5 μm). Earth-emitted radiation would then be deduced by differencing the two channels. Estimate the maximum distance z from the entrance aperture to the primary mirror to ensure operation well into the Fresnel regime at all wavelengths of interest.

SOLUTION

We will use as our design criterion $d(2/\lambda z)^{1/2} = 10$, where now d is the diameter of the circular aperture. Thus,

$$z_{max} = \frac{d^2}{50\lambda} = \frac{(10{,}000\ \mu m)^2}{(50)\ (100\ \mu m)} = 20{,}000\ \mu m = 2.0\ cm$$

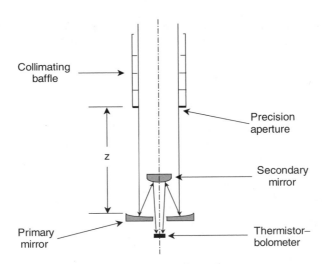

Figure 5.7 A scanning thermistor–bolometer radiometer for monitoring the earth's radiative energy balance

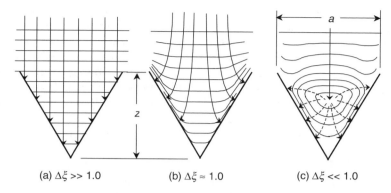

(a) $\Delta\xi \gg 1.0$ (b) $\Delta\xi \approx 1.0$ (c) $\Delta\xi \ll 1.0$

Figure 5.8 Radiation normally incident to a long V-groove for three values of $\Delta\xi = a(2/\lambda z)^{1/2}$: (a) $\Delta\xi \gg 1$, (b) $\Delta\xi \approx 1$, and (c) $\Delta\xi \ll 1$

5.3 CORNER EFFECTS

Consider the long triangular cavity, or "V-groove," formed by two perfectly conducting planes meeting at half-angle $\theta = \tan^{-1}(a/2z)$, as shown in Figure 5.8. Now suppose that a plane wave of monochromatic electromagnetic radiation is normally incident to the cavity opening. How is radiation reflected from the cavity? Three extremes are illustrated in Figure 5.8. Figure 5.8a depicts a situation in which the wavelength of the incident radiation is short compared to the dimensions of the cavity, Figure 5.8c depicts a situation in which the wavelength is long compared to the dimensions of the cavity, and Figure 5.8b depicts an intermediate situation in which the wavelength of the incident radiation is on the order of the dimensions of the cavity.

In Figure 5.8a the "rays" and wave fronts form a rectangular grid, with the rays extending, for all practical purposes, all the way to the surface. Essentially all of the electromagnetic energy in the cavity is tied up in traveling waves. Because the wall is a polished metal the rays would undergo specular reflections and eventually be reflected from the cavity. The wave view of thermal radiation comes into play only at the surface itself, as described in Chapter 4. Geometric optics can be used to a high degree of accuracy to describe the situation depicted in Figure 5.8a.

In Figure 5.8b only a relatively small number of wavelengths of the incident radiation will "fit" into the cavity, and, therefore, a significant fraction of the electromagnetic energy within the cavity is stored in a complex standing-wave structure. The wave fronts are distorted near the walls such that the "rays" follow curved paths there. Maxwell's equations must be solved to accurately describe the resulting potential gradients and the ultimate disposition of incident electromagnetic energy, and geometric optics can provide only an estimate of the pattern of rays reflected from the cavity.

In Figure 5.8c only a fraction of a wavelength of the incident monochromatic electromagnetic wave fits into the cavity. In this case the electrical and magnetic fields oscillate in a "bubble" structure whose complex shape is determined by the solution to Maxwell's equations subject to the zero-electric-field boundary condition imposed by the perfectly conducting walls.* The continuous curves now represent equipotential surfaces, and energy is passed back and forth between the bubble and the walls through the resulting potential gradient. The amount of reflection from the cavity would be highly dependent on wavelength, and the directionality of scattered radiation would approach a cylindrical distribution. Geometric optics could not be used to describe this situation.

It is clear that the situation depicted in Figure 5.8c will occur at all wavelengths and in all triangular cavities regardless of their width and depth. This is because as we move deeper into such cavities the available space becomes more crowded until the situation evolves from that of Figure 5.8a, through that of Figure 5.8b, and eventually to that of Figure 5.8c. This means that Planck's assumption, cited at the beginning of this chapter, can never be completely valid in enclosures having corners. Fortunately, however, in most situations of practical engineering interest the energy involved in the aberrant behavior depicted in Figures 5.8b and c is negligibly small.

Finally, the reader is reminded that the geometry described in Figure 5.8, involving ideal plane surfaces formed by perfectly conducting walls, is highly idealized: corners in real enclosures are composed of textured imperfect conductors and nonconductors. However, while the situation hypothesized in Figure 5.8 does not represent these more realistic situations, the figure and its discussion are useful for understanding the inherent complexity of corner effects in thermal radiation.

Example Problem 5.2

In Section 4.9 surface topography is defined in terms of the mechanical roughness σ_m, a concept used to explain off-specular peaking when $\lambda \ll \sigma_m$. Model a roughened surface as a series of parallel triangular grooves for which $a = z = \sigma_m$, as shown in Figure 5.9. What value of σ_m should be used in conjunction with a green ($\lambda = 0.55$ μm) light source in an experiment intended to test the explanation of off-specular peaking offered in Section 4.8?

SOLUTION

We will use as our criterion $\sigma_m(2/\lambda\sigma_m)^{1/2} = 10$. Thus, $\sigma_m = 50\lambda = (50)(0.55 \ \mu\text{m}) = 27.5 \ \mu$m.

---

* Even though the cavity is an infinitely long groove, under certain circumstances a series of these "bubbles" could form, each having a finite dimension along the length of the groove. The length of the bubbles would depend on the wavelength of the incident monochromatic radiation.

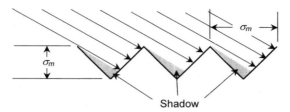

Figure 5.9   Hypothetical surface topography for testing the explanation for off-specular peaking offered in Section 4.8

5.4   POLARIZATION EFFECTS

Thermal radiation emanating from most natural sources is randomly polarized, as illustrated in Figure 5.10a. Familiar examples of sources of randomly polarized thermal radiation include the sun, combustion processes, and most heated surfaces. Exceptions of the latter would be surfaces composed of a material having a regular crystalline structure or whose surface has been micro-machined to produce a pattern of parallel lines of alternating index of refraction. The reader is referred back to Section 1.9 and especially to Figure 1.7 for a brief introduction to the nature of polarized radiation. Note that the terms "randomly polarized" and "unpolarized" are used interchangeably.

Randomly polarized electromagnetic radiation may be polarized by passing it through a polarizing filter, such as the lens of a pair of Polaroid sunglasses. The polarizing filter depicted in Figure 5.10b consists of a transparent substrate onto which has been deposited a pattern of closely spaced parallel rows of a material having a complex index of refraction different from that of the substrate. When randomly polarized light such as that depicted in Figure 5.10a is incident to the filter, only the component at right angles to the pattern is able to pass through without being attenuated, as depicted in Figure 5.10c. Attenuation of unpolarized light by a

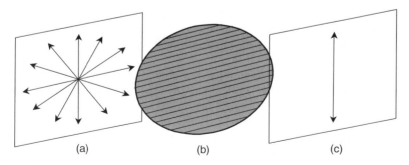

Figure 5.10   Illustration of (a) randomly polarized light, (b) a polarizing filter, and (c) light that has been vertically polarized by the polarizing filter (the electric field vector is represented by double-ended arrows)

single commercial polarizing filter is typically on the order of 60 to 70 percent, depending on the wavelength. If two polarizing filters are rotated by 90 deg with respect to each other (i.e., "crossed"), very little light can pass through. An attenuation of up to 99.99 percent can be achieved by crossing two commercial polarizing filters. Thus, a device can be built that permits the continuous variation of the intensity of light passing through the two filters by rotating one of them with respect to the other (see Team Project TP 5.2). In general, the transmittance of a polarizing filter depends on wavelength because of the variation with wavelength of the complex index of refraction.

While most natural light sources are unpolarized, many polarizing mechanisms exist in nature. For example, sunlight may be polarized by passage through the atmosphere, and especially through clouds. This is because asymmetrical molecules such as H_2O may seek a preferential orientation for one reason or another, for example, in the presence of a sufficiently strong electric field such as might exist in a cloud. Electromagnetic radiation passing through a region of aligned asymmetrical molecules will be polarized because of the difference in dipole moment of the molecule along its long and short axes. For this reason sunlight reaching the ground typically has a different polarization than sunlight reflected back into space, a fact that may be exploited as an atmospheric sounding tool.

Recall that Fresnel's equation,

$$p'_\lambda = \frac{1}{2}(\rho'_{p,\lambda} + \rho'_{s,\lambda})$$

$$= \frac{1}{2}\left[\frac{\tan^2(\theta - \theta_t)}{\tan^2(\theta + \theta_t)} + \frac{\sin^2(\theta - \theta_t)}{\sin^2(\theta + \theta_t)}\right] \qquad (4.26)$$

$$= \frac{1}{2}\frac{\sin^2(\theta - \theta_t)}{\sin^2(\theta + \theta_t)}\left[1 + \frac{\cos^2(\theta + \theta_t)}{\cos^2(\theta - \theta_t)}\right]$$

states that the reflectivity of unpolarized light at the interface between two media whose indices of refraction are n_1 and n_2 is the average of the reflectivities of the transverse-magnetic (p) and transverse-electric (s) polarized components. Then, if $n_2/n_1 = \tan\theta$, the transverse-magnetic polarized component of reflectivity is zero and only transverse-electric polarized radiation is reflected. The angle $\theta = \tan^{-1}(n_2/n_1)$ is called the *Brewster angle*, and the phenomenon it represents is the basis of a laboratory polarizer.

Most real surfaces of engineering interest will polarize reflected radiation to a certain extent. This effect is usually unimportant in radiation heat transfer modeling except, once again, in the design of optical instrumentation, where it can have a significant effect on the spectral distribution of heat flux to the detector. Particular care should be exercised when designing with mirrors. In this case, the degree of polarization of the reflected radiation at a given angle will be wavelength dependent to a degree that depends on the structure of the mirror. This problem is minimized in the case of symmetrical mirrors, as in the Cassegrain telescopes of Figures 5.7 and 5.12, but could be significant in the case of asymmetrical mirror systems.

Team Projects

TP 5.1 In the discussion of Figure 5.3 it is suggested that each slit in a diffraction grating behaves like a cylindrical line source whose intensity varies periodically with time at the frequency of the incident light, and that all of the sources are in phase (please see Discussion Points DP 5.5 and DP 5.6). Consider Young's two-slit experiment. First, see if you can reproduce it. You will need to build a nearly monochromatic light source and find a way to create two parallel slits of the proper width and spacing. Perhaps reference to Young's original work would be helpful here. Then after successfully reproducing Young's experiment, see if you can use a variation on geometrical optics to explain the results. Do this by computing the path lengths of the rays arriving at the same point on the screen from the two slits. Then add the amplitudes of the two rays, taking into account their relative phases. This should be done for a suitable number and spacing of points on the screen. Present the results as a plot of this sum against position on the screen.

TP 5.2 Buy a pair (or borrow your roommate's pair) of "inexpensive" Polaroid sunglasses. Remove the lenses, trim them, and mount them in concentric cardboard tubes as shown in Figure 5.11. A mailing tube for maps and posters, usually available at the local post office, works well in this application because it consists of two snugly fitting cardboard tubes.

The device shown in Figure 5.11 could in principle be used for viewing the sun during a solar eclipse, assuming the quality of the lenses is sufficiently high. *Users are cautioned to first adjust the two lenses so that they are "crossed," for example, by viewing the circumsolar sky, before attempting to view the disk of the sun.*

Discussion Points

DP 5.1 Five conditions are listed in Section 5.1, any one of which could lead to a violation of the assumptions permitting a geometric optics description of radiation heat transfer. Discuss each of these and justify its inclusion on the list.

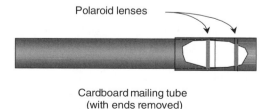

Figure 5.11   An inexpensive rotating polarizer for viewing the sun during an eclipse

DP 5.2   Briefly refer to the description of the "net-exchange" method given in Chapter 8 of this book. What are the assumptions and restrictions inherent to this method?

DP 5.3   Based on what you have read in this chapter and elsewhere in this book, when may we use the geometric model and when must we consider wave-like behavior?

DP 5.4   The terms "coherent" and "incoherent" are used in this chapter. Based on the use of the terms, what may we deduce their definitions to be? What well-known property of laser light depends on its high level of coherence? Please explain the role of coherence in producing this property.

DP 5.5   In the description of Figure 5.3 it is suggested that each slit in a diffraction grating behaves like a cylindrical line source whose intensity varies periodically with time at the frequency of the incident light, and that all of the sources are in phase. Does this mean that the incident light source must be coherent for there to be a diffraction pattern? Please explain your answer.

DP 5.6   In the description of Figure 5.3 it is suggested that each slit in a diffraction grating behaves like a cylindrical line source whose intensity varies periodically with time at the frequency of the incident light, and that all of the sources are in phase (please see Discussion Point DP 5.5). Based on this analogy, how might you expect the diffraction pattern to change if the light illuminating the grating was incident at an angle?

DP 5.7   Did you ever look out the window of an airplane and see the shadow of the plane on the ground or on a cloud? If so, you may have also noticed a rainbow of light surrounding the shadow. How do you explain this phenomenon?

DP 5.8   Where do you believe the factor of "2" comes from in the definition of $\Delta \xi$? [*Hint:* See Problem P 5.1.]

DP 5.9   The diffraction patterns in Figure 5.6 suggest Fourier integral approximations of the "box" function in the figure. What does this suggest to you? (See Problem P 5.4.)

DP 5.10   Consider the scanning radiometer in Figure 5.7. Suppose we were able to redesign the optical train (the Cassegrain telescope) so that the edge of the secondary mirror was in the plane of, and, therefore, helped form, the precision aperture, as shown in Figure 5.12. Do you believe this would be a fundamental improvement in the design of the instrument? Explain your answer.

DP 5.11   Consider the system of parallel triangular grooves depicted in Figure 5.9. Suppose monochromatic light incident to this grating at an angle is reflected onto a screen. How do you believe the light pattern on the screen would vary with position along the screen? Why? [*Hint:* See "diffraction gratings."]

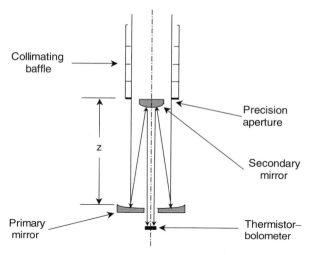

Figure 5.12   A possible modification of the scanning thermistor–bolometer radiometer in Figure 5.7

DP 5.12   Explain the relationship between polarization of light and the electric dipole moment of molecules, implied in the paragraph above Equation 4.26 in this chapter.

DP 5.13   Prepare to lead a discussion on the topic, "How a pattern of parallel lines of differing index of refraction polarizes light." What is the essential role of the index of refraction?

DP 5.14   Elaborate on the last sentence in the chapter. Why is a symmetric-mirror optical train superior to an asymmetrical-mirror optical train for passing polarized light without distortion?

Problems

P 5.1   Show that Equation 5.1 can be written [3]

$$\frac{i_\lambda(\lambda, z, x)}{i_{\lambda,0}} = \frac{1}{2}\,[C(\xi_2) - C(\xi_1)]^2 + \frac{1}{2}[S(\xi_2) - S(\xi_1)]^2 \qquad (5.2)$$

where

$$C(\xi) \equiv \int_0^\xi \cos\left(\frac{1}{2}\pi u^2\right) du \qquad (5.3)$$

and

$$S(\xi) \equiv \int_0^\xi \sin\left(\frac{1}{2}\pi u^2\right) du \qquad (5.4)$$

under the change of variables

$$\xi_1 = \frac{1}{2}\,\Delta\xi\left(\frac{\xi_0}{\Delta\xi} - 1\right) \qquad (5.5)$$

and

$$\xi_2 = \frac{1}{2}\,\Delta\xi\left(\frac{\xi_0}{\Delta\xi} + 1\right) \qquad (5.6)$$

In Equations 5.5 and 5.6, $\xi_0 = -2x(2/\lambda z)^{1/2}$ and $\Delta\xi = a(2/\lambda z)^{1/2}$.

P 5.2   Create a spreadsheet that uses Equations 5.2 through 5.6 to reproduce Figure 5.6. Plot the figure for $-100 \le x \le 100\ \mu$m.

P 5.3   Rework Example Problem 5.1 for the geometry suggested in Discussion Point DP 5.10. Let the diameter of the secondary mirror be 0.5 cm.

P 5.4   Use a Fourier integral approach to approximate the "box" function in Figure 5.6. Consider the possibility that there may be a relationship between the diffraction patterns in the figure and the Fourier integral approximation.

P 5.5   Verify that the parallel component of the directional–hemispherical, spectral reflectivity is zero and that all of the reflected radiation is polarized normal to the plane of incidence if $n_2/n_1 = \tan\theta$ at the interface between two dielectrics whose indices of refraction are n_1 and n_2.

P 5.6   What is the Brewster angle for window glass in the visible part of the thermal spectrum?

REFERENCES

1. Planck, M., *The Theory of Heat Radiation,* Dover Publications, Inc., New York, 1959.
2. Goodman, J. W., *Introduction to Fourier Optics,* McGraw-Hill, New York, 1968, pp. 70–72.
3. Haskell, R. E., "A simple experiment on Fresnel diffraction," *American Journal of Physics,* Vol. 38, No. 8, August 1970, pp. 1039–1042.

6

RADIATION IN A PARTICIPATING MEDIUM

To this point we have confined our study of thermal radiation to its interaction with surfaces. Implicit to the development has been the idea that these surfaces communicate radiatively with their surroundings through a vacuum or some other medium that does not participate in the radiative process. We now consider the case of a participating medium; that is, a volume that absorbs and emits and may also scatter thermal radiation. An important special case is gaseous radiation, and especially radiation in the atmosphere, although the material in this chapter is equally applicable to radiation traversing window glass, lenses, and certain filters.

6.1 MOTIVATION FOR THE STUDY OF RADIATION IN A PARTICIPATING MEDIUM

So far we have considered only situations in which radiation interacts with a surface.* It has been assumed that for all practical purposes the medium lying beyond the surface is a vacuum. This is an acceptable approximation in many situations of engineering interest. Well-known exceptions are glazed solar collectors, which take advantage of the ability of ordinary glass to freely transmit most of the solar spectrum while appearing opaque to infrared radiation, and combustion processes, in which gaseous radiation is often the dominant mechanism of heat transfer. Also, in many

---

* Max Planck [1] reminds us that "we frequently speak of the surface of a body as radiating heat to the surroundings, but this form of expression does not imply that the surface actually emits heat rays. Strictly speaking, the surface of a body never emits rays, but rather it allows part of the rays coming from the interior to pass through." Thus, it is perhaps more correct to say that so far we have considered only situations in which radiation crosses the boundary separating a participating medium from a nonparticipating medium.

common situations involving the atmosphere and optical components such as lenses and filters, the radiation emitted, scattered and absorbed by volume elements can be significant. Examples include the study of atmospheric energetics (e.g., the so-called "greenhouse" effect) and prediction of the infrared signature of aircraft and their exhaust plumes. In the case of atmospheric energetics we are interested both in radiative processes within the atmosphere, including the effects of clouds and aerosols, and the design and analysis of radiometric instruments for measuring the earth's radiative energy budget.

6.2 EMISSION FROM GASES AND (SEMI-)TRANSPARENT SOLIDS AND LIQUIDS

We define the *single-species spectral absorption coefficient,* $\kappa_{\lambda,n}(\lambda, \mathbf{s})$, where $\mathbf{s}$ is a position vector, such that a differential volume element dV located at location $\mathbf{s}$ and containing only species n will emit a power of

$$d^2 Q_{e,\lambda} = 4\pi \kappa_{\lambda,n}(\lambda, \mathbf{s})\, i_{b\lambda}(\lambda, T)\, dV\, d\lambda \quad \text{(W)} \tag{6.1}$$

where, in general, $T = T(\mathbf{s})$. If the medium is a pure gas such as water vapor (H_2O) or carbon dioxide (CO_2), then the subscript n identifies the single gaseous species. In this case, the value of the absorption coefficient is proportional to the mass (or number) density of the gas. If the medium is a liquid or solid then the subscript n simply represents the homogeneous medium such as water or glass. Equation 6.1 can be taken as the definition of the single-species spectral absorption coefficient. This may seem strange since the defining relationship describes emission rather than absorption; however, the reason for this will become clear in the next section.

Consideration of Equation 6.1 reveals that $\kappa_{\lambda,n}$ has the dimensions of reciprocal length L^{-1}. Note that Equation 6.1 embodies the assumption that emission is *isotropic;* that is, that the volume element emits equally in all directions over the 4π solid angle. The quantity $d^2 Q_{e,\lambda}$ can be thought of as the *radiative strength* of the volume element dV in wavelength interval $d\lambda$ about wavelength λ.

6.3 ABSORPTION BY GASES AND (SEMI-)TRANSPARENT SOLIDS AND LIQUIDS

Consider a narrow beam of radiation subtending solid angle $d\omega$ about direction θ, ϕ and passing a distance ds along an arbitrary coordinate axis s through a medium containing only species n whose local single-species spectral absorption coefficient is $\kappa_{\lambda,n}(\lambda, \mathbf{s})$, as illustrated in Figure 6.1. Let the beam be monochromatic, having energy only in wavelength interval $d\lambda$ about wavelength λ. Then the power absorbed by the volume element $dV = A(\mathbf{s})ds$ due to this one beam is

$$d^3 Q_{a,\lambda} = -di_{\lambda,a}(\lambda, \mathbf{s})\, A(\mathbf{s})\, d\omega\, d\lambda \quad \text{(W)} \tag{6.2}$$

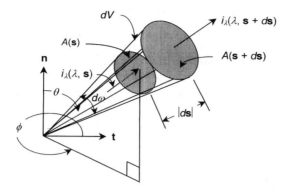

Figure 6.1   Attenuation of monochromatic radiation by absorption

where

$$di_{\lambda,a}(\lambda, \mathbf{s}) = i_\lambda(\lambda, \mathbf{s} + d\mathbf{s}) - i_\lambda(\lambda, \mathbf{s}) = -\kappa_{\lambda,n}(\lambda, \mathbf{s}) \, i_\lambda(\lambda, \mathbf{s}) \, ds \qquad (6.3)$$

Equation 6.3 is known as *Beer's law* (although it should probably be called *Beer's hypothesis*). Beer's law (rather than Equation 6.1) is often taken as the definition of the single-species spectral absorption coefficient, $\kappa_{\lambda,n}(\lambda, \mathbf{s})$. In fact, the value of $\kappa_{\lambda,n}$ is often obtained experimentally by measuring the attenuation of a spectral beam over a known path length and applying Beer's law to the result. With the introduction of Equation 6.3, Equation 6.2 may be written

$$d^3 Q_{a,\lambda} = \kappa_{\lambda,n}(\lambda, \mathbf{s}) \, i_\lambda(\lambda, \mathbf{s}) \, d\omega \, dV \, d\lambda \qquad (6.4)$$

Now suppose the volume element dV is completely surrounded by blackbodies at temperature T. Then, when sufficient time has passed for the volume element to reach thermal equilibrium, we have

$$d^2 Q_{a,\lambda} = d^2 Q_{e\lambda} \qquad (6.5)$$

where now $d^2 Q_{a,\lambda}$ is the power absorbed by dV due to radiation incident from the surrounding 4π-space,

$$d^2 Q_{a,\lambda} = \int_{4\pi} \kappa_{\lambda,n}(\lambda, \mathbf{s}) \, i_{b\lambda}[\lambda, T(\mathbf{s})] \, dV \, d\lambda \, d\omega$$
$$= 4\pi \kappa_{\lambda,n}(\lambda, \mathbf{s}) i_{b\lambda}(\lambda, T) dV \, d\lambda \qquad (6.6)$$

Comparing this result for $d^2 Q_{a,\lambda}$ with the expression for $d^2 Q_{e,\lambda}$, Equation 6.1, it is clear that $\kappa_{\lambda,n}(\lambda, \mathbf{s})$ must be the same quantity in the two expressions. Thus, Kirchhoff's law also holds for the spectral absorption coefficient, and so the same symbol is used both for emission and absorption.

6.4   THE BAND-AVERAGED INTENSITY AND SPECTRAL EMISSION COEFFICIENT

Consider a beam of *band-averaged spectral intensity* $i_{\Delta\lambda,0}$ passing in the positive-s direction along the arbitrary coordinate axis s from the point $s = 0$, as depicted in Figure 6.2. What is the intensity of the beam at a point s if the beam path is through species n only and the *single-species band-averaged spectral emission coefficient* is $\kappa_{\Delta\lambda,n}(\Delta\lambda, s)$? By "band-averaged spectral" it is meant that the relevant quantities have been suitably averaged over the finite wavelength interval of width $\Delta\lambda = \lambda_2 - \lambda_1$; that is,

$$i_{\Delta\lambda} \equiv \frac{1}{\Delta\lambda} \int_{\lambda_1}^{\lambda_2} i_\lambda(\lambda)\, d\lambda \tag{6.7}$$

and

$$\kappa_{\Delta\lambda,n}(\Delta\lambda, s) \equiv \frac{\displaystyle\int_{\lambda_1}^{\lambda_2} \kappa_{\lambda,n}(\lambda, s)\, i_\lambda(\lambda, T)\, d\lambda}{\displaystyle\int_{\lambda_1}^{\lambda_2} i_\lambda(\lambda, T)\, d\lambda} = \frac{\displaystyle\int_{\lambda_1}^{\lambda_2} \kappa_{\lambda,n}(\lambda, s)\, i_\lambda(\lambda, T)\, d\lambda}{i_{\Delta\lambda}(\Delta\lambda, T)\, \Delta\lambda} \tag{6.8}$$

Note that in the limit as $\Delta\lambda$ approaches zero, these quantities converge to their spectral forms so that the symbol $\Delta\lambda$ can be replaced by the symbol λ everywhere in the following development. In general, band-averaged spectral behavior is of more practical interest than spectral behavior in engineering applications, and thus is emphasized in this book.

In practice, a level of approximation is often acceptable in which the band-averaged spectral absorption coefficient is assumed to be uniform over wavelength interval $\Delta\lambda$ with a value given by

$$\kappa_{\Delta\lambda,n}(\Delta\lambda, s) \cong \frac{1}{\Delta\lambda} \int_{\lambda_1}^{\lambda_2} \kappa_{\lambda,n}(\lambda, s)\, d\lambda \tag{6.8a}$$

The approximation represented by Equation 6.8a is accurate only if wavelength intervals are chosen in which the variation of the spectral absorption coefficient $\kappa_{\lambda,n}(\lambda,s)$ is relatively minor. This approach is especially appropriate for use with *band models* for the spectral absorption coefficient, which vary relatively slowly

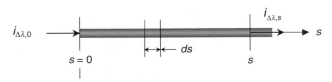

Figure 6.2   A beam of radiation propagating through an absorbing and emitting medium in the positive-s direction

with wavelength, but generally cannot be used with *line-by-line models,* for which the spectral absorption coefficient exhibits strong wavelength dependence. Line-by-line and band models are described in Section 6.8. Because the radiative transport equations become linear under the approximation represented by Equation 6.8a, it permits significant simplification to be achieved in any modeling effort based on Equation 6.3.

In many if not most cases we are interested in the band-averaged spectral absorption coefficient for a mixture of gases. In this case the subscript "*n*" carried by the symbol for the band-averaged spectral absorption coefficient is suppressed. For a given spectral interval $\Delta\lambda$ it may be that only a single species in a gaseous mixture is radiatively active; that is, its single-species, band-averaged spectral absorption coefficient may dominate over those for the other species present. However, the value of its single-species band-averaged spectral absorption coefficient generally will be different from the value for the pure gas. In the development that follows the subscript "*n*" has been suppressed and the resulting equations apply equally well to pure gases, semitransparent liquids and solids, and mixtures of gases. In other words, where appropriate it will be assumed that the contributions of the various species present have been suitably combined to produce a mixture band-averaged spectral absorption coefficient. This procedure is carried out particularly well and efficiently by the statistical band models described in Section 6.9.

6.5 RADIATION SOURCES AND SINKS WITHIN A PURELY ABSORBING, EMITTING MEDIUM

For the situation depicted in Figure 6.2, as the beam propagates along the s axis it suffers both attenuation (by absorption) and augmentation (by emission), so that for a volume element of differential length ds we have

$$di_{\Delta\lambda} = di_{\Delta\lambda,a} + di_{\Delta\lambda,e} = di_{\text{sink},\Delta\lambda} + di_{\text{source},\Delta\lambda} \qquad (6.9)$$

In Equation 6.9, $di_{\Delta\lambda,a}$ is the differential attenuation due to absorption along $ds,$ and so may be thought of as a *sink* of radiation, while $di_{\Delta\lambda,e}$ is the differential augmentation along ds due to emission and so may be thought of as a *source* of radiation. In the case of a narrow beam of solid angle $d\omega$ and cross-sectional area $dA,$ Equation 6.1 can be written

$$\frac{d^3 Q_e}{d\omega\,\Delta A\,d\lambda} = di_{\Delta\lambda,e} = \kappa_{\Delta\lambda} i_{b,\Delta\lambda}[\Delta\lambda,T(s)]\,ds \qquad (6.10)$$

Then, with the introduction of Equation 6.10 and the appropriate form of Beer's law, Equation 6.3, Equation 6.9 may be written

$$di_{\Delta\lambda} = -\kappa_{\Delta\lambda} i_{\Delta\lambda}\,(\Delta\lambda,\,s)\,ds + \kappa_{\Delta\lambda} i_{b,\Delta\lambda}[\Delta\lambda,\,T(s)]\,ds \qquad (6.11)$$

At this point it is convenient to combine the spectral absorption coefficient with the differential space coordinate ds to form the *differential optical coordinate* $d\tau_{\Delta\lambda} = \kappa_{\Delta\lambda}ds$. Then Equation 6.11 can be rewritten as

$$\frac{di_{\Delta\lambda}}{d\tau_{\Delta\lambda}} = -i_{\Delta\lambda}(\Delta\lambda, \tau_{\Delta\lambda}) + i_{b,\Delta\lambda}[\Delta\lambda, T(\tau_{\Delta\lambda})] \tag{6.12}$$

Note that this result tacitly adopts the approximation represented by Equation 6.8a. Equation 6.12 is the *equation of radiative transfer* for radiation propagating in the solid angle $d\omega$ in direction θ, ϕ along the arbitrary optical axis $\tau_{\Delta\lambda}$ through a purely absorbing, emitting medium.

Solution of the equation of radiative transfer in an absorbing, emitting medium may be obtained by multiplying through by the integrating factor e^{τ} to obtain

$$e^{\tau}\frac{di_{\Delta\lambda}}{d\tau} = \frac{d(i_{\Delta\lambda}e^{\tau})}{d\tau} - i_{\Delta\lambda}e^{\tau} = -i_{\Delta\lambda}e^{\tau} + i_{b,\Delta\lambda}[\Delta\lambda, T(\tau)]e^{\tau} \tag{6.13}$$

In Equation 6.13 and the following development the subscript $\Delta\lambda$ carried by τ has been suppressed in the interest of notational simplicity. Equation 6.13 is then multiplied through by $d\tau$ and integrated from $\tau = 0$ to τ, yielding

$$\int_{0}^{\tau} d[i_{\Delta\lambda}(\Delta\lambda, \tau')\, e^{\tau'}] = \int_{0}^{\tau} i_{b,\Delta\lambda}[\Delta\lambda, T(\tau')]\, e^{\tau'}d\tau' \tag{6.14}$$

or

$$i_{\Delta\lambda}(\Delta\lambda, \tau)\, e^{\tau} = i_{\Delta\lambda,0} + \int_{0}^{\tau} i_{b,\Delta\lambda}[\Delta\lambda, T(\tau')]\, e^{\tau'}d\tau' \tag{6.15}$$

Finally, dividing Equation 6.15 through by e^{τ} we obtain

$$i_{\Delta\lambda}(\Delta\lambda, \tau) = i_{\Delta\lambda,0}e^{-\tau} + \int_{0}^{\tau} i_{b,\Delta\lambda}[\Delta\lambda, T(\tau')]\, e^{\tau'-\tau}d\tau' \tag{6.16}$$

The first term on the right-hand side of Equation 6.16 represents the attenuation of band-averaged spectral radiation due to absorption along the path from $\tau = 0$ to τ, and the second term represents augmentation of the band-averaged spectral radiation by emission along the path, reduced by absorption between the emission point τ' and the point τ.

In principle, for the case of a three-dimensional domain filled with an absorbing, emitting medium, Equation 6.12 can be written for all solid angles $d\omega$, i.e., for all directions θ,ϕ, making up the 4π-space surrounding every point in the domain. Then the intensity at a given point is the algebraic sum of the intensities obtained by solving the resulting equations of transfer subject to the boundary condition $i_{\Delta\lambda}(0)$ as indicated above. An approximation of this approach, which would require that the temperature distribution be known throughout the domain, is the basis of the discrete-ordinate method developed in Chapter 7.

6.6 OPTICAL REGIMES

It is often convenient to classify a particular problem into one of three optical regimes based on the value of the band-averaged optical coordinate $\tau_{\Delta\lambda}$ integrated over a distance s,

$$\tau_{\Delta\lambda} \equiv \int_0^s d\tau_{\Delta\lambda} = \int_0^s \kappa_{\Delta\lambda}[\Delta\lambda, T(s')]\, ds' \tag{6.17}$$

The utility of the so-called *band-averaged optical thickness* is that its mean (or extreme) value in a problem domain often dictates the approach used to attack the problem.

$\tau_{\Delta\lambda} \gg 1$. A medium is said to be *optically thick* in a given wavelength interval $\Delta\lambda$ whenever $\tau_{\Delta\lambda} \gg 1$. In an optically thick medium, radiation exchange occurs only among neighboring volume elements. This is the *diffusion limit* in which the governing radiative transport equations are differential equations.

$\tau_{\Delta\lambda} \ll 1$. A medium is said to be *optically thin* in a given wavelength interval $\Delta\lambda$ whenever $\tau_{\Delta\lambda} \ll 1$. Radiation emitted within an optically thin medium travels to the bounding walls of the enclosure surrounding the medium without being absorbed. The optically thin assumption usually leads to significant simplifications in the treatment of a radiation problem.

$\tau_{\Delta\lambda} \approx 1$. If the optical depth of a medium is on the order of unity, the medium is neither optically thick nor optically thin. In this case, radiation is exchanged among all volume elements of the medium and the governing equations of transfer are integro-differential equations.

6.7 TRANSMITTANCE AND ABSORPTANCE OVER AN OPTICAL PATH

It is frequently of interest to know the transmitted or absorbed fraction for radiation incident to a specified thickness of an absorbing medium. Now the concepts of transmissivity and absorptivity have already been introduced in Chapter 3 as properties of surfaces. The same terms could be, and indeed often are, applied to attenuation by radiatively participating volumes. However, to clearly differentiate between the absorption of thermal radiation by surfaces, on the one hand, and by volumes, on the other, in this book the alternative terms *transmittance t* and *absorptance a* are used for volumes.

Consider once again Figure 6.2, which shows a beam of band-averaged spectral intensity $i_{\Delta\lambda,0}$ at $s = 0$ passing along an optical path in the positive-s direction. Then if the band-averaged spectral intensity at s is $i_{\Delta\lambda,s}$, the *band-averaged spectral transmittance* of the optical path of length s is defined as

$$t_{\Delta\lambda}(\Delta\lambda, s) \equiv \frac{i_{\Delta\lambda,s}}{i_{\Delta\lambda,0}} \qquad (6.18)$$

Beer's law, Equation 6.3, can be adapted to the case of band-averaged intensity, rearranged, and integrated over the path length s, to yield

$$\int_{s=0}^{s} \frac{di_{\Delta\lambda}(\Delta\lambda, s')}{i_{\Delta\lambda}(\Delta\lambda, s')} = -\int_{s=0}^{s} \kappa_{\Delta\lambda}(\Delta\lambda, s') \, ds' \qquad (6.19)$$

or

$$\ln\{[i_{\Delta\lambda}(\Delta\lambda, s')]\}\Big|_{i_{\Delta\lambda,0}}^{i_{\Delta\lambda,s}} = \ln\left(\frac{i_{\Delta\lambda,s}}{i_{\Delta\lambda,0}}\right) = -\tau_{\Delta\lambda}(\Delta\lambda, s) \qquad (6.20)$$

As in the case of Equation 6.12—indeed, as in the case of all expressions involving the band-averaged optical coordinate $\tau_{\Delta\lambda}$—Equations 6.19 and 6.20 incorporate the approximation represented by Equation 6.8a. Exponentiating Equation 6.20 and introducing the result into Equation 6.18, there results

$$t_{\Delta\lambda}(\Delta\lambda, s) = e^{-\tau_{\Delta\lambda}} \qquad (6.21)$$

The *band-averaged spectral absorptance* is then defined as

$$a_{\Delta\lambda}(\Delta\lambda, s) \equiv \frac{i_{\Delta\lambda,0} - i_{\Delta\lambda,s}}{i_{\Delta\lambda,0}} = 1 - t_{\Delta\lambda}(\Delta\lambda, s) \qquad (6.22)$$

6.8 EMISSION AND ABSORPTION MECHANISMS IN GASES

A gas molecule stores energy in one of four modes: electronic, translational, vibrational, and rotational.* The electronic storage mode, already discussed briefly in Section 1.6 and illustrated in Figure 1.5, leads to both *line* and *continuum* absorption and emission. The absorption (emission) spectrum of the atmosphere, or of any other mixture of gases for that matter, consists of a distribution of spectral lines superimposed on a background continuum, as illustrated in Figure 1.6.

Figure 6.3 shows the spectral absorption coefficient of a typical spectral line plotted as a function of wavelength. We define the *line strength* as

$$S \equiv \int_{0}^{\infty} \kappa_\lambda(\lambda) \, d\lambda \qquad (6.23)$$

---

* Molecules also store energy in their interatomic bonds as well as in various nuclear forms, but these are unimportant for thermal radiation, whose energy is generally too low to dissociate molecules.

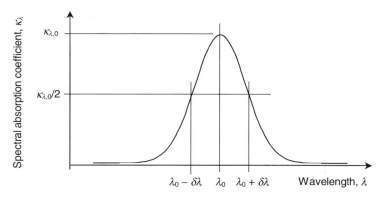

Figure 6.3   A typical line profile and its nomenclature

so that

$$\kappa_\lambda(\lambda) = S\, g_\lambda(\lambda - \lambda_0) \tag{6.24}$$

where the *shape factor* $g_\lambda(\lambda - \lambda_0)$ carries the units of inverse wavelength and has the property

$$\int_{-\infty}^{\infty} g_\lambda(\lambda - \lambda_0)\, d(\lambda - \lambda_0) = 1.0 \tag{6.25}$$

The *line half-width* $\delta\lambda$ corresponds to the two values of wavelength for which the spectral absorption coefficient is one-half of its peak value; that is,

$$\kappa_\lambda(\lambda_0 \pm \delta\lambda) = \frac{\kappa_\lambda(\lambda_0)}{2} \tag{6.26}$$

From Equation 1.14,

$$\nu = \frac{E_a - E_b}{h} \tag{1.14}$$

where $\nu = c/\lambda$, E_a, and E_b are fixed quantum energy states, and c and h are constants, it might be expected that

$$g(\lambda - \lambda_0) = \delta(\lambda - \lambda_0) \tag{6.27}$$

where $\delta(\lambda - \lambda_0)$ is the Dirac delta function. Several mechanisms are responsible for the deviation of the profile of spectral lines from the behavior suggested by Equation 6.27. In addition to the so-called "natural" profile that results from the probabilistic nature of quantum mechanics (i.e., E_a and E_b are not really fixed), several additional mechanisms exist for *line broadening*. Chief among these is *collision* broadening, in which spectral lines are broadened by molecular collisions that distort the molecular structure, and *Doppler broadening,* which occurs because some molecules have a component of velocity toward the observer while others have a component away from the observer. Collision broadening is sensitive to the local number density, or pressure, in the gas, while Doppler broadening is sensitive to the local temperature. Thus, careful measurement of the profiles of certain spectral lines offers opportunities for pressure and temperature sounding in the atmosphere and in other gas mixtures.

When spectral lines are sufficiently closely spaced that they overlap,* they are said to form a *band.* Band structure is especially prevalent in the infrared region of the spectrum, which is dominated by *rotational* or *vibrational* bands associated with water vapor (H_2O) and carbon dioxide (CO_2). A molecule consisting of n atoms has $3n$ degrees of freedom, or nonelectronic energy storage modes. Three of these are translational modes along the three principal axes. Then, for example, a linear diatomic molecule such as O_2 has six degrees of freedom, of which two are rotational, while the triatomic molecule CO_2 has nine degrees of freedom: four bending modes (of which two are identical, or *degenerate*) and two rotational modes. Figure 6.4 shows the structure of the water vapor molecule and its available vibrational and rotational degrees of freedom. Several important bands in the infrared are associated with vibrational and rotational modes of carbon dioxide and water vapor, both of which are "greenhouse" gases.

The translational degree of freedom does not in itself lead to emission from a molecule, but translational energy can be converted into any of the other three forms by collisions. In addition, side bands associated with frequency modulation may be created when collisions associated with the translational modes distort the molecular structure during emission.

---

* Even when broadened lines do not actually overlap, their spacing may be smaller than the spectral resolution of the spectrometer observing them. In this case, they appear as bands (cf., the "bands" in Figure 1.6).

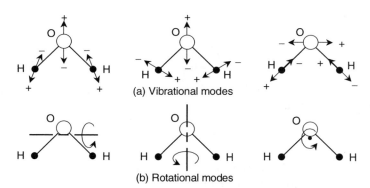

(a) Vibrational modes

(b) Rotational modes

Figure 6.4   Molecular structure and available (a) vibrational and (b) rotational degrees of freedom of water vapor

6.9   SPECTRAL ABSORPTION COEFFICIENT MODELS

The models available for computing spectral absorption coefficients may be divided into two broad classes: *line-by-line models* and *band models*. Line-by-line models seek to represent the electromagnetic spectrum as it would be viewed by a spectrometer. That is, semi-empirical coefficients, based on quantum theory and spectroscopy, are provided that can be used to reassemble the actual line spectrum of a mixture of gases to a high degree of spectral resolution. These are data-intensive models requiring relatively large amounts of computer resources. Their chief advantage is that they can have very fine spectral resolution, which may be required in certain applications such as the design of infrared weapons systems and related instrumentation. Examples of available line-by-line models are LOWTRAN, MODTRAN, HITRAN, and FASCODE [2, 3] to compute increasingly high-resolution spectral absorption coefficients in mixtures of gases whose temperature and species partial pressures are specified.

In radiation heat transfer calculations it is usually adequate, and it is certainly more cost effective in terms of computer resources, to use band-averaged spectral absorption coefficients. These so-called *band models* use a much smaller set of semi-empirical constants to loft curves through the actual line structure. This approach is especially accurate in regions of the spectrum featuring groups of lines that, because of their close spacing and breadth, overlap to form bands rather than identifiable lines (the so-called "weak-line" limit). More to the point, due to the integral nature of radiation heat transfer in which the final product is often the total heat transfer to a surface, the spectral integral of a quantity over the range of a band model can approach the same value as the spectral integral of the quantity over the same spectral range using a high-resolution line-by-line model. In view of their relative simplicity, which translates into greater computational speed and efficiency, band models are the logical choice for many if not most heat transfer calculations.

A band model may be thought of as a low-pass filter that slides along a line spectrum ingesting individual lines $i_{\lambda,L}(\lambda)$ on the front end and spitting out a continuous curve $i_{\lambda,C}(\lambda)$ on the back end such that

$$i_{\lambda,C}(\lambda) = \frac{1}{\lambda} \int_0^{\lambda} i_{\lambda,L}(\lambda')\, d\lambda' \tag{6.28}$$

The concept inherent in Equation 6.28 is illustrated in Figure 6.5. In fact, any practical instrument intended to measure the emission or absorption spectrum of a gas will have a finite spectral resolution. Therefore, it will tend to integrate the energy entering its aperture over a band whose width is equal to its spectral resolution, thereby blurring the spectral sharpness of the spectrum. Stated another way, a spectrometer of spectral resolution $\Delta\lambda$, when viewing a line spectrum, will sense a value of band-averaged intensity over a wavelength interval $\Delta\lambda$ equal to

$$i_{\Delta\lambda} = \frac{1}{\Delta\lambda} \int_{\lambda}^{\lambda+\Delta\lambda} i_{\lambda,L}(\lambda')\, d\lambda' \tag{6.29}$$

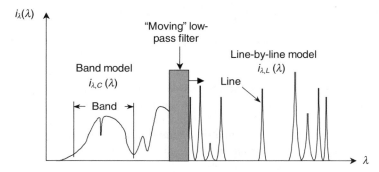

Figure 6.5 Illustration of the relationship between a band model and a line-by-line model

Statistical band models comprise an important class of band models. They are based on quantum mechanical ideas but rely heavily on semi-empirical constants to capture the average behavior of the spectrum over relatively broad spectral intervals. A well-known statistical band model is the so-called NASA band model [4]. A recent enhancement to the NASA band model [5, 6] uses the same equations as those in the model but applies them over much narrower wavelength intervals, thereby providing significantly improved spectral resolution. An application of the enhanced model is documented in Reference 7.

6.10 SCATTERING BY GASES AND (SEMI-)TRANSPARENT SOLIDS AND LIQUIDS

Scattering in participating media is analogous to reflection in surface radiation. The main difference is that scattering occurs in 4π-space while reflection is limited to 2π-space. Scattering may be by molecules or by particles. The *Rayleigh scattering* model for molecular scattering is the topic of Section 6.14. According to Rayleigh's theory, the scattered energy in any direction varies as the fourth power of the frequency ν, or as the inverse of the fourth power of the wavelength λ, of the scattered light. Rayleigh's theory applies when the wavelength of the scattered light is long compared to the size of the scattering centers. This is generally true for gaseous molecules and visible and infrared radiation. The "fourth-power law" means that blue light is scattered more efficiently than red light by molecules in the atmosphere. This explains why the daytime sky appears blue, and why the solar disk appears reddish-orange at sunset and sunrise when the path length through the atmosphere (the "atmospheric mass") is longest.

Mie scattering theory, which is the topic of Section 6.15, describes the interaction between particles and incident radiation when the scattering particles are not small compared to the wavelength of the light being scattered. Mie scattering theory treats the scattering centers as semi-transparent bodies, usually spherical in shape, and scattering may be thought of as a combination of reflection, refraction, and diffraction.

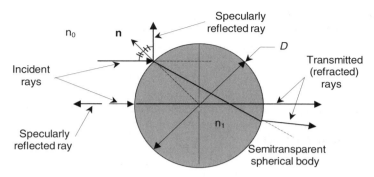

Figure 6.6   Illustration of scattering for the case of a spherical particle with $n_1 > n_0$ and $\pi D \gg \lambda$

When the scattering body is sufficiently large compared to the wavelength of the light being scattered, refraction and reflection dominate and it is possible to treat the radiation as rays. In this case, the situation is as depicted in Figure 6.6, with individual rays being reflected from the surface and refracted within the body, following the law of specular reflection and Snell's law, Equations 4.12 and 4.13, respectively. Results obtained from a geometric ray trace are comparable to those obtained by Mie scattering theory if the wavelength of the incident radiation is sufficiently short compared to the diameter of the sphere. In this case, it would be relatively easy to determine the scattering pattern (called the *phase function*) for a Mie scattering center using the Monte Carlo ray-trace method developed in Part III of the book.

When scattering occurs the scattered radiation continues through the medium, as illustrated in Figure 6.7. The attenuation of an incident beam is given by an expres-

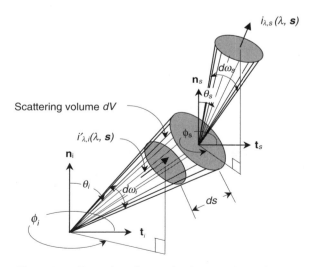

Figure 6.7   Illustration of scattering from a volume element

sion similar to Beer's law. However, it must be kept in mind that radiation is scattered out of the beam rather than absorbed; scattered radiation continues to propagate through the medium. Then we write for the beam attenuation by scattering

$$di_{\lambda,s}(\lambda, s) = -\sigma_\lambda(\lambda, s) \, i_{\lambda,i}(\lambda, s) \, ds \tag{6.30}$$

where σ_λ is the *spectral scattering coefficient,* defined by Equation 6.30.

In general, attenuation of a beam in wavelength interval $\Delta\lambda$ will be due both to absorption and scattering; that is,

$$di_{\text{sink},\Delta\lambda} = di_{\Delta\lambda,a} + di_{\Delta\lambda,s} \tag{6.31}$$

or

$$
\begin{aligned}
di_{\text{sink},\Delta\lambda}(\Delta\lambda, S) &= -[\kappa_{\Delta\lambda}(\Delta\lambda, s) + \sigma_{\Delta\lambda}(\Delta\lambda, s)] \, i_{\Delta\lambda,i}(\Delta\lambda, s) \, ds \\
&= -\beta_{\Delta\lambda}(\Delta\lambda, s) \, i_{\Delta\lambda,i}(\Delta\lambda, s) \, ds
\end{aligned}
\tag{6.32}
$$

where $\beta_{\Delta\lambda}$ is the *band-averaged spectral extinction coefficient.* Note that a band-averaged version of Equation 6.30 can be written using equations analogous to Equations 6.8 and 6.8a. Equations 6.31 through 6.35 are then subject to the same level of approximation as are Equations 6.12 through 6.16.

Rearranging Equation 6.32, we obtain

$$\frac{di_{\text{sink},\Delta\lambda}}{i_{\Delta\lambda}} = -\beta_{\Delta\lambda} \, ds \tag{6.33}$$

and integrating over the path length s yields

$$\int_0^s \frac{di_{\text{sink},\Delta\lambda}}{i_{\Delta\lambda}} = \ln\left[\frac{i_{\text{sink},\Delta\lambda}(s)}{i_{\Delta\lambda,0}}\right] = -\int_0^s \beta_{\Delta\lambda}(s') \, ds' = -\tau_{\Delta\lambda} \tag{6.34}$$

Finally, exponentiating Equation 6.34 and rearranging the result yields

$$i_{\text{sink},\Delta\lambda}(\Delta\lambda, s) = i_{\Delta\lambda,0} \, e^{-\tau_{\Delta\lambda}} \tag{6.35}$$

The quantity $\tau_{\Delta\lambda}$, which accounts for both absorption and emission, is the more general case of the band-averaged optical coordinate defined by Equation 6.17; the classifications in Section 6.6 apply to this quantity when both absorption and scattering are present.

6.11   THE SCATTERING PHASE FUNCTION, Φ

The *local band-averaged spectral scattering phase function,* $\Phi_{\Delta\lambda}(\Delta\lambda, \mathbf{s}, \theta_i, \phi_i, \theta_s, \phi_s)$, is defined as an intensity ratio. Specifically, it is the local ratio of the band-averaged spectral intensity of radiation scattered in direction θ_s, ϕ_s, at position **s** due to a beam incident from direction θ_i, ϕ_i, to the band-averaged spectral intensity that would be scattered in direction θ_s, ϕ_s if scattering were isotropic (i.e., uniform in all directions). Then following this definition the change in band-averaged spectral intensity due to scattering in direction θ_s, ϕ_s of a beam $i_{\Delta\lambda,i}(\Delta\lambda, \mathbf{s}, \theta_i, \phi_i)$ incident from direction θ_i, ϕ_i is

$$di_{\Delta\lambda,s}(\Delta\lambda, \mathbf{s}, \theta_i, \phi_i, \theta_s, \phi_s) = \sigma_{\Delta\lambda} i_{\Delta\lambda,i}(\Delta\lambda, \mathbf{s}, \theta_i, \phi_i)$$

$$\times \frac{\Phi_{\Delta\lambda}(\Delta\lambda, \mathbf{s}, \theta_i, \phi_i, \theta_s, \phi_s)}{4\pi} ds \qquad (6.36)$$

where ds is once again the differential path length along arbitrary ordinate s. Note that the scattering phase function is normalized such that

$$\frac{1}{4\pi} \int_{4\pi} \Phi_{\Delta\lambda}(\Delta\lambda, \mathbf{s}, \theta_i, \phi_i, \theta_s, \phi_s) \, d\omega_i = 1.0 \qquad (6.37)$$

6.12   THE RADIATION SOURCE FUNCTION

When scattering is present in a participating medium, a volume element will act as a source of both emitted radiation, as described in Section 6.2, and scattered radiation, as described in Section 6.11. This latter contribution will be due to radiation incident to the volume element that is scattered without being absorbed within the volume element. It is noted that scattering may be accompanied by a wavelength shift. This could be an important effect in some applications, such as spectroscopy, but it is usually unimportant in radiation heat transfer, especially in the context of a band-averaged analysis.

It is convenient to include scattered radiation along with emitted radiation in a source term similar to Equation 6.1; that is,

$$dQ_{\text{source},\Delta\lambda} = dQ_{e,\Delta\lambda} + dQ_{s,\Delta\lambda} = 4\pi\kappa_{\Delta\lambda}(\Delta\lambda, \mathbf{s}) \, i_{b,\Delta\lambda}(\Delta\lambda, T) \, \Delta\lambda \, dV$$

$$+ \int_{4\pi}\int_{4\pi} \sigma_{\Delta\lambda}(\Delta\lambda, \mathbf{s}) i_{\Delta\lambda,i}(\Delta\lambda, \mathbf{s}) \, \Delta\lambda \qquad (6.38)$$

$$\times \frac{\Phi_{\Delta\lambda}(\Delta\lambda, \mathbf{s}, \theta_i, \phi_i, \theta_s, \phi_s)}{4\pi} \, d\omega_i \, d\omega_s \, dV$$

Equation 6.38 can be differentiated with respect to the cross-sectional area normal to the path s and the scattered solid angle $d\omega_s$ to obtain the change in band-averaged source intensity for a path of differential length ds, yielding

$$di_{\text{source},\Delta\lambda}(\Delta\lambda, \mathbf{s}, \theta_s, \phi_s) = \frac{d^3 Q_{\text{source},\Delta\lambda}}{dA\, d\omega_s} = \kappa_{\Delta\lambda}(\Delta\lambda, \mathbf{s})\, i_{b,\Delta\lambda}(\Delta\lambda, T)\, \Delta\lambda\, ds$$

$$+ \frac{1}{4\pi} \int_{4\pi} \sigma_{\Delta\lambda}(\Delta\lambda, \mathbf{s})\, i_{\Delta\lambda,i}(\Delta\lambda, \mathbf{s})\, \Delta\lambda\, \Phi_{\Delta\lambda} \qquad (6.39)$$

$$\times (\Delta\lambda, \mathbf{s}, \theta_i, \phi_i, \theta_s, \phi_s)\, d\omega_i\, ds$$

In Equation 6.39, the first term on the right-hand side represents emission in direction θ_s, ϕ_s and the second term represents radiation scattered in direction θ_s, ϕ_s due to radiation incident to the volume element from 4π-space. The subscript s in Equations 6.38 and 6.39 coincidentally represents both the scattered direction and the direction along the path s since the two directions are the same in these equations.

Finally, we can now reinterpret the second equality in Equation 6.9 to include emission, absorption, and scattering; that is,

$$di_{\Delta\lambda} = di_{\text{sink},\Delta\lambda} + di_{\text{source},\Delta\lambda} \qquad (6.40)$$

where now $di_{\text{sink},\Delta\lambda}$ is given by Equation 6.32 and $di_{\text{source},\Delta\lambda}$ is given by Equation 6.39.

6.13 THE EQUATIONS OF RADIATIVE TRANSFER

Radiation in a participating medium can be classified into three broad categories: (1) radiation in a purely absorbing, emitting medium (no scattering); (2) radiation in a purely scattering medium (no absorption or emission); and (3) radiation in an absorbing, emitting, and scattering medium. Further, when scattering is present, as in categories 2 and 3, scattering can be *isotropic* (independent of direction) or *anisotropic* (directionally dependent). In this section we present the *equations of radiative transfer* governing the three categories. Derivation of these equations is a straightforward application of Equations 6.32, 6.39, and 6.40.

6.13.1 Radiation in a Purely Absorbing, Emitting Medium (No Scattering)

The equation of transfer describing radiation in a purely absorbing, emitting medium is Equation 6.12 rewritten in terms of the band-averaged spectral absorption coefficient; that is,

$$\frac{di_{\Delta\lambda}}{ds} = -\kappa_{\Delta\lambda} i_{\Delta\lambda}(\Delta\lambda, x) + \kappa_{\Delta\lambda} i_{b,\Delta\lambda}[\Delta\lambda, T(s)] \qquad (6.12a)$$

6.13.2 Radiation in a Purely Scattering Medium (No Absorption or Emission)

We know that radiation is continuously emitted by any medium whose temperature exceeds absolute zero (Prevost's law). However, in many situations the amount of emission and absorption is negligible compared to the amount of scattering, in which case the first term on the right-hand side of Equation 6.37 can be neglected in favor of the second term. An important example is the scattering of sunlight in the atmosphere by molecules and aerosols. In this limit the equation of transfer in the positive-s direction is

$$\frac{di_{\Delta\lambda,s}}{ds} = -\sigma_{\Delta\lambda}i_{\Delta\lambda}(\Delta\lambda, \mathbf{s})$$

$$+\frac{1}{4\pi}\int_{4\pi}\sigma_{\Delta\lambda}(\Delta\lambda, \mathbf{s})\,i_{\Delta\lambda,i}(\Delta\lambda, \mathbf{s})\,\Phi_{\Delta\lambda}(\Delta\lambda,\mathbf{s},\theta_i,\theta_s,\phi_s)\,d\omega_i \qquad (6.41)$$

The first term on the right-hand side of Equation 6.41 represents the contribution to the divergence of intensity in the positive-s direction due to scattering out of the beam incident from the negative-s direction; and the second term represents the contribution due to scattering of radiation, incident to the volume element from 4π-space, in the positive-s ($= \theta_s,\phi_s$) direction. Some radiation incident from the negative-s direction will be *forward scattered* in the positive-s direction. This contribution is included in the second term. As was the case for Equation 6.12a, a version of Equation 6.41 can be written along all possible ordinates of 4π-space. Scattering may be anisotropic, and so more than three ordinate directions may be needed to describe the directionality of scattering.

6.13.3 Radiation in an Absorbing, Emitting, and Scattering Medium

In the most general case of a radiatively participating medium, significant contributions are made to the local divergence of intensity by absorption, emission, and scattering. In this case, the equation of transfer becomes

$$\frac{di_{\Delta\lambda}}{ds} = -\beta_{\Delta\lambda}i_{\Delta\lambda}(\Delta\lambda, \mathbf{s}) + \kappa_{\Delta\lambda}i_{b,\Delta\lambda}(\Delta\lambda, T)$$

$$+\frac{1}{4\pi}\int_{4\pi}\sigma_{\Delta\lambda}(\Delta\lambda, \mathbf{s})\,i_{\Delta\lambda,i}(\Delta\lambda, \mathbf{s})\,\Phi_{\Delta\lambda}(\Delta\lambda,\mathbf{s},\theta_i,\phi_i,\theta_s,\phi_s)\,d\omega_i \qquad (6.42)$$

In Equation 6.42, the first term on the right-hand side represents the contribution to the divergence of the intensity in the positive-s direction due to a combination of absorption and scattering out of the beam, the second term represents the contribution due to local emission, and the third term represents the contribution due to scattering into the beam from the surroundings.

Traditional strategies for the solution of Equations 6.12a, 6.41, and 6.42 are surveyed in Chapter 7, while the most flexible and robust strategy, the Monte Carlo ray-trace method, is the subject of Part III of this book.

6.14 RAYLEIGH SCATTERING

Rayleigh scattering describes scattering by molecules whose size is small compared to the wavelength of the scattered radiation. Rayleigh scattering is relatively easy to explain in terms of the material in Chapter 1, even if the underlying mathematical development is somewhat tedious. Recall that our perspective in Chapter 1 was that of emission from a single atom within a regular crystalline array. Emission occurred when the atom was jostled and thus distorted by its neighbors so that a periodically oscillating electric dipole was created. Then it was argued that a time-varying electric dipole moment behaves like a dipole antenna, radiating electromagnetic energy into the surroundings at the frequency of oscillation of the electric dipole, as developed in Appendix A. Rayleigh scattering is a similar phenomenon except that now the molecule is polarized by an incident electric field $\mathbf{E}_i$ rather than by its rowdy neighbors. Because of its relatively long wavelength the incident field is spatially uniform in the vicinity of the molecule. The polarized molecule, oscillating at the frequency of the incident radiation, then acts as a source of electromagnetic radiation at that frequency, just as in the case of the atom in Chapter 1.

In Section 1.12 we introduced the idea of the *electric dipole moment* of an atom or molecule, and in Appendix A its scalar magnitude is given as $\mathscr{P} = q\delta$ for the geometry of Figure 1.11. However, the electric dipole moment $\mathscr{P}$ is generally a vector quantity defined

$$\mathscr{P} = (q\delta)_x \, \mathbf{i} + (q\delta)_y \, \mathbf{j} + (q\delta)_z \, \mathbf{k} = \bar{\alpha} \mathbf{E}_i \tag{6.43}$$

In Equation 6.43, q is the magnitude of the two charges separated by a distance δ along the specified coordinate axis and $\bar{\alpha}$ is the *polarizability* of the atom or molecule. The polarizability is a tensor quantity because an electric field can produce an electric dipole moment with a component normal to the direction of the applied field in the case of an asymmetric molecule.

We consider the special case of a symmetrical, or isotropic, molecule for which

$$\mathscr{P} = \alpha E_i \tag{6.44}$$

that is, for which the polarizability is a scalar so that polarization occurs only in the direction of the incident electric field. First, let us suppose that linearly polarized radiation with $\mathbf{E}_i = |E_i| e^{-2\pi i c_0 t/\lambda} \, \mathbf{k}$ is incident along the y axis to the molecule shown in Figure 6.8. As derived in Appendix A, the magnitudes of the electric and magnetic fields at a point $\mathbf{s}$ in the far field of a molecular source stimulated by an incident electric field of wavelength λ are

$$E_\theta (\lambda, r, \theta) = \frac{\pi \mathscr{P}}{\varepsilon_0 r \lambda^2} \sin(\theta) \, e^{-2\pi i r/\lambda} \tag{6.45}$$

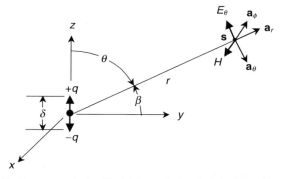

Figure 6.8   The geometry for Rayleigh scattering from an isotropic molecule

and

$$H_\phi(\lambda,r,\theta) = \frac{\pi c_0 \mathcal{P}}{r\lambda^2} \sin(\theta)\, e^{-2\pi i r/\lambda} \tag{6.46}$$

where r is the distance to the field point **s** and θ is the angle between the axis of polarization of the molecule and the line connecting the molecule and the field point. The line r connecting the molecule and the field point **s** lies in the y, z plane, sometimes called the *scattering plane,* as illustrated in Figure 6.8. In Equations 6.45 and 6.46 the scalar quantities E_θ, H_ϕ, and $\mathcal{P}$ are the magnitudes of the corresponding vector quantities. The radiated electric field vector is in the same plane as that of the polarizing electric field vector, and the radiated magnetic field vector is orthogonal to both the direction of propagation (along r) and the electric field vector.

In terms of the angle $\beta = \pi/2 - \theta$, Equations 6.45 and 6.46 may be written

$$E_\theta(\lambda,r,\beta) = \frac{\pi \mathcal{P}}{\varepsilon_0 r\lambda^2} \cos(\beta)\, e^{-2\pi i r/\lambda} \tag{6.47}$$

and

$$H_\phi(\lambda,r,\beta) = \frac{\pi c_0 \mathcal{P}}{r\lambda^2} \cos(\beta)\, e^{-2\pi i r/\lambda} \tag{6.48}$$

Finally, with the introduction of Equation 6.44, the radiated electric and magnetic fields may be expressed in terms of the incident electric field; that is,

$$E_\theta(\lambda,r,\beta) = \frac{\pi \alpha E_i}{\varepsilon_0 r\lambda^2} \cos(\beta)\, e^{-2\pi i r/\lambda} \tag{6.49}$$

and

$$H_\phi(\lambda,r,\beta) = \frac{c_0 \pi \alpha E_i}{r\lambda^2} \cos(\beta)\, e^{-2\pi i r/\lambda} \tag{6.50}$$

Inspection of Equations 6.49 and 6.50 reveals that

$$H_\phi = \frac{E_\theta}{Z_0} \tag{6.51}$$

where $Z_0 = 1/c_0\varepsilon_0$ (Ω) is the *characteristic impedance of free space.* Thus, the power flux at field point **s** propagating along the line directed from the molecule to the field point is

$$P_{r,\lambda}(\lambda,r,\beta) = E_\theta H_\phi^* = \frac{E_\theta E_\theta^*}{Z_0} = \frac{c_0}{\varepsilon_0} \left(\frac{\pi \alpha E_i}{r\lambda^2} \right)^2 \cos^2 \beta \tag{6.52}$$

We next consider randomly polarized, or "natural," electromagnetic radiation incident along the y axis to a molecule at the origin of the x,y,z-coordinate system shown in Figure 6.8. Orientation of the coordinate system r, θ, ϕ whose origin is at field point **s** is indicated by the orthogonal set of unit vectors $\mathbf{a}_r, \mathbf{a}_\theta, \mathbf{a}_\phi$. The unit vector $\mathbf{a}_r$ is colinear with the line r extending from the radiating molecule to the field point **s,** the unit vector $\mathbf{a}_\theta$ lies in the y, z plane and is orthogonal to $\mathbf{a}_r$ and oriented in the direction of angle θ, and the unit vector $\mathbf{a}_\phi$ completes the right-handed system; that is,

$$\mathbf{a}_\phi = \mathbf{a}_r \times \mathbf{a}_\theta \tag{6.53}$$

Because the incident radiation is now unpolarized its electric field can be arbitrarily resolved into two equal-strength components along the x and z axes such that $E_{i,x} = E_{iz} = E_i/2$. Then in addition to Equations 6.49 and 6.50, which still apply with E_i replaced by $E_i/2$, we have

$$E_\phi(\lambda,r) = \frac{\pi \alpha E_i}{2\varepsilon_0 r\lambda^2}\, e^{-2\pi i r/\lambda} \tag{6.54}$$

and

$$H_\theta(\lambda,r) = \frac{c_0 \pi \alpha E_i}{2r\lambda^2}\, e^{-2\pi i r/\lambda} \tag{6.55}$$

since $\theta = \pi/2$ when scattering is normal to the y, z plane. Then the spectral flux propagating away from the molecule along r at **s** is

$$P_{r,\lambda} = \frac{1}{Z_0}(E_\theta E_\theta^* + E_\theta E_\theta^*) \quad (\text{W/m}^2) \tag{6.56}$$

Introducing Equations 6.49 (with E_i replaced by $E_i/2$) and 6.54, Equation 6.56 becomes

$$P_{r,\lambda} (\lambda,r,\beta) = \frac{c_0}{\varepsilon_0} \left(\frac{\alpha \pi E_i}{2r\lambda^2} \right)^2 (\cos^2\beta + 1) \tag{6.57}$$

The average of this radiated flux over 4π-space is

$$\langle P_{r,\lambda} (\lambda,r) \rangle = \frac{1}{4\pi} \int_{4\pi} P(r,\beta)\, d\omega = \frac{4}{3} \frac{c_0}{\varepsilon_0} \left(\frac{\alpha \pi E_i}{2r\lambda^2} \right)^2 \tag{6.58}$$

Thus, *the scattering phase function for Rayleigh scattering of natural radiation by an isotropic molecule* is

$$\Phi_\lambda(\lambda,\beta) = \frac{P_r}{\langle P_r \rangle} = \frac{3}{4} (\cos^2\beta + 1) \tag{6.59}$$

Molecules in the atmosphere are separated by distances much larger than their physical dimensions and are randomly oriented and spaced. Therefore, scattering may be considered *incoherent* and it is reasonable to assume that Equation 6.59 represents the behavior of a finite volume element filled with a statistically representative number of molecules.

We now consider a small volume whose characteristic dimension Δ is large compared to the molecular dipole moment length δ but small compared to the wavelength of the incident radiation λ and, especially, compared to the distance r. Let the volume Δ^3 contain N randomly distributed and oriented molecules of sufficient number so that the molecular number density $\aleph$ is uniform and continuous throughout the volume. It is convenient to express the incident electric field in terms of flux,

$$P_{\Delta,\lambda} = \frac{E_i^2}{2Z_0} = \frac{c_0 \varepsilon_0}{2} E_i^2 \tag{6.60}$$

Then, for incoherent scattering from the volume element containing N molecules, Equation 6.57 can be written

$$P_{r,\lambda} (\lambda,r,\beta) = \frac{N}{2\varepsilon_0^2} \left(\frac{\alpha \pi}{r\lambda^2} \right)^2 P_{\Delta,\lambda} (\cos^2\beta + 1) \tag{6.61}$$

Now the two spectral fluxes $P_{r,\lambda}$ and $P_{\Delta,\lambda}$ in Equation 6.61 can both be converted into intensities by dividing each by the appropriate solid angle. Dividing both sides of Equation 6.61 by A/r^2 and multiplying the right-hand side by Δ^2/Δ^2 yields

$$i_{r,\lambda} \equiv \frac{P_{r,\lambda} (\lambda,r,\beta)}{A/r^2} = \frac{N}{2\varepsilon_0^2} \left(\frac{\alpha \pi}{\lambda^2} \right)^2 \frac{i_{\Delta,\lambda}}{\Delta^2} (\cos^2\beta + 1) \tag{6.62}$$

where

$$i_{r,\lambda} = \frac{P_{r,\lambda}}{A/r^2} \qquad (6.63)$$

and

$$i_{\Delta,\lambda} = \frac{P_{\Delta,\lambda}}{A/\Delta^2} \qquad (6.64)$$

In Equations 6.63 and 6.64 the area A is the invariant cross-section of the beam propagating from the volume Δ^3 to the field point **s.**

We now compare Equation 6.62 with Equation 6.34. If in the latter equation the differential path length ds is replaced by the short but finite path length Δ, then the intensity defined by the equation is finite and is the same quantity as the intensity defined by Equation 6.62. When Equation 6.62 is equated to the suitably modified Equation 6.36 and Equation 6.58 is introduced, there results

$$\frac{N}{2\varepsilon_0^2}\left(\frac{\alpha\pi}{\lambda^2}\right)^2 \frac{i_{\Delta,\lambda}}{\Delta^2}(\cos^2\beta + 1) = \sigma_\lambda i_{\Delta,\lambda}\frac{3}{16\pi}(\cos^2\beta+1)\Delta \qquad (6.65)$$

or, after simplification,

$$\sigma_\lambda = \frac{8}{3}\frac{\pi^3}{\varepsilon_0^2}\aleph\,\alpha^2\lambda^{-4} \qquad (6.66)$$

Finally, with the introduction of the Lorentz–Lorenz relation*

$$\alpha\aleph = 3\varepsilon_0\frac{n^2-1}{n^2+2} \qquad (6.67)$$

we have

$$\sigma_\lambda = \frac{24\pi^3}{\aleph\lambda^4}\left(\frac{n^2-1}{n^2+2}\right)^2 \qquad (6.68)$$

where n is the (real) index of refraction.

6.15 MIE SCATTERING

Scattering centers in the atmosphere include individual molecules, aerosols, and precipitation such as ice crystals and raindrops. While real scattering centers are generally inhomogeneous and irregular in shape, they are usually modeled as being ho-

---

* Coincidentally, this relationship was discovered independently and almost simultaneously by two investigators with remarkably similar names [8].

mogeneous and regular in shape to permit a closed-form solution for the wavelength dependence of their scattering phase function. In 1908 Gustav Mie [9] published a seminal article in which he formulated the problem of scattering of electromagnetic radiation by particles whose complex index of refraction differs from that of the propagating medium. The other important scattering model, attributed to Lord Rayleigh, is limited to scattering from particles, such as individual molecules, whose size is very small compared to the wavelength of the scattered radiation. The Mie scattering model on the other hand can be used to describe scattering for a wide range of values of the size parameter $a = 2\pi r/\lambda$, including molecular scattering. While the Mie model can be used to characterize scattering from irregularly shaped scattering volumes, its use is typically limited to describing scattering from homogeneous spherical particles. Finally, the Mie scattering model can also be used to describe scattering from particles whose characteristic dimension is large compared to the wavelength of the scattering radiation; however, this can also be accomplished—often more directly—using the MCRT approach developed in Part III of this book.

The Mie scattering model is based on solving Maxwell's equations in the interior of the scattering volume and in the surrounding medium, and then matching the two solutions at the interface. Even for homogeneous spherical scattering volumes the formulation is very tedious, involving infinite series of products of periodic functions whose eigenvalues are roots of equations involving Bessel and Hankel functions. Program UNO listed in Appendix B is based on a version of the Mie theory that incorporates certain simplifications that limit the validity of results obtained to values of the size parameter $a = 2\pi r/\lambda$ of less than 250. The program is valid for most atmospheric radiation applications of practical interest, including those involving the scattering of visible and infrared radiation by atmospheric aerosols and precipitation. Readers interested in details about formulating the Mie scattering model are referred to Lenoble's book on atmospheric radiation [10].

Figure 6.9 shows the scattering phase function due to scattering of unpolarized 0.55-μm radiation by a 2.0-μm-diameter spherical aerosol whose complex index of

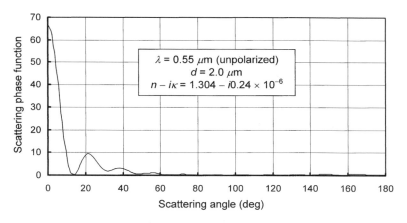

Figure 6.9 Scattering of unpolarized 0.55-μm radiation by a 2.0-μm-diameter spherical aerosol whose complex index of refraction is $1.304 - i0.24 \times 10^{-6}$

refraction is $1.304 - i0.24 \times 10^{-6}$. The phase function is defined such that a value of 1.0 in a given direction corresponds to diffuse scattering without absorption. We see in this example that radiation is strongly scattered in the forward direction ($\theta = 0$) with very little absorption.

Team Project

TP 6.1 Build a simple apparatus for testing, at least qualitatively, the numerical results obtained in Problem P 6.6. Use a clear glass marble whose diameter is near the value specified in the problem, and then scale the remaining dimensions appropriately. (White light may be used rather than green light.)

Discussion Points

DP 6.1 Do you believe that there exists a participating medium for which emission is not isotropic? Can you think of an example? Please justify your answer.

DP 6.2 What distinction do we make between physical law and hypothesis? For example, we might question the practice of referring to Beer's law (or even to Fourier's law) as a law. Why is this?

DP 6.3 Describe a possible experiment for measuring the single-species spectral absorption coefficient. Include in your discussion how you would interpret the data.

DP 6.4 In what situations of practical engineering interest do you believe that it might be acceptable to deal with band-averaged spectral emission coefficients rather than spectral coefficients? In what situations are spectral coefficients required?

DP 6.5 Why do you think the values of single-species absorption coefficients are different in a mixture of gases from their values for the isolated species?

DP 6.6 Think of some examples of when, for a given wavelength interval, the single-species absorption coefficient dominates. Why might one species be much more radiatively active than other species present in a given wavelength interval?

DP 6.7 List several practical situations in which a medium can be considered optically thin. List several practical situations in which a medium can be considered optically thick. Can you think of situations in which the medium is neither optically thin nor optically thick?

DP 6.8 How might collision broadening be used to sound the earth's atmosphere from space to measure the vertical profile of species number density?

DP 6.9 How might Doppler broadening be used to sound the earth's atmosphere from space to measure the vertical profile of temperature?

DP 6.10 Create a version of Figure 6.4 for carbon dioxide (CO_2).

DP 6.11 Go to the literature and find information about "greenhouse" gases. Which gases are considered greenhouse gases? In what wavelength bands do they absorb? What are the absorption mechanisms? Finally, see if you can find a plot of the solar spectrum as viewed through the earth's atmosphere.

DP 6.12 "Side bands associated with frequency modulation" are mentioned in the last sentence of Section 6.8. What is being referred to here? Can you find any reference to or examples of this phenomenon in the literature?

DP 6.13 Visit the websites referenced in connection with LOWTRAN, MOD-TRAN, HITRAN, and FASCODE referred to in Section 6.9. How are they described? What are their limitations? Are these databases and codes generally available to the public? How would you obtain a copy? What would it cost?

DP 6.14 Prepare a brief discussion that completes the chain of logic, begun at the end of the first paragraph in Section 6.10, concerning blue skies and red sunsets.

DP 6.15 How is the spectral scattering coefficient similar to the spectral absorption coefficient, and how is it different?

DP 6.16 Does Rayleigh scattering enhance or diminish the global radiation field? Where in the theoretical development of Section 6.14 is this question answered?

DP 6.17 Prepare a discussion of the statement in Section 6.14 that "the polarizability is a tensor quantity because an electric field can produce an electric dipole moment with a component normal to the direction of the applied field in the case of an asymmetric molecule."

DP 6.18 What is the physical significance of the characteristic impedance of free space, Z_0, defined in the line following Equation 6.51?

DP 6.19 Create a figure illustrating the relative sizes described in the paragraph above Equation 6.60.

Problems

P 6.1 A sheet of 2.0-mm-thick glass whose spectral transmissivity is given in Figure 6.10 is used as the cover of a solar collector. A diffuse solar flux of 350 W/m² is incident on one side, and the other side faces a black collector plate maintained at 340 K. Excluding the solar input, the effective blackbody temperature of the surroundings on the outside of the glazing is

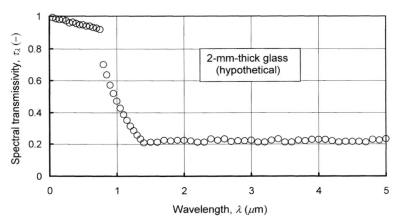

Figure 6.10 Spectral transmissivity of a 2.0-mm-thick sheet of window glass (hypothetical)

300 K. Estimate the equilibrium temperature of the sheet of glass. (You may neglect conduction and convection. Also, assume $\rho_\lambda = 0$ for all wavelengths.)

P 6.2 Categorize the following media as optically thick, optically thin, or optically moderate in thickness (neither thick nor thin). You will have to do some library or Internet research to find the required spectral absorption coefficients.

  (a) Atmospheric air, $\lambda = 9.5\ \mu m$, $s = 10\ km$

  (b) Atmospheric air, $\lambda = 0.55\ \mu m$, $s = 10\ km$

  (c) Atmospheric air, $\lambda = 3.5\ \mu m$, $s = 1.0\ km$

  (d) Window glass, $\lambda = 5\ \mu m$, $s = 2.5\ mm$

  (e) Window glass, $\lambda = 0.5\ \mu m$, $s = 2.5\ mm$

  (f) Pure water, $\lambda = 9.5\ \mu m$, $s = 10\ m$

  (g) Pure water, $\lambda = 0.55\ \mu m$, $s = 10\ m$

P 6.3 Estimate the spectral absorption coefficient for a WG295 filter at 0.4 μm based on the data in Table 3.3. What would be the spectral transmissivity at 0.4 μm of a WG295 filter 1.0 cm thick?

P 6.4 A spherical glass ball 1.0 cm in diameter is irradiated with a beam of randomly polarized green light ($\lambda = 0.55\ \mu m$) also 1.0 cm in diameter, as shown in Figure 6.11. The index of refraction of the glass is 1.5. Use a geometric ray trace to construct a scatter diagram of 10,000 rays incident on a

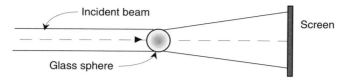

Figure 6.11 A spherical ball irradiated by a uniform cylindrical beam

screen oriented normal to the optical axis 10 cm behind the center of the sphere.

P 6.5 For the situation described in Problem P 6.4, devise a phase function generator based on a geometrical ray trace that produces the zenith (θ_s) and azimuth (ϕ_s) scattering angles with respect to the axis of illumination **n** and an arbitrary tangent vector **t** normal to **n**. The input to the generator should be two random numbers uniformly distributed between zero and unity. The geometry is illustrated in Figure 6.12. Use only the *result* from the ray trace in Problem P 6.4; that is, do not use the known geometry and index of refraction.

P 6.6 Now set up a cubic volume 1.0 m on a side filled with 5000 randomly positioned spherical scattering centers of the type described in Problem P 6.4. Orient the cubic volume as illustrated in Figure 6.13. Trace 100,000 rays through the cubic volume, with all rays normally incident to the face at $x = -0.5$ m. Present the results in terms of scatter diagrams that show where the 100,000 rays exit the six faces of the cubic volume.

P 6.7 Use Program Uno in Appendix B to create plots of the scattering phase function versus scattering angle for values of the size parameter $2\pi r/\lambda$ of 0.01, 0.1, 1, 10, and 40 using the same value of the complex index of refraction n $- i\kappa$ as in the example of Figure 6.9. Discuss the results obtained. What trends with relative particle size do you see? Compare the result for the smallest particle size with the Rayleigh scattering phase function, Equation 6.59. What may we conclude from this comparison?

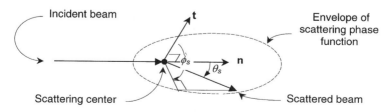

Figure 6.12 Geometry for Problem P 6.7

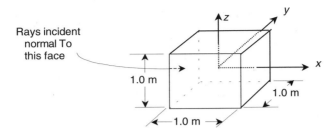

Figure 6.13 A cubic volume 1.0 m on a side filled with 5000 spherical scattering centers of the type described in Problem P 6.6

REFERENCES

1. Planck, M., *The Theory of Heat Radiation,* Dover Publications, Inc., New York, 1959.
2. *http://www.ontar.com/publications.html*
3. *http://www.ontar.com/faq.html*
4. Ludwig, C. B., W. Malkmus, J. E. Reardon, and J. A. L. Thompson, *Handbook of Infrared Radiation from Combustion Gases,* Marshall Spaceflight Center, National Aeronautics and Space Administration Special Publication, NASA SP-3080, 1973.
5. Nelson, E. L., *Development of an Infrared Gaseous Radiation Band Model Based on NASA SP-3080 for Computational Fluid Dynamic Code Validation Applications,* Master-of-Science Thesis, Department of Mechanical Engineering, Virginia Polytechnic Institute and State University, Blacksburg, VA, August 1992.
6. Villeneuve, P. V., *A Parametric Study of the Validity of the Weak-line and Strong-line Limits of Infrared Band Absorption,* Master-of-Science Thesis, Department of Mechanical Engineering, Virginia Polytechnic Institute and State University, Blacksburg, VA, August 1992.
7. Nelson, E. L., J. R. Mahan, L. D. Birckelbaw, J. A. Turk, D. A. Wardwell, and C. E. Hange, *Temperature, Pressure, and Infrared Image Survey of an Axisymmetric Heated Exhaust Plume,* National Aeronautics and Space Administration Technical Memorandum 110382, February 1996.
8. Born, M., and E. Wolf, *Principles of Optics,* Pergamon Press, London, 1959, p. 86 (footnote).
9. Mie, G., "Optics of turbid media," *Annalen der Physik,* Vol. 25, No. 3, 1908, pp. 377–445.
10. Lenoble, J., *Atmospheric Radiative Transfer,* A Deepak Publishing, Hampton, VA, 1993.

PART II

TRADITIONAL METHODS OF RADIATION HEAT TRANSFER ANALYSIS

SOLUTION OF THE EQUATION OF RADIATIVE TRANSFER

In Chapter 6 we learned that radiation through a participating medium is described by an equation of radiative transfer that equates the local gradient of intensity in a given direction to the difference between a local source term arising from emission and net scattering in that direction and local attenuation due to absorption. We now turn our attention to strategies for solution of the equation of radiative transfer. The basic ideas are illustrated by considering classical one-dimensional problems for which analytical or approximate solutions may be readily found. Then two popular approaches suitable for the solution of general multidimensional problems, the differential, or P_N, approximation and the discrete–ordinate, or S_n, approximation, are presented. Finally, the chapter ends with a brief survey of recent improvements to and applications of these two methods.

7.1 ANALYTICAL SOLUTION OF THE EQUATION OF RADIATIVE TRANSFER IN A PURELY ABSORBING, EMITTING, ONE-DIMENSIONAL MEDIUM

Relatively few problems of practical interest permit exact, closed-form, analytical solution of the equation of radiative transfer. However, our interest in such problems is nonetheless justified because of their pedagogical importance. In this section we consider radiative transfer through a purely absorbing, emitting medium bounded by infinite, parallel, plane walls.

7.1.1 Radiative Transfer Through a Purely Absorbing, Emitting Gray Medium Bounded by Infinite, Plane, Parallel Black Walls

Consider a system, illustrated in Figure 7.1, consisting of two infinite, parallel, plane walls separated by a distance L and bounding a uniform, absorbing, emitting

185

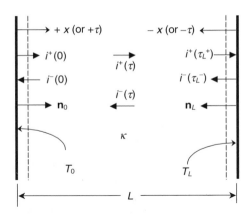

Figure 7.1 Geometry for radiative transfer through a medium bounded by two infinite, parallel, plane walls

medium. To work our way gradually into the topic we begin our study by at first letting the medium be gray and the walls be black; later we will relax these restrictions and consider a nongray medium and diffuse, nongray walls.

It is convenient for mathematical purposes to consider the intensity passing in any direction s to be the difference between a component $i^+(s)$ propagating in the positive-s direction and a component $i^-(s)$ propagating in the negative-s direction, i.e., $i(s) = i^+(s) - i^-(s)$. Then assuming that both absorption and emission in the participating medium are isotropic, Equation 6.12a becomes

$$\frac{di^+}{ds} = -\kappa i^+(s) + \kappa i_b[T(s)] \tag{7.1}$$

for radiation propagating in the positive-s direction, and

$$-\frac{di^-}{ds} = \kappa i^-(s) + \kappa i_b[T(s)] \tag{7.2}$$

for radiation propagating in the negative-s direction. If we further specify that the temperature is uniform at value T in the participating medium, then $i_b(T) = \sigma T^4/\pi$.

For any point between the two bounding planes it is convenient to consider that all beams of intensity i^+ passing through the point propagate into the 2π-space to the right of the point, while all beams of intensity i^- passing through the point propagate into the 2π-space to the left of the point, as illustrated in Figure 7.2a. Now consider a given beam whose path s makes angle θ with respect to the x axis, and let this beam intersect two imaginary parallel planes separated by a distance dx. Then the corresponding distance ds traveled by the beam between the two planes is $dx/\cos\theta$, as illustrated in Figure 7.2b. With this change of variables Equations 7.1 and 7.2 can be written

$$\cos\theta^+ \frac{di^+}{dx} = -\kappa i^+\left(\frac{x}{\cos\theta^+}\right) + \frac{\kappa\sigma T^4}{\pi} \tag{7.3}$$

and

$$-\cos\theta^-\frac{di^-}{dx} = -\kappa i^-\left(\frac{x}{\cos\theta^-}\right) + \frac{\kappa\sigma T^4}{\pi} \tag{7.4}$$

where θ^+ in Equation 7.3 is measured with respect to the normal $\mathbf{n}_0$ to the left-hand bounding plane at $x = 0$, and θ^- in Equation 7.4 is measured with respect to the normal $\mathbf{n}_L$ to the right-hand bounding plane at $x = L$. Finally, with the introduction of the optical coordinate $\tau = \kappa x$, Equations 7.3 and 7.4 become

$$\mu^+\frac{di^+}{d\tau} = -i^+\left(\frac{\tau}{\mu^+}\right) + \frac{\sigma T^4}{\pi} \tag{7.5}$$

and

$$-\mu^-\frac{di^-}{d\tau} = -i^-\left(\frac{\tau}{\mu^-}\right) + \frac{\sigma T^4}{\pi} \tag{7.6}$$

where now $\mu = \cos\theta$.

The exact solution to Equations 7.5 and 7.6 with a uniform temperature is relatively easy to obtain; however, before continuing with the exact solution it is inter-

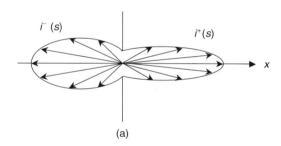

(a)

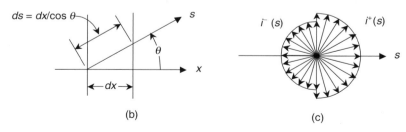

(b)   (c)

Figure 7.2   (a) Intensity distribution in the positive-x and negative-x hemispheres; (b) the relationship between the axis of propagation s, the distance x from the left-hand wall, and the angle θ between s and x; and (c) the assumed intensity distribution in the two-flux model

esting to consider an approximate solution of considerable pedagogical importance. In the so-called *two-flux approximation* the intensity is assumed to be independent of direction within each of the two hemispheres on either side of a plane passing normal to the x axis, as illustrated in Figure 7.2c. Subject to this assumption Equations 7.5 and 7.6 can be multiplied through by $d\omega$ and integrated over the appropriate 2π-space, obtaining

$$\frac{1}{2}\frac{di^+(\tau)}{d\tau} = -i^+(\tau) + \sigma T^4 \tag{7.5a}$$

and

$$-\frac{1}{2}\frac{di^-(\tau)}{d\tau} = -i^-(\tau) + \sigma T^4 \tag{7.6a}$$

We will return to this approximation presently, but first let us complete the exact solution of Equations 7.5 and 7.6.

With the change of variable $I = i - \sigma T^4/\pi$, Equations 7.5 and 7.6 can be written

$$\mu^+\frac{dI^+}{d\tau} = -I^+\left(\frac{\tau}{\mu^+}\right) \tag{7.7}$$

and

$$-\mu^-\frac{dI^-}{d\tau} = -I^-\left(\frac{\tau}{\mu^-}\right) \tag{7.8}$$

Equations 7.7 and 7.8 are variable–separable and subject to the initial conditions $I^+ = I^+(0)$ at $\tau = 0$ and $I^- = I^-(\tau_L/\mu^-)$ at $\tau = \tau_L$. Their solutions are

$$\frac{I^+(\tau/\mu^+)}{I^+(0)} = \frac{i^+(\tau/\mu^+) - \sigma T^4/\pi}{i^+(0) - \sigma T^4/\pi} = e^{-\tau/\mu+} \tag{7.9}$$

and

$$\frac{I^-(\tau/\mu^-)}{I^-(\tau_L/\mu^-)} = \frac{i^-(\tau/\mu^-) - \sigma T^4/\pi}{i^-(\tau_L/\mu^-) - \sigma T^4/\pi} = e^{-(\tau_L-\tau)/\mu^-} \tag{7.10}$$

where $i^+(0) = \sigma T_0^4/\pi$ and $i^-(\tau_L/\mu^-) = \sigma T_L^4/\pi$. Finally, the net flux crossing an imaginary plane located at optical coordinate τ between the two bounding planes is

$$q''(\tau) = q''^+(\tau) - q''^+(\tau) = \int_{2\pi}i^+\left(\frac{\tau}{\mu^+}\right)\mu^+\,d\omega^+ - \int_{2\pi}i^-\left(\frac{\tau}{\mu^-}\right)\mu^-\,d\omega^-$$

$$= 2(\sigma T_0^4 - \sigma T^4)\,E_3(\tau) - 2(\sigma T_L^4 - \sigma T^4)\,E_3(\tau_L - \tau) \tag{7.11}$$

where E_3 is the *exponential integral function of the third kind,* defined

$$E_3(\tau) = \int_0^1 e^{-\tau/\mu}\, \mu\, d\mu \qquad (7.12)$$

The exponential integral of the third kind is a member of the family of exponential integrals of the *n*th kind, defined

$$E_n(\tau) = \int_0^1 e^{-\tau/\mu}\mu^{n-2}\, d\mu \qquad (7.13)$$

Properties of exponential integral functions are given in Chandrasekhar [1] and tables of their values may be found in Kourganoff [2].

Now returning to Equations 7.5a and 7.6a, the net flux subject to the two-flux approximation is given by

$$q''(\tau) = (\sigma T_0^4 - \sigma T^4)\, e^{-2\tau} - (\sigma T_L^4 - \sigma T^4)\, e^{-2(\tau_L - \tau)} \qquad (7.11a)$$

The analysis represented by Equations 7.3 through 7.11 may at first seem to be of only academic interest because of the relatively rare occurrence of a uniform-temperature participating medium. In fact, only two situations lead to a uniform temperature distribution within an absorbing, emitting medium bounded by infinite, parallel, black walls maintained at T_0 and T_L. First, in the absence of other modes of heat transfer (*radiative equilibrium*) a uniform temperature distribution can occur in the medium only if the medium is optically thin, in which case it can be shown that

$$T = [\tfrac{1}{2}(T_0^4 + T_L^4)]^{1/4} \qquad \text{(optically thin)} \qquad (7.14)$$

In this case, the net flux crossing an imaginary plane parallel to the bounding planes and located at any optical coordinate τ is independent of τ and is given by

$$q'' = \sigma(T_0^4 - T_L^4) \qquad \text{(optically thin)} \qquad (7.15)$$

Equation 7.15 follows directly from Equation 7.11 since by definition an absorbing, emitting medium is optically thin in the limit as $\tau_L \geq \tau$ approaches zero.

The other situation in which the temperature distribution within an absorbing, emitting medium bounded by two infinite, parallel black walls can be uniform occurs in the presence of other modes of heat transfer. A local energy balance then gives

$$\dot{q}(\tau) = \frac{dq''(\tau)}{d\tau} \qquad (7.16)$$

where the source term $\dot{q}$ represents the local heat release due to a combination of other modes of heat transfer and volumetric heat generation. With the introduction of Equation 7.11 we find that for the temperature distribution between the two bounding walls to be uniform at T, the local source term must vary with τ according to

$$\dot{q}(\tau) = \frac{dq''(\tau)}{d\tau} = -2[E_2(\tau)(\sigma T_0^4 - \sigma T^4) - E_2(\tau_L - \tau)(\sigma T_L^4 - \sigma T^4)] \quad (7.17)$$

where E_2 is the exponential integral function of the second kind. Clearly, this is a very special condition that would not be expected to occur very often in practice. In spite of this limitation, however, the development leading to Equation 7.11 has an important practical significance. Suppose the temperature profile $T(\tau)$ in a gray absorbing, emitting medium between two parallel black walls is known. Then it is always possible to subdivide the domain between the plane walls into isothermal zones of appropriate width such that the resulting stepwise temperature profile adequately approximates the actual temperature profile. The result obtained for radiative exchange can be expected to approach the exact result in the limit as the number of zones becomes large. The analysis leading to the wall heat fluxes consistent with this approximation, which is a special case of Hottel's zone method [3], is illustrated in the following example problem.

Example Problem 7.1

Suppose the two infinite, parallel black walls shown in Figure 7.1 are spaced a distance $L = 1.0$ m apart and are maintained at temperatures $T_0 = 900$ K and $T_L = 300$ K, while the temperature in the intervening absorbing, emitting medium varies according to $T(x) = 700 - 300x/L$. Let the absorbing–emitting medium be gray with an absorption coefficient of $\kappa = 0.15$ m^{-1}. What is the net radiative heat flux distribution in W/m^2 in the space bounded by the two surfaces?

SOLUTION

We divide the domain between the infinite parallel walls into 10 zones whose temperatures are uniform at a value equal to the mean of the actual temperature distribution in the zone. These temperatures are given in Table 7.1. Then generalizing Equations 7.9 and 7.10, the intensity leaving the nth zone in the positive- and negative-x direction is

$$i^+\left(\frac{\tau_n}{\mu^+}\right) = \frac{\sigma T_0^4}{\pi} e^{-n\Delta\tau/\mu^+} + \sum_{j=1}^{n} [e^{-(n-j)\Delta\tau/\mu^+} - e^{-(n-j+1)\Delta\tau/\mu^+}] \frac{\sigma T_j^4}{\pi} \quad (7.18)$$

Table 7.1 Temperatures and heat fluxes for the ten zones in Example Problem 7.1

n	x_n (m)	T_n (K)	$q''^{+}(n)$ (kW/m^2)	$q''^{+}(n)$ (kW/m^2)	$q''(n)$ (kW/m^2)
0	0	900	37.2	1.86	35.3
1	0.1	685	36.5	1.56	34.9
2	0.2	655	35.8	1.31	34.4
3	0.3	625	35.0	1.11	33.9
4	0.4	595	34.3	0.940	33.3
5	0.5	565	33.5	0.803	32.7
6	0.6	535	32.7	0.694	32.0
7	0.7	505	32.0	0.609	31.3
8	0.8	475	31.2	0.543	30.7
9	0.9	445	30.5	0.495	30.0
10	1.0	415	29.7	0.459	29.3

and

$$i^{-}\left(\frac{\tau_{n-1}}{\mu^{-}}\right) = \frac{\sigma T_L^4}{\pi} e^{-(11 - n)\Delta\tau/\mu^{-}}$$

$$+ \sum_{j=n}^{10} [e^{-(j - n)\Delta\tau/\mu^{-}} - e^{-(j-n+1)\Delta\tau/\mu^{-}}] \frac{\sigma T_j^4}{\pi} \qquad (7.19)$$

Note that these expressions are general, i.e., they apply for any number N of zones, if the numbers 10 and 11 in Equation 7.19 are replaced by N and $N + 1$, respectively. Expressions for the flux leaving zone n in the positive- and negative-x directions, respectively, can be obtained by introducing Equations 7.18 and 7.19 into Equation 7.11; that is,

$$q''^{+}(\tau_n) = [2n^2 E_3 (\Delta\tau, n)] \sigma T_0^4$$

$$+ \sum_{j=1}^{n} [2(n - j)^2 E_3 (\Delta\tau, n - j) - 2(n - j + 1)^2 \qquad (7.20)$$

$$E_3(\Delta\tau, n - j + 1)] \sigma T_j^4$$

and

$$q''^{-}(\tau_{n-1}) = [2(11 - n)^2 E_3 (\Delta\tau, 11 - n)] \sigma T_L^4$$

$$+ \sum_{j=n}^{10} [2(j - n)^2 E_3 (\Delta\tau, j - n) - 2(j - n + 1)^2 \qquad (7.21)$$

$$E_3(\Delta\tau, j - n + 1)] \sigma T_j^4$$

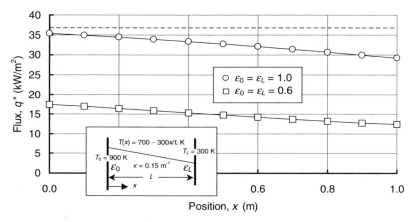

Figure 7.3   Heat flux profiles for Example Problems 7.1 and 7.2

where the *fractional exponential integral of the third kind* $E_3(\Delta\tau, n)$ is defined

$$E_3(\Delta\tau, n) = \int_0^{1/n} e^{\Delta\tau/\mu}\, \mu\, d\mu \tag{7.22}$$

with $2(0)^2 E_3(\Delta\tau, 0) = 1$. Thus, when $n = 1$, $\tau_{n-1} = \tau_0 = 0$ is the optical coordinate of the left-hand wall at $x = 0$, and when $n = 10$, $\tau_{10} = \tau_L = 0.15$ is the optical coordinate of the right-hand wall at $x = 1.0$ m. Table 7.1 gives the values of $q''^{+}(\tau)$, $q''^{-}(\tau)$, and $q''(\tau)$ at the imaginary planes dividing the domain into zones, and Figure 7.3 gives these results in graphical form.

The net flux from the left-hand plane is 35.3 kW/m², and the net heat flux from the right-hand plane is -29.3 kW/m². The horizontal dashed line in Figure 7.3 represents the value of the heat flux crossing any plane parallel to the two bounding planes that would occur in the absence of the intervening participating medium. Therefore, the net effect of the intervening medium in this example is to absorb radiation emitted by the two bounding planes, thereby reducing the net heat transfer between the two planes.

7.1.2   Radiative Transfer Through a Purely Absorbing, Emitting Gray Medium Bounded by Infinite, Plane, Parallel, Gray, Diffuse Walls

The foregoing example was significantly simplified by the fact that the bounding walls were black. If, instead, the walls had been gray and diffuse, the initial conditions for Equations 7.9 and 7.10, $i^{+}(0)$ and $i^{-}(\tau_L/\mu^{-})$, would have been

$$i^{+}(0) = \frac{\varepsilon_0\, \sigma T_0^4}{\pi} + 2(1-\varepsilon_0)\int_0^1 i^{-}(0)\mu^{-}\, d\mu^{-} \tag{7.23}$$

and

$$i^-\left(\frac{\tau_L}{\mu^-}\right) = \frac{\varepsilon_L \sigma T_L^4}{\pi} + 2(1-\varepsilon_L) \int_0^1 i^+\left(\frac{\tau_L}{\mu^+}\right)\mu^+ d\mu^+ \tag{7.24}$$

where $\varepsilon_0 (= \alpha_0)$ and $\varepsilon_L (= \alpha_L)$ are the emissivity of the two gray walls. Introducing Equations 7.23 and 7.24 into Equations 7.9 and 7.10 and evaluating the resulting equations at $\tau = \tau_L$ and $\tau = 0$, respectively, yields

$$i^+\left(\frac{\tau_L}{\mu^+}\right) = \frac{\sigma T^4}{\pi} + \left[\frac{\varepsilon_0 \sigma T_0^4}{\pi} + 2(1-\varepsilon_0)\int_0^1 i^-(0)\,\mu^- d\mu^- - \frac{\sigma T^4}{\pi}\right]e^{-\tau_L/\mu^+} \tag{7.25}$$

and

$$i^-(0) = \frac{\sigma T^4}{\pi}$$
$$+ \left[\frac{\varepsilon_L \sigma T_L^4}{\pi} + 2(1-\varepsilon_L)\int_0^1 i^+\left(\frac{\tau_L}{\mu^+}\right)\mu^+ d\mu^+ - \frac{\sigma T^4}{\pi}\right]e^{-\tau_L/\mu^-} \tag{7.26}$$

Equations 7.23 through 7.26 can now be solved simultaneously for $i^+(0)$ and $i^-(\tau_L/\mu^-)$, yielding

$$i^+(0) = A_0\frac{\sigma T^4}{\pi} + B_0\frac{\sigma T_0^4}{\pi} + C_0\frac{\sigma T_L^4}{\pi} + D \tag{7.27}$$

and

$$i^-\left(\frac{\tau_L}{\mu^-}\right) = A_L\frac{\sigma T^4}{\pi} + B_L\frac{\sigma T_L^4}{\pi} + C_L\frac{\sigma T_0^4}{\pi} + D \tag{7.28}$$

where

$$A_{0,L} = \frac{(1-\varepsilon_{0,L})\{1 - 2E_3(\tau_L) - 4(1-\varepsilon_{0,L})[E_3(\tau_L)]^2\}}{1 - 4(1-\varepsilon_0)(1-\varepsilon_L)[E_3(\tau_L)]^2} \tag{7.29}$$

$$B_{0,L} = \frac{\varepsilon_{0,L}}{1 - 4(1-\varepsilon_0)(1-\varepsilon_L)[E_3(\tau_L)]^2} \tag{7.30}$$

$$C_{0,L} = \frac{2(1-\varepsilon_{0,L})\varepsilon_{L,0}E_3(\tau_L)}{1 - 4(1-\varepsilon_0)(1-\varepsilon_L)[E_3(\tau_L)]^2} \tag{7.31}$$

and

$$D = \frac{2(1-\varepsilon_0)(1-\varepsilon_L)E_3(\tau_L)}{1 - 4(1-\varepsilon_0)(1-\varepsilon_L)[E_3(\tau_L)]^2} \tag{7.32}$$

Once again, the situation of a uniform-temperature absorbing, emitting layer is in itself of limited practical importance; however, Hottel's idea of approximating a nonisothermal layer as a collection of isothermal layers whose individual temperatures are the average of the temperature distribution over their widths remains viable. This idea is illustrated for the case of gray, diffuse bounding walls in Example Problem 7.2.

Example Problem 7.2

Suppose in Example Problem 7.1 that both bounding walls have an emissivity of 0.6. How does this change the results for the net heat flux distribution between the two walls?

SOLUTION

The solution domain is once again subdivided into $N = 10$ equal-thickness zones, each 0.1 m thick. Then we have for the plane at τ_n separating the nth zone from the $n + 1$st zone

$$q_n'' = q_n''^+ - q_n''^- = 2\pi \left[\int_0^1 i^+ \left(\frac{\tau_n}{\mu^+} \right) \mu^+ d\mu^+ - \int_0^1 i^- \left(\frac{\tau_n}{\mu^-} \right) \mu^- d\mu^- \right] \quad (7.33)$$

where

$$i^+ \left(\frac{\tau_n}{\mu^+} \right) = \frac{\sigma T_n^4}{\pi} + \left[i^+ \left(\frac{\tau_{n-1}}{\mu^+} \right) - \frac{\sigma T_n^4}{\pi} \right] e^{-\Delta\tau_n/\mu+} \quad (7.34)$$

and

$$i^- \left(\frac{\tau_n}{\mu^-} \right) = \frac{\sigma T_{n+1}^4}{\pi} + \left[i^- \left(\frac{\tau_{n+1}}{\mu^-} \right) - \frac{\sigma T_{n+1}^4}{\pi} \right] e^{-\Delta\tau_{n+1}/\mu^-} \quad (7.35)$$

Taking into account the recursive nature of Equations 7.34 and 7.35, they may be written

$$i_n^+ = \sum_{j=1}^n [(1 - e^{-\Delta\tau/\mu+})e^{-(j-1)\Delta\tau/\mu+}] \frac{\sigma T_{n+1-j}^4}{\pi}$$

$$+ [(e^{-n\Delta\tau/\mu+})\varepsilon_0] \frac{\sigma T_0^4}{\pi} + [2(1 - \varepsilon_0)e^{-n\Delta\tau/\mu+}] \int_0^1 i_0^- (\mu^-)\mu^- d\mu^- \quad (7.36)$$

and

$$i_{n-1}^- = \sum_{j=n}^{10}[(1 - e^{-\Delta\tau/\mu^-})e^{-(j-n)\Delta\tau/\mu^-}]\frac{\sigma T_j^4}{\pi} + [(e^{-(11-n)\Delta\tau/\mu^-})\varepsilon_L]\frac{\sigma T_L^4}{\pi}$$
$$+ [2(1 - \varepsilon_L)e^{-(11-n)\Delta\tau/\mu^-}]\int_0^1 i_{10}^+(\mu^+)\mu^+ d\mu^+ \tag{7.37}$$

If Equations 7.36 and 7.37 are now introduced into the integrals in Equation 7.33, there result

$$q_n^+ = 2\sum_{j=1}^{n}[(j-1)^2 E_3(\Delta\tau, j-1) - j^2 E_3(\Delta\tau, j)]\sigma T_{n+1-j}^4$$
$$+ 2n^2\varepsilon_0 E_3(\Delta\tau, n)\sigma T_0^4 + 2n^2(1 - \varepsilon_0)E_3(\Delta\tau, n)q_0^- \tag{7.38}$$

and

$$q_{n-1}^- = 2\sum_{j=n}^{10}[(j-1)^2 E_3(\Delta\tau, -1) - (j-n+1)^2 E_3(\Delta\tau, j-n+1)]\sigma T_j^4$$
$$+ 2n^2\varepsilon_L E_3(\Delta\tau, n)\sigma T_L^4 + 2n^2(1 - \varepsilon_L)E_3(\Delta\tau, n)q_{10}^+ \tag{7.39}$$

Particular notice should be taken of the subscripts on i^+ and i^- in Equations 7.37 and 7.39. The possibility for confusion exists because τ_n is the optical coordinate of the plane between zone n and zone $n + 1$, while $q_n^{''-}$ is the left-propagating flux component at τ_n. Thus, i_{10}^- and $q_{10}^{''-}$ are the boundary conditions and so are not represented by Equations 7.37 and 7.39. Then, for example, when using Equation 7.33 to compute $q_8^{''}$, we set $n = 8$ in Equation 7.38 and $n = 9$ in Equation 7.39.

The two unknown boundary conditions may be found by evaluating Equation 7.38 for $n = 10$ and Equation 7.39 for $n = 1$, and then solving the two resulting equations for q_0^- and q_{10}^+. When this is done there results

$$q_{10}^+ = \frac{A^+ + B_0 + (A^- + B_L)C_0}{1 - C_0 C_L} \tag{7.40}$$

and

$$q_0^- = \frac{A^- + B_L + (A^+ + B_0)C_L}{1 - C_0 C_L} \tag{7.41}$$

where

$$A^+ = 2\sum_{j=1}^{10}[(j-1)^2 E_3(\Delta\tau, j-1) - j^2 E_3(\Delta\tau, j)]\sigma T_{11-j}^4 \tag{7.42}$$

Table 7.2 Temperatures and heat fluxes for the ten zones in Example Problem 7.2

n (−)	x_n (m)	T_n (K)	$q''^{+}(n)$ (kW/m^2)	$q''^{-}(n)$ (kW/m^2)	$q''(n)$ (kW/m^2)
0	0	900	25.6	8.08	17.5
1	0.1	685	25.2	8.21	17.0
2	0.2	655	24.8	8.34	16.4
3	0.3	625	24.3	8.47	15.9
4	0.4	595	23.9	8.57	15.3
5	0.5	565	23.4	8.66	14.7
6	0.6	535	22.9	8.73	14.2
7	0.7	505	22.4	8.77	13.7
8	0.8	475	22.0	8.77	13.2
9	0.9	445	21.5	8.72	12.8
10	1.0	415	21.0	8.61	12.4

$$B_0 = 200 \, \varepsilon_0 \, E_3(\Delta\tau, 10)\sigma T_0^4 \tag{7.43}$$

$$C_0 = 200(1 - \varepsilon_0)E_3(\Delta\tau, 10) \tag{7.44}$$

$$A^- = 2\sum_{j=1}^{10} [(j-1)^2 E_3(\Delta\tau, j-1) - j^2 E_3(\Delta\tau, j)]\sigma T_j^4 \tag{7.45}$$

$$B_L = 200 \, \varepsilon_L \, E_3(\Delta\tau, 10)\sigma T_L^4 \tag{7.46}$$

and

$$C_L = 200(1 - \varepsilon_L)E_3(\Delta\tau, 10) \tag{7.47}$$

The heat flux distribution given in Table 7.2 and shown in Figure 7.3 results when the numerical values of L, Δx, ε_0, ε_L, κ, T_0, T_L, and the temperature distribution in the absorbing, emitting medium are introduced for the current example. The not unexpected effect of partially reflecting walls is to lower the heat flux throughout the solution domain in comparison to the results for black walls given in Example Problem 7.1.

7.1.3 Radiative Transfer Through a Purely Absorbing, Emitting Nongray Medium Bounded by Infinite, Plane, Parallel, Nongray, Diffuse Walls

We now consider the more realistic situation for which the purely absorbing, emitting medium and the bounding walls in Figure 7.1 have wavelength-dependent properties. In this case Equations 7.5 and 7.6 become

$$\mu^+ \frac{di_{\Delta\lambda}^+}{dx} = -\kappa_{\Delta\lambda} i_{\Delta\lambda}^+ \left(\frac{x}{\mu^+}\right) + i_{b,\Delta\lambda}(\Delta\lambda, T) \tag{7.48}$$

and

$$-\mu^- \frac{di_{\Delta\lambda}^-}{dx} = -\kappa_{\Delta\lambda} i_{\Delta\lambda}^- \left(\frac{x}{\mu^-}\right) + i_{b,\Delta\lambda}(\Delta\lambda, T) \qquad (7.49)$$

where now the band-averaged spectral absorption coefficient $\kappa_{\Delta\lambda}$ has been separated from the coordinate axis x. This was done because the solutions to Equations 7.48 and 7.49, once obtained, need to be summed over the wavelength bands at common values of x rather than at common values of $\tau_{\Delta\lambda} = \kappa_{\Delta\lambda} x$. In writing Equations 7.48 and 7.49 the medium has once again been assumed to be isothermal at temperature T. Then the general solutions to Equations 7.48 and 7.49 are

$$i_{\Delta\lambda}^+ \left(\Delta\lambda, \frac{x}{\mu^+}\right) = i_{b,\Delta\lambda}(\Delta\lambda, T) + [i_{\Delta\lambda}^+ (\Delta\lambda, 0) - i_{b,\Delta\lambda}(\Delta\lambda, T)]e^{-\kappa_{\Delta\lambda}x/\mu^+} \qquad (7.50)$$

and

$$i_{\Delta\lambda}^- \left(\Delta\lambda, \frac{x}{\mu^-}\right) = i_{b,\Delta\lambda} (\Delta\lambda, T)$$

$$+ [i_{\Delta\lambda}^- \left(\Delta\lambda, \frac{x}{\mu^-}\right) - i_{b,\Delta\lambda}(\Delta\lambda, T)]e^{-\kappa_{\Delta\lambda}(L-x)/\mu^-} \qquad (7.51)$$

The initial conditions to Equations 7.50 and 7.51 are similar in form to those given by Equations 7.23 and 7.24; that is,

$$i_{\Delta\lambda}^+(0) = \varepsilon_{\Delta\lambda,0} i_{b,\Delta\lambda}(\Delta\lambda, T_0) + 2(1 - \varepsilon_{\Delta\lambda,0}) \int_0^1 i_{\Delta\lambda}^-(0)\mu^- d\mu^- \qquad (7.52)$$

and

$$i_{\Delta\lambda}^- \left(\frac{L}{\mu^-}\right) = \varepsilon_{\Delta\lambda,L} i_{b,\Delta\lambda}(\Delta\lambda, T_L) + 2(1 - \varepsilon_{\Delta\lambda,L}) \int_0^1 i_{\Delta\lambda}^+ \left(\frac{L}{\mu^+}\right)\mu^+ d\mu^+ \qquad (7.53)$$

When Equations 7.52 and 7.53 are introduced into Equations 7.50 and 7.51 and the results evaluated at the right- and left-hand walls, respectively, there result

$$i_{\Delta\lambda}^+ \left(\Delta\lambda, \frac{L}{\mu^+}\right) = i_{b,\Delta\lambda}(\Delta\lambda, T)$$

$$+ [\varepsilon_{\Delta\lambda,0} i_{b,\Delta\lambda} (\Delta\lambda, T_0) \qquad (7.54)$$

$$+ 2(1 - \varepsilon_{\Delta\lambda,0}) \int_0^1 i_{\Delta\lambda}^-(0)\mu^- d\mu^- - i_{b,\Delta\lambda}(\Delta\lambda, T)]e^{-\kappa_{\Delta\lambda}L/\mu^+}$$

and

$$
\begin{aligned}
i_{\Delta\lambda}^-(\Delta\lambda,0) = &\; i_{b,\Delta\lambda}(\Delta\lambda,T) \\
&+ \Bigg[\varepsilon_{\Delta\lambda,L} i_{b,\Delta\lambda}(\Delta\lambda,T_L) + 2(1 - \varepsilon_{\Delta\lambda,L}) \\
&\times \int_0^1 i_{\Delta\lambda}^+\!\left(\Delta\lambda,\frac{L}{\mu^+}\right)\mu^+ d\mu^+ - i_{b,\Delta\lambda}(\Delta\lambda,T) \Bigg] e^{-\kappa_{\Delta\lambda}L/\mu^-}
\end{aligned}
\tag{7.55}
$$

Equations 7.52 through 7.55 can now be solved simultaneously for the initial conditions $i_{\Delta\lambda}^+(\Delta\lambda, 0)$ and $i_{\Delta\lambda}^-(\Delta\lambda, L/\mu^-)$, obtaining

$$
\begin{aligned}
i_{\Delta\lambda}^+(0) = &\; A_{\Delta\lambda,0} i_{b,\Delta\lambda}(\Delta\lambda,T) + B_{\Delta\lambda,0} i_{b,\Delta\lambda}(\Delta\lambda,T_0) \\
&+ C_{\Delta\lambda,0} i_{b,\Delta\lambda}(\Delta\lambda,T_L) + D_{\Delta\lambda}
\end{aligned}
\tag{7.56}
$$

and

$$
\begin{aligned}
i_{\Delta\lambda}^-\!\left(\Delta\lambda,\frac{L}{\mu^-}\right) = &\; A_{\Delta\lambda,L} i_{b,\Delta\lambda}(\Delta\lambda,T) + B_{\Delta\lambda,L} i_{b,\Delta\lambda}(\Delta\lambda,T_L) \\
&+ C_{\Delta\lambda,L} i_{b,\Delta\lambda}(\Delta\lambda,T_0) + D_{\Delta\lambda}
\end{aligned}
\tag{7.57}
$$

where

$$
A_{\Delta\lambda,0,L} = \frac{(1 - \varepsilon_{\Delta\lambda,0,L})\{1 - 2E_3(\tau_{\Delta\lambda,L}) - 4(1 - \varepsilon_{\Delta\lambda,0,L})[E_3(\tau_{\Delta\lambda,L})]^2\}}{1 - 4(1 - \varepsilon_{\Delta\lambda,0})(1 - \varepsilon_{\Delta\lambda,L})[E_3(\tau_{\Delta\lambda,L})]^2}
\tag{7.58}
$$

$$
B_{\Delta\lambda,0,L} = \frac{\varepsilon_{\Delta\lambda,0,L}}{1 - 4(1 - \varepsilon_{\Delta\lambda,0})(1 - \varepsilon_{\Delta\lambda,L})[E_3(\tau_{\Delta\lambda,L})]^2}
\tag{7.59}
$$

$$
C_{\Delta\lambda,0,L} = \frac{2(1 - \varepsilon_{\Delta\lambda,0,L})\varepsilon_{\Delta\lambda,L,0} E_3(\tau_{\Delta\lambda,L})}{1 - 4(1 - \varepsilon_{\Delta\lambda,0})(1 - \varepsilon_{\Delta\lambda,L})[E_3(\tau_{\Delta\lambda,L})]^2}
\tag{7.60}
$$

and

$$
D_{\Delta\lambda} = \frac{2(1 - \varepsilon_{\Delta\lambda,0})(1 - \varepsilon_{\Delta\lambda,L})E_3(\tau_{\Delta\lambda,L})}{1 - 4(1 - \varepsilon_{\Delta\lambda,0})(1 - \varepsilon_{\Delta\lambda,L})[E_3(\tau_{\Delta\lambda,L})]^2}
\tag{7.61}
$$

In Equations 7.58 through 7.61, $\tau_{\Delta\lambda,L} = \kappa_{\Delta\lambda}L$.

Once again, the interesting case is the one for which the temperature varies with position between the two bounding walls. In this case the right- and left-propagating components of heat flux crossing the plane between zone n and zone $n + 1$ are given by relations analogous to Equations 7.38 and 7.39; that is

$$
\begin{aligned}
q_{\Delta\lambda,n}^+ = &\; 2\pi \sum_{j=1}^n [(j - 1)^2 E_3(\Delta\tau_{\Delta\lambda}, j - 1) - j^2 E_3(\Delta\tau_{\Delta\lambda,j})] i_{b,\Delta\lambda}(\Delta\lambda,T_{n+1-j}) \\
&+ 2\pi n^2 \varepsilon_{\Delta\lambda,0} E_3(\Delta\tau_{\Delta\lambda}, n) i_{b,\Delta\lambda}(\Delta\lambda,T_0) + 2n^2 (1 - \varepsilon_0)E_3(\Delta\tau_{\Delta\lambda},n)q_{\Delta\lambda,0}^-
\end{aligned}
\tag{7.62}
$$

and

$$
\begin{aligned}
q_{\Delta\lambda,n-1}^- = 2\pi \sum_{j=n}^{10} [&(j-1)^2 \, E_3(\Delta\tau_{\Delta\lambda}, j-1) \\
&- (j-n+1)^2 \, E_3(\Delta\tau_{\Delta\lambda}, j-n+1)] \, i_{b,\Delta\lambda}(\Delta\lambda, T_j) \\
&+ 2\pi n^2 \varepsilon_{\Delta\lambda,L} E_3(\Delta\tau_{\Delta\lambda}, n) \, i_{b,\Delta\lambda}(\Delta\lambda, T_L) \\
&+ 2n^2 \, (1 - \varepsilon_{\Delta\lambda,L}) E_3(\Delta\tau_{\Delta\lambda}, n) \, q_{\Delta\lambda,10}^+
\end{aligned}
\tag{7.63}
$$

where $\Delta\tau_{\Delta\lambda} = \kappa_{\Delta\lambda}\Delta x$ and where the boundary conditions $q_{\Delta\lambda,0}^-$ and $q_{\Delta\lambda,10}^+$ are obtained in a manner analogous to that leading to Equations 7.40 and 7.41. The band-averaged spectral intensities and absorption coefficients in the above equations are defined by Equations 6.7 and 6.8.

Finally, the total local net heat flux crossing the plane between zone n and zone $n + 1$ is

$$
q_n = \sum_{k=1}^{K} (q_{\Delta\lambda_k,n}^+ - q_{\Delta\lambda_k,n}^-)
\tag{7.64}
$$

where $k = 1, 2, \ldots, K$ is the index of the K wavelength bands.

In the most general situation the temperature profile is, in fact, unknown, in which case equations must be available expressing conservation of energy and, in the case of a flowing medium, conservation of mass and momentum. Additional relations describing conservation of chemical species and thermodynamic state will be required in the presence of chemical reactions. All of these equations must then be solved simultaneously to obtain the temperature and intensity distributions.

7.2 ANALYTICAL SOLUTION OF THE EQUATION OF RADIATIVE TRANSFER IN A PURELY SCATTERING ONE-DIMENSIONAL MEDIUM

An important example of one-dimensional radiative transfer through a purely scattering medium is the propagation of solar radiation through the atmosphere. In this case the wavelength interval occupied by most of the solar flux is sufficiently removed from the important atmospheric absorption and emission bands that these phenomena can for the most part be neglected. Further, because of the atmosphere's thinness relative to the radius of the earth, adequate accuracy can often be obtained by considering the atmosphere, or at least sufficiently thin slices of it, to be *plane and parallel,* meaning that the scattered intensity is uniform within parallel planes.

7.2.1 Isotropic Scattering

In the case of isotropic scattering ($\Phi = 1$) in a gray atmosphere, the intensity in the upwelling (positive-z) direction is

$$i^+(\tau, \mu^+) = i^+(0, \mu^+) e^{-\tau/\mu^+} + \int_0^\tau \bar{i}(\tau') e^{-(\tau-\tau')/\mu^+} d\left(\frac{\tau'}{\mu^+}\right) \qquad (7.65)$$

and the intensity in the downwelling (negative-z) direction is

$$i^-(\tau, \mu^-) = i^-(\tau_L, \mu^-) e^{-(\tau_L - \tau)/\mu^-} + \int_{\tau_L}^\tau \bar{i}(\tau') e^{-(\tau' - \tau)/\mu^-} d\left(\frac{\tau'}{\mu^-}\right) \qquad (7.66)$$

where now $\tau = \sigma z$ and

$$\bar{i}(\tau) \equiv \frac{1}{4\pi} \int_{4\pi} i(\tau, \omega_i)\, d\omega_i = \frac{1}{2}\left[\int_0^1 i^+(\tau, \mu^+)\, d\mu^+ + \int_0^1 i^-(\tau, \mu^-)\, d\mu^-\right] \qquad (7.67)$$

Equations 7.65 and 7.66 follow immediately by application of an integrating factor to solve Equation 6.41 specialized to the case of plane, parallel radiation. The reader desiring details concerning the use of integrating factor to solve first-order differential equations is invited to review the derivation of Equation 6.16.

In terms of the two components of intensity described by Equations 7.65 and 7.66, the net flux propagating in the positive-z direction can be written

$$
\begin{aligned}
q'' = q''^+ - q''^- &= 2\pi\left[\int_0^1 i^+(\tau, \mu^+)\mu^+ d\mu^+ - \int_0^1 i^-(\tau, \mu^-)\, \mu^- d\mu^-\right]\\
&= 2\pi\left[\int_0^1 i^+(0, \mu^+) e^{-\tau/\mu^+}\, \mu^+ d\mu^+\right.\\
&\quad + \int_0^\tau i(\tau') \int_0^1 e^{-(\tau-\tau')/\mu^+}\, d\mu^+ d\tau'\\
&\quad - \int_0^1 i^-(\tau_L, \mu^-) e^{-(\tau_L - \tau)/\mu^-}\, \mu^- d\mu^-\\
&\quad \left. - \int_{\tau_L}^\tau \bar{i}(\tau') \int_0^1 e^{-(\tau' - \tau)/\mu^-}\, d\mu^- d\tau'\right]
\end{aligned}
\qquad (7.68)
$$

There being no sources or sinks of radiation in a purely scattering medium, then in the absence of other modes of heat transfer the flux q'' must be uniform throughout the region, i.e., $dq''/d\tau = 0$ (radiative equilibrium). Therefore, its value can be determined by evaluating Equation 7.68 for any value of the optical coordinate τ. It is convenient to choose $\tau = \tau_L/2$, in which case

$$
q'' = 2E_3\left(\frac{\tau_L}{2}\right)(\sigma T_0^4 - \sigma T_L^4)
$$

$$
+ 2\pi\left[\int_0^{\tau_L/2} \bar{i}(\tau')E_2\left(\frac{\tau_L}{2} - \tau'\right) d\tau' - \int_{\tau_L}^{\tau_L/2} \bar{i}(\tau')E_2\left(\tau' - \frac{\tau_L}{2}\right) d\tau'\right] \tag{7.69}
$$

if the two bounding walls are black and maintained at temperatures T_0 and T_L. An integral equation for the local mean intensity $\bar{i}(\tau)$ defined by Equation 7.68 can be obtained by substituting Equations 7.65 and 7.66 into Equation 7.67, yielding

$$
\bar{i}(\tau) = \frac{1}{2\pi}[E_2(\tau) \sigma T_0^4 + E_2(\tau_L - \tau)\sigma T_L^4]
$$

$$
+ \frac{1}{2}\left[\int_0^\tau \bar{i}(\tau')E_1(\tau - \tau') d(\tau - \tau') - \int_0^{\tau_L-\tau} \bar{i}(\tau')E_1(\tau' - \tau) d(\tau' - \tau)\right] \tag{7.70}
$$

An exact analytical solution of Equation 7.70 has proven to be elusive; however, a reasonable first approximation for small values of optical depth may be obtained by assuming that $\bar{i}(\tau)$ is uniform throughout the layer. Subject to this approximation, Equation 7.69 becomes

$$
q'' \cong 2E_3\left(\frac{\tau_L}{2}\right)(\sigma T_0^4 - \sigma T_L^4) \tag{7.71}
$$

This result is exact in the limit as τ_L approaches zero, where $q'' = \sigma T_0^4 - \sigma T_L^4$; however, it underestimates the heat flux to an increasing degree as optical thickness increases, as indicated in Table 7.3 and Figure 7.4. Therefore, for values of optical

Table 7.3 Comparison of $2E_3(\tau_L/2)$ with the exact variation of $f(\tau_L)$ [4]

τ_L	$f(\tau_L)$ [4]	$2E_3$ $(\tau_L/2)$	Error (%)
0	1.0	1.0	0
0.1	0.9157	0.9098	−0.64
0.2	0.8491	0.8326	−1.94
0.3	0.7934	0.7646	−3.63
0.4	0.7458	0.7038	−5.63
0.5	0.7040	0.6494	−7.76
0.6	0.6672	0.6000	−10.1
0.8	0.6046	0.5146	−14.9
1.0	0.5532	0.4432	−19.9

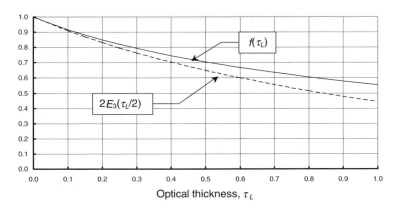

Figure 7.4   Comparison of $2E_3(\tau_L/2)$ with the exact variation of $f(\tau_L)$ [4]

depth greater than about 0.5, where the error is 7.76 percent, it is recommended that a better approximation be obtained for $\bar{i}(\tau)$. The actual expression for q'' and any appropriate approximation will have the general form

$$q'' = f(\tau_L)(\sigma T_0^4 - \sigma T_L^4) \tag{7.72}$$

7.2.2 Diffuse, Gray Walls

A somewhat more interesting situation arises when the bounding planes are diffuse and gray, in which case the boundary conditions become

$$q_0''^+ = \varepsilon_0 \sigma T_0^4 + (1 - \varepsilon_0)q_0''^- \tag{7.73}$$

and

$$q_L''^- = \varepsilon_L \sigma T_L^4 + (1 - \varepsilon_L)q_L''^+ \tag{7.74}$$

Then, noting that $q'' = q_0''^+ - q_0''^- = q_L''^+ - q_L''^-$, we have

$$q_0''^+ = \sigma T_0^4 - \left(\frac{1 - \varepsilon_0}{\varepsilon_0}\right)q'' \tag{7.75}$$

and

$$q_L''^- = \sigma T_L^4 + \left(\frac{1 - \varepsilon_L}{\varepsilon_L}\right)q'' \tag{7.76}$$

Replacing σT_0^4 and σT_L^4 in Equation 7.72 with the new boundary conditions given by Equations 7.75 and 7.76 and solving the resulting expression for q'' yields

$$q'' = \frac{f(\tau_L)}{1 + (1/\varepsilon_0 + 1/\varepsilon_L - 2)f(\tau_L)} (\sigma T_0^4 - \sigma T_L^4) \tag{7.77}$$

It should be pointed out that the approximation represented by Equation 7.71 improves significantly when the walls are gray. For example, if both walls have an emissivity of 0.7, the heat flux error using the approximation is only 5 percent when $\tau_L = 0.5$.

7.2.3 Nongray Walls and Media

Extension of these results to nongray walls and scattering media parallels the foregoing development while drawing on ideas leading to Equation 7.64. The details are left as a student exercise.

7.3 SOLUTION OF THE EQUATION OF RADIATIVE TRANSFER IN A ONE-DIMENSIONAL ABSORBING, EMITTING, AND SCATTERING MEDIUM

We now consider the problem of one-dimensional radiation in a plane layer of a gray absorbing, emitting, and scattering medium separated by parallel walls spaced a distance L apart. The equations of transfer governing this case are

$$\mu^+ \frac{di^+}{dx} = -(\kappa + \sigma)i^+(x, \mu^+) + \kappa i_b(T)$$

$$+ \frac{\sigma}{2} \left[\int_0^1 i^+(x, \mu'^+)\Phi(\mu'^+, \mu^+) \, d\mu'^+ \right. \tag{7.78}$$

$$\left. + \int_0^1 i^-(x, \mu'^-)\Phi(\mu'^-, \mu^+) \, d\mu'^- \right]$$

and

$$-\mu^- \frac{di^-}{dx} = -(\kappa + \sigma)i^-(x, \mu^-) + \kappa i_b(T)$$

$$+ \frac{\sigma}{2} \left[\int_0^1 i^-(x, \mu'^-)\Phi(\mu'^-, \mu^-) \, d\mu'^- \right. \tag{7.79}$$

$$\left. + \int_0^1 i^+(x, \mu'^+)\Phi(\mu'^+, \mu^-) \, d\mu'^+ \right]$$

7.3.1 Isotropic Scattering

In writing Equations 7.78 and 7.79 it has been assumed that the medium is uniform, i.e., that κ, σ, and Φ are independent of x. For the moment we will assume further that scattering is isotropic, which means that the scattering phase function is independent of both the incident and scattered directions and so has a value of 1.0. Taking this into account and introducing the *single-scattering albedo* $\Omega = \sigma/(\kappa + \sigma)$, Equations 7.78 and 7.79 can be written

$$\mu^+ \frac{di^+}{d\tau} = -i^+(\tau, \mu^+) + (1 - \Omega)i_b[T(\tau)]$$

$$+ \frac{\Omega}{2}\left[\int_0^1 i^+(\tau, \mu^+)\, d\mu^+ + \int_0^1 i^-(\tau, \mu^-)\, d\mu^-\right] \tag{7.80}$$

and

$$-\mu^- \frac{di^-}{d\tau} = -i^-(\tau, \mu^-) + (1 - \Omega)i_b[T(\tau)]$$

$$+ \frac{\Omega}{2}\left[\int_0^1 i^-(\tau, \mu^-)\, d\mu^- + \int_0^1 i^+(\tau, \mu^+)\, d\mu^+\right] \tag{7.81}$$

where now $\tau = (\kappa + \sigma)x$. Solution of Equations 7.80 and 7.81 proceeds in a manner similar to that demonstrated in Section 7.2. Application of the integrating factor $e^{-\tau}$ leads to

$$i^+\left(\frac{\tau}{\mu^+}\right) = i^+(0)e^{-\tau/\mu^+} + (1 - \Omega)\int_0^\tau i_b[T(\tau')]e^{(\tau' - \tau)/\mu^+}\frac{d\tau'}{\mu^+}$$

$$+ \Omega\int_0^\tau \bar{i}(\tau')e^{(\tau' - \tau)/\mu^+}\frac{d\tau'}{\mu^+} \tag{7.82}$$

and

$$i^-\left(\frac{\tau}{\mu^-}\right) = i^-\left(\frac{\tau_L}{\mu^+}\right)e^{-(\tau_L - \tau)/\mu^-} + (1 - \Omega)\int_{\tau_L}^\tau i_b[T(\tau')]e^{(\tau - \tau')/\mu^-}\frac{d\tau'}{\mu^+}$$

$$+ \Omega\int_{\tau_L}^\tau \bar{i}(\tau')e^{(\tau - \tau')/\mu^-}\frac{d\tau'}{\mu^-} \tag{7.83}$$

Equations 7.82 and 7.83 can be used to derive an integral equation for the function $\bar{i}(\tau)$ in a manner identical to the derivation of Equation 7.70. When this is done assuming diffuse boundary conditions $i^+(0)$ and $i^-(\tau_L)$ there results

$$\bar{i}(\tau) = \frac{1}{2}\left(i^+(0)E_2(\tau) + i^-(\tau_L)E_2(\tau_L - \tau)\right.$$

$$+ (1 - \Omega)\left\{\int_0^\tau i_b[T(\tau')]E_1(\tau - \tau')\,d\tau' - \int_\tau^{\tau_L} i_b[T(\tau')]E_1(\tau' - \tau)\,d\tau'\right\} \quad (7.84)$$

$$+ \Omega\left[\int_0^\tau \bar{i}(\tau')E_1(\tau - \tau')\,d\tau' - \int_\tau^{\tau_L} \bar{i}(\tau')E_1(\tau' - \tau)\,d\tau'\right]\right)$$

Then, in principle, when the boundary conditions and the temperature distribution in the medium are specified, Equation 7.84 can be used to obtain the source function $\bar{i}(\tau)$.

If Equations 7.80 and 7.81 are multiplied by $d\omega$ and integrated over their respective 2π-spaces and the results are added together, we obtain

$$\frac{dq''}{d\tau} = \frac{dq''^+}{d\tau} - \frac{dq''^-}{d\tau} = -\bar{i}(\tau) + 4\pi(1 - \Omega)i_b[T(\tau)] + 4\pi\Omega\bar{i}(\tau) \quad (7.85)$$

7.3.2 Radiative Equilibrium ($dq''/d\tau = 0$)

In the case of radiative equilibrium ($dq''/d\tau = 0$) we have from Equation 7.85

$$\bar{i}(\tau) = i_b[T(\tau)] = \frac{\sigma T^4(\tau)}{\pi} \quad (7.86)$$

An integral equation for the corresponding temperature distribution in the medium can be derived by replacing $\bar{i}(\tau)$ with $\sigma T^4(\tau)/\pi$ in Equation 7.84, obtaining

$$\sigma T^4(\tau) = \frac{1}{2}\left[E_2(\tau)\sigma T_0^4 + E_2(\tau_L - \tau)\sigma T_L^4\right.$$

$$\left. + \int_0^\tau \sigma T^4(\tau')E_1(\tau - \tau')\,d\tau' + \int_\tau^{\tau_L} \sigma T^4(\tau')E_1(\tau' - \tau)\,d\tau'\right] \quad (7.87)$$

7.3.3 Uniform-Temperature Participating Medium Between Parallel Black Walls ($dq''/d\tau \neq 0$)

Recall that the temperature in the medium can be uniform at T only in the presence of appropriately distributed sources or sinks of heat; that is, the system cannot be in radiative equilibrium. These sources or sinks could be due to conduction, convection, chemical processes such as combustion, and so forth. In the case of a uniform-temperature, absorbing, emitting, and scattering medium bounded by black walls maintained at T_0 and T_L, Equations 7.82 and 7.83 become

$$
i^+\left(\frac{\tau}{\mu^+}\right) = \frac{\sigma T_0^4}{\pi} e^{-\tau/\mu^+} + (1 - \Omega)\frac{\sigma T^4}{\pi}\int_0^\tau e^{(\tau'-\tau)/\mu^+}\frac{d\tau'}{\mu^+}
$$
$$
+ \Omega\int_0^\tau \bar{i}(\tau')e^{(\tau'-\tau)/\mu^+}\frac{d\tau'}{\mu^+}
$$

(7.88)

and

$$
i^-\left(\frac{\tau}{\mu^-}\right) = \frac{\sigma T_L^4}{\pi} e^{-(\tau_L-\tau)/\mu^-} + (1 - \Omega)\frac{\sigma T^4}{\pi}\int_{\tau_L}^\tau e^{(\tau-\tau')/\mu^-}\frac{d\tau'}{u^-}
$$
$$
+ \Omega\int_{\tau_L}^\tau \bar{i}(\tau')e^{(\tau-\tau')/\mu^-}\frac{d\tau'}{\mu^-}
$$

(7.89)

and Equation 7.84 becomes

$$
\bar{i}(\tau) = \frac{1}{2}\left\{E_2(\tau)\frac{\sigma T_0^4}{\pi} + E_2(\tau_L - \tau)\frac{\sigma T_L^4}{\pi}\right.
$$
$$
+ (1 - \Omega)\frac{\sigma T^4}{\pi}[E_2(\tau_L - \tau) - E_2(\tau)]
$$

(7.90)

$$
\left.+\Omega\left[\int_0^\tau \bar{i}(\tau')E_1(\tau - \tau')\,d\tau' - \int_\tau^{\tau_L}\bar{i}(\tau')E_1(\tau' - \tau)\,d\tau'\right]\right\}
$$

Finally, the net heat flux at any plane τ is

$$
q''(\tau) = 2\pi\int_0^1 i^+\left(\frac{\tau}{\mu^+}\right)\mu^+\,d\mu^+ - 2\pi\int_0^1 i^-\left(\frac{\tau}{\mu^-}\right)\mu^-\,d\mu^-
$$
$$
= 2E_3(\tau)\sigma T_0^4 - 2E_3(\tau_L - \tau)\sigma T_L^4
$$
$$
+ 2(1 - \Omega)\sigma T^4 [E_3(\tau_L - \tau) - E_3(\tau)]
$$

(7.91)

$$
+ 2\pi\Omega\left[\int_0^\tau \bar{i}(\tau')E_2(\tau - \tau')\,d\tau' + \int_\tau^{\tau_L}\bar{i}(\tau')E_2(\tau' - \tau)\,d\tau'\right]
$$

The strategy for computing the local net heat flux is to first solve Equation 7.90 for $\bar{i}(\tau)$, using a numerical approach or one of the approximate methods discussed in the next section. Then the variation of $\bar{i}(\tau)$ with τ is substituted into Equation 7.91.

7.3.4   Uniform-Temperature Participating Medium Between Parallel, Diffuse, Gray Walls ($dq''/d\tau \neq 0$)

In the case of diffuse, gray walls the quantities σT_0^4 and σT_L^4 in Equations 7.90 and 7.91 are replaced with the boundary conditions defined by Equations 7.75 and 7.76

with $q'' = q''(\tau)$ evaluated at $\tau = 0$ or $\tau = \tau_L$ as appropriate. This introduces the further complication that, when solved, Equation 7.91 gives $q''(\tau)$ in terms of the two unknown boundary conditions $q''(0)$ and $q''(\tau_L)$. The result must be evaluated at the two boundaries and the two resulting algebraic equations solved simultaneously to recover the boundary conditions.

7.4 SOLUTION OF THE EQUATION OF RADIATIVE TRANSFER IN MULTIDIMENSIONAL SPACE

The material in the preceding sections is useful for understanding the basic physics and trends associated with the equation of transfer and its solution; however, most problems of practical interest involve two- and three-dimensional solution domains in which scattering is usually not isotropic, reflection is seldom diffuse, and the geometry is often irregular. Exact, closed-form, analytical solution of the integrodifferential equations describing these real-world situations is rarely possible. This has led to the development of two broad categories of approaches for modeling thermal radiation in participating media: (1) approximate formulations in which either (a) the intensity is approximated as a truncated series of appropriate transcendental functions or (b) the integrals are approximated as appropriately weighted sums, and (2) the more or less "exact" Monte Carlo ray-trace method.

The Monte Carlo ray-trace (MCRT) method, which is treated in detail in Part III of this book, has much to recommend for its flexibility and accuracy. In fact, it is standard practice for authors to publish results from a Monte Carlo-based analysis along side those obtained from the alternative method they are promoting. This is testimony to both the relative ease of formulating an MCRT model and the wide acceptance of the inherent accuracy of such models. In the past the MCRT method has been criticized for its excessive use of computer resources; however, this criticism is rapidly losing its credibility in face of the almost daily arrival of bigger, faster, less-expensive computers.

A notable advantage of the Monte Carlo ray-trace method is its unique ability to conform to irregular geometries while dealing with directional emission, absorption, reflection, and scattering. An underappreciated charm of the method is its intuitive appeal and the attendant relative ease with which the underlying principles may be understood and applied. However, perhaps the greatest strength of the MCRT method is that surface-to-surface radiation, surface-to-volume-element radiation, and volume-element-to-volume-element radiation are treated in the same manner, so that issues of matching surface radiation to radiation in a surrounding participating medium are resolved automatically. While this book emphasizes the MCRT method for radiation heat transfer modeling, two of the more popular alternatives to the method are presented below in the interest of completeness.

7.4.1 The Differential, or P_N, Approximation

The differential, or P_N, approximation is based on the idea that the intensity in a participating medium can be represented as a rapidly converging series whose terms are based on orthogonal spherical harmonics. Any number of terms in the series can, in principle, be retained, but experience has shown that a small number is often adequate. We begin our study of the P_N approximation by developing the P_1 approximation, so called because only the zeroth and first moments of the intensity are retained in the series.

The three-dimensional version of the equation of radiative transfer for a purely absorbing, emitting medium may be written in rectangular coordinates as

$$\ell \frac{\partial i}{\partial_x} + m \frac{\partial i}{\partial y} + n \frac{\partial i}{\partial z} = -\kappa[i - i_b(T)] \tag{7.92}$$

where, in general, $T = T(x,y,z)$ and ℓ, m, and n are the direction cosines

$$\ell = \cos\alpha = \sin\theta \cos\phi$$
$$m = \cos\beta = \sin\theta \sin\phi \tag{7.93}$$
$$n = \cos\gamma = \cos\theta$$

The angles appearing in Equations 7.93 are illustrated in Figure 7.5. We approximate the deviation of local intensity from its local mean value in terms of the local gradients along the principal axes of its mean value; that is,

$$i(x, y, z, \theta, \phi) - i_0(x, y, z) \cong -\frac{1}{\kappa}\left(\ell \frac{\partial i_0}{\partial x} + m \frac{\partial i_0}{\partial y} + n \frac{\partial i_0}{\partial z}\right) \tag{7.94}$$

where

$$i_0(x, y, z) = \frac{1}{4\pi} \int_0^{2\pi} \int_0^{2\pi} i(x, y, z, \theta, \phi) \sin\theta \, d\theta \, d\phi \tag{7.95}$$

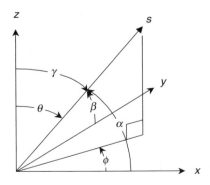

Figure 7.5 Geometry showing relationship between the direction cosine angles α, β, and γ and the angles θ and ϕ

The appearance of the absorption coefficient κ in Equation 7.94 serves both to convert the physical distances into the more meaningful optical distances and to preserve the dimensional integrity of the relation. The P_1 approximation for intensity $i(x, y, z, \theta, \phi)$ is obtained by adding $i_0(x,y,z)$ to both sides of Equation 7.94.

A three-step process leads to a differential equation for the local mean intensity, i_0. First, Equation 7.92 is multiplied through by $d\omega$ and the result integrated over 4π-space, yielding

$$\frac{\partial q''_x}{\partial x} + \frac{\partial q''_y}{\partial y} + \frac{\partial q''_z}{\partial z} = -4\pi\kappa[i_0 - i_b(T)] \tag{7.96}$$

The reader is reminded that here and elsewhere q'' represents a net flux, i.e., the difference between the fluxes into the two hemispheres on either side of a plane passing through a point. The next step is to multiply Equation 7.92 through successively by each of the three direction cosines ℓ, m, and n, and by $d\omega$, and then to integrate the result in each case over 4π-space. Because the direction cosines are members of an orthogonal set, this process produces three equations relating the local gradients of the mean intensity to the local components of heat flux in the same direction; that is,

$$\frac{4\pi}{3} \frac{\partial i_0}{\partial x} = -\kappa q''_x \tag{7.97}$$

$$\frac{4\pi}{3} \frac{\partial i_0}{\partial y} = -\kappa q''_y \tag{7.98}$$

and

$$\frac{4\pi}{3} \frac{\partial i_0}{\partial z} = -\kappa q''_z \tag{7.99}$$

The final step is to eliminate the components of q'' among Equations 7.96 through 7.99. When this is done there results

$$\frac{\partial^2 i_0}{\partial x^2} + \frac{\partial^2 i_0}{\partial y^2} + \frac{\partial^2 i_0}{\partial z^2} = 3\kappa^2 [i_0 - i_b(T)] \tag{7.100}$$

Equation 7.100 is the governing equation for the P_1 approximation.

Equation 7.100 is subject to two boundary conditions in each dimension. In the case of diffuse boundaries it must be true that

$$i_w(x, y, z) = \varepsilon_w i_b(T_w) + (1 - \varepsilon_w)\frac{1}{\pi}\int_{2\pi} (\ell, m, n)i(x, y, z, \theta, \phi)\, d\omega \tag{7.101}$$

where the choice of the direction cosine ℓ, m, or n as a factor under the integral depends on the orientation of the wall with respect to the coordinate axes. Also, the intensity under the integral is evaluated at a specific value of one of the three space co-ordinates, e.g., $x = 0$, $x = L$, etc. Equation 7.101 is not in a convenient form because the formulation is reticent on the angular distribution of intensity needed to evaluate the integral. However, the integral represents the incident flux q_w'', which can be ex-pressed as the difference between right- and left-propagating components as was done elsewhere in this chapter. For example, $q_x'' = q_x''^+ - q_x''^-$, where, with the intro-duction of Equation 7.94 for $i(x, y, z, \theta, \phi)$,

$$q_x''^+ \cong \int_{-\pi/2}^{\pi/2} \int_0^\pi \ell \left[i_0 - \frac{1}{\kappa} \left(\ell \frac{\partial i_0}{\partial x} + m \frac{\partial i_0}{\partial y} + n \frac{\partial i_0}{\partial z} \right) \right] \sin\theta \, d\theta \, d\phi$$

$$= \pi i_0 - \frac{2\pi}{3\kappa} \frac{\partial i_0}{\partial x} \tag{7.102}$$

and

$$q_x''^- \cong \int_{-\pi/2}^{3\pi/2} \int_0^\pi \ell \left[i_0 - \frac{1}{\kappa} \left(\ell \frac{\partial i_0}{\partial x} + m \frac{\partial i_0}{\partial y} + n \frac{\partial i_0}{\partial z} \right) \right] \sin\theta \, d\theta \, d\phi$$

$$= \pi i_0 + \frac{2\pi}{3\kappa} \frac{\partial i_0}{\partial x} \tag{7.103}$$

When Equations 7.102 and 7.103 are introduced into Equation 7.101 (recognizing that the integral is a heat flux), there results

$$i_0 \pm \frac{2}{3\kappa} \left(\frac{2 - \varepsilon_w}{\varepsilon_w} \right) \frac{\partial i_0}{\partial x} = i_b(T_w) \tag{7.104}$$

where the plus (+) sign holds for the wall at x (or y or z) $= 0$ and the minus ($-$) sign holds for x (or y or z) $= L_x$ (or L_y or L_z). Equations 7.104 exist for each coordinate axis, with the subscript w indicating the appropriate wall.

Equation 7.100 is a diffusion equation reminiscent of those encountered in con-duction heat transfer, except that here the temperature T has been replaced by i_0. Its solution subject to appropriate boundary conditions would proceed routinely using one of the several finite difference schemes available to solve other diffusion equa-tions. This is a significant advantage of the P_N approximation when radiative trans-port occurs in the presence of other modes of heat transfer and fluid flow, whose models also consist of coupled systems of differential equations. After Equation 7.100 has been solved for the spatial distribution of i_0, the local intensity distribution can be found from Equation 7.94, and the three components of flux can be found from Equations 7.97 through 7.99.

It was not essential to invoke the idea of moments in the foregoing derivation of the P_1 approximation. However, the quantity $4\pi i_0$ is, in fact, the zeroth moment of the intensity and the quantity q'' is the first moment of the intensity with respect to the appropriate coordinate axis. That is,

$$M_0(x, y, z) = 4\pi i_0(x, y, z) = \int_{4\pi} i(x, y, z, \theta, \phi) \, d\omega \tag{7.105}$$

$$M_{1,\ell}(x, y, z) = q_x'' = \int_{4\pi} \ell i(x, y, z, \theta, \phi)\, d\omega = -\frac{4\pi}{3\kappa}\frac{\partial i_0}{\partial x} \qquad (7.106)$$

and so forth. While it was unnecessary to view the P_N approximation in these terms when $N = 1$, for higher orders of the approximation it is convenient to think in terms of moments of the intensity. Then, while the P_1 approximation required only the zeroth moment and three first moments of intensity for a total of four moments, the P_3 approximation requires 18 moments. While it is generally true that the accuracy of the P_N approximation increases with N, according to Siegel and Howell [5] even-ordered approximations do not sufficiently increase the accuracy of the approximation over the next lower odd-ordered approximation to justify their increase in effort. Furthermore, while solutions based on the P_3 approximation can be significantly more accurate than those based on the P_1 approximation, relatively little additional gain in accuracy is obtained using the P_5 approximation. Therefore, given the considerable increase in complexity associated with increasing order, the P_3 approximation may be considered optimum for most engineering applications.

Readers interested in a more general development that emphasizes the P_3 approximation and that includes scattering with spatially and directionally dependent radiative properties of the bounding walls and the participating medium are encouraged to consult the book by Siegel and Howell [5] or the article by Mengüç and Viskanta [6].

7.4.2 The Discrete-Ordinate Approximation

A second increasingly popular approach to solution of the equations of radiative transfer is the discrete-ordinate, or S_n, approximation. This approach first came to the attention of the radiation heat transfer community in 1960 with the republication by Dover of Chandrasekhar's 1950 book on radiative transfer [1]. An influential early description of this approach, applied to neutron transport theory, is the oft-cited article by Carlson and Lathrop writing in the *Computing Methods in Reactor Physics* [7]; and a contemporary application to the equations of radiative transfer is described in Love's 1968 textbook [8]. More recent developments may be found in Siegel and Howell [5], Modest [9], and in articles by Fiveland [10] and others.

In the discrete–ordinate, or S_n, approximation the integrals appearing in the equations of radiative transfer are approximated as weighted sums. Consider the equation of radiative transfer written for a rectangular three-dimensional enclosure containing a gray, absorbing, emitting, and scattering medium,

$$\ell\frac{\partial i(x, y, z, \omega)}{\partial x} + m\frac{\partial i(x, y, z, \omega)}{\partial y} + n\frac{\partial i(x, y, z, \omega)}{\partial z}$$

$$= -\beta i(x, y, z, \omega) + \kappa i_b[T(x, y, z)] + \frac{\sigma}{4\pi}\int_{4\pi} i(x, y, z, \omega')\,\Phi(\omega', \omega)\, d\omega' \qquad (7.107)$$

As in Equation 7.92, here the direction cosines ℓ, m, and n are defined by Equations

7.93 through 7.95 and range in value between -1 and 1. In the method of discrete ordinates we approximate the integral in Equation 7.107 at a node point x,y,z as

$$\int_{4\pi} i(x, y, z, \omega') \, \Phi(\omega', \omega) \, d\omega' \cong \sum_{m'=1}^{M} w_{m'} i_{m'} \Phi_{m',m} \qquad (7.108)$$

where the $w_{m'}$, are the M *weights,* whose values are usually chosen to maximize the accuracy of the approximation for a given number M of *ordinates.* The term "ordinate" here refers to a direction, or axis, in 4π-space. For every ordinate passing in one direction a second ordinate passes in the opposite direction, reminiscent of the two-flux approximation described briefly in Section 4.4.1. Therefore, the ordinates appear in pairs and so M must be an even number. The prime (') associated with the symbol m identifies m' as the index of an incident ordinate, while m without the prime is the index of an ordinate exiting from the node point.

The M direction cosines associated with the M weights and the M values of intensity are determined as follows. Consider the first octant of a unit sphere, shown in Figure 7.6, whose center is at the origin of the orthogonal ℓ, m, n-coordinate system. Let some number n of cutting planes pass normal to each of the three coordinate axes at $\ell = \pm\ell_1, \ell = \pm\ell_2, \ldots, m = \pm m_1, m = \pm m_2, \ldots, n = \pm n_1, n = \pm n_2$, and so forth (in the figure $n = 4$, with two cutting planes intersecting each direction cosine axis at positive values and two at negative values). Then the curves formed by the intersection of these cutting planes with the unit sphere intersect at $M = n(n + 2)$ points, where n is an even number. In Figure 7.6, $M = 24$, or three per octant. The ordinates

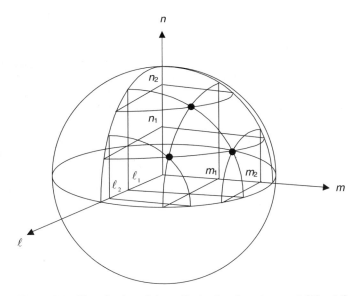

Figure 7.6   Visualization of the ordinate directions for $n = 4$ ($M = 24$)

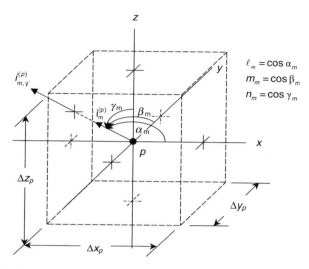

Figure 7.7 Volume element $\Delta x_p \Delta y_p \Delta z_p$ showing nomenclature for Equation 7.110

referred to are then the axes formed by extending lines from the node point in directions $\pm \ell_i$, $\pm m_j$, $\pm n_k$, where i, j, and k are permitted to range over all positive integers from 1 to $n/2$. The approximation issuing from this process is then referred to as the S_n approximation.

It is convenient, although by no means essential, to select the points on the imaginary unit sphere in Figure 7.6 such that the weights are those associated with Gauss quadrature. When this is done the approximation represented by Equation 7.108 is exact as long as the function represented by the product $i(x, y, z, \omega)\Phi(\omega', \omega)$ is a polynomial of degree less than or equal to $2M$. When the approximation represented by Equation 7.108 is introduced into Equation 7.107 there results

$$\ell_m \frac{\partial i_m}{\partial x} + m_m \frac{\partial i_m}{\partial y} + n_m \frac{\partial i_m}{\partial y} = -\beta i_m + \kappa i_b[T(x, y, z)]$$

$$\hspace{4cm}\text{(7.109)}$$

$$+ \frac{\sigma}{4\pi} \sum_{m'=1}^{M} w_{m'} i_{m'} \Phi_{m',m} \qquad m = 1, 2, \ldots, M$$

Equation 7.109 represents M first-order partial differential equations in M intensities. In principle, each of these M equations requires three boundary conditions, one for each coordinate direction x, y, and z. In practice, the nature of these boundary conditions depends on the approach used to solve the equations. The volume element approach outlined below is based on an article by Fiveland [10].

We consider each node p of the solution domain to be at the center of a rectangular volume element of dimensions $\Delta x_p \Delta y_p \Delta z_p$, as illustrated in Figure 7.7. Then integrating Equation 7.109 over the volume element surrounding node p yields

$$\ell_m \Delta y_p \Delta z_p (i_{m,x^+}^{(p)} - i_{m,x^-}^{(p)}) + m_m \Delta x_p \Delta z_p (i_{m,y^+}^{(p)} - i_{m,y^-}^{(p)}) + n_m \Delta x_p \Delta y_p (i_{m,z^+}^{(p)} - i_{m,z^-}^{(p)})$$

$$= -\beta \, \Delta x_p \Delta y_p \Delta z_p i_m^{(p)} + \kappa \, \Delta x_p \Delta y_p \Delta z_p i_b^{(p)} \quad (7.110)$$

$$+ \frac{\sigma}{4\pi} \Delta x_p \Delta y_p \Delta z_p \sum_{m'=1}^{M} w_{m'} i_m^{(p)} \, \Phi_{m',m}$$

where it is understood that Equation 7.110 represents an approximation whose accuracy generally improves as the size of the volume element is reduced until convergence is achieved. Of course, the form of the discrete-ordinate phase function $\Phi_{m',m}$ depends on the scattering model used. For the case of isotropic scattering it is equal to unity. Fiveland suggests the reasonable model

$$\Phi(\varphi) = 1 + a \cos\varphi \quad (7.111)$$

where φ is the cone angle measured from the ordinate direction before scattering and a is a parameter that can take any value between -1 and $+1$. Then, if $a = 0$, scattering is isotropic; if $a = 1$, scattering is 100-percent forward; and if $a = -1$, scattering is 100-percent backward. If this model is adopted, then

$$\Phi_{m',m} = 1.0 + a \, (\ell_m \ell_{m'} + m_m m_{m'} + n_m n_{m'}) \quad (7.112)$$

For the case of specified wall temperatures the boundary conditions for Equations 7.110 are

at $x = 0$ (for which $\ell_{m'} > 0$);

$$i_m = \varepsilon_{x_0} i_b(T_{x_0}) + \frac{1 - \varepsilon_{x_0}}{\pi} \sum_{m'-} w_{m'} \left| \ell_{m'} \right| i_{m'} \quad (7.113)$$

at $x = L_x$ (for which $\ell_{m'} < 0$);

$$i_m = \varepsilon_{L_x} i_b \, (T_{L_x}) + \frac{1 - \varepsilon_{L_x}}{\pi} \sum_{m'+} w_{m'} \ell_{m'} i_{m'} \quad (7.114)$$

with similar expressions for the y and z directions. In Equation 7.113 the symbol $m' -$ on the summation indicates summation only over indices m' for which $\ell_{m'}$ is negative, and in Equation 7.114 the symbol $m' +$ on the summation indicates summation only over indices m' for which $\ell_{m'}$ is positive.

Because the boundary conditions contain the unknown intensities, it is necessary to solve the system iteratively. In the first iteration we assume that the boundaries are black. Then, beginning with the first volume element (for which $p = 1$) in the corner, where $x = y = z = 0$ and assuming that the variation of intensity along an ordinate within the volume element is linear, we can write for each value of m for which the direction cosines are positive

$$i_m^{(1)} = \tfrac{1}{2}(i_{m,x^-}^{(1)} + i_{m,x^+}^{(1)}) = \tfrac{1}{2}(i_{m,y^-}^{(1)} + i_{m,y^+}^{(1)}) = \tfrac{1}{2}(i_{m,z^-}^{(1)} + i_{m,z^+}^{(1)}) \qquad (7.115)$$

Then, since $i_{m,x^-}^{(1)}$, $i_{m,y^-}^{(1)}$, and $i_{m,z^-}^{(1)}$ are known boundary conditions, Equations 7.115 can be used to eliminate $i_{m,x^+}^{(1)}$, $i_{m,y^+}^{(1)}$, and $i_{m,z^+}^{(1)}$ from Equation 7.110. The resulting equation can then be solved for $i_m^{(1)}$ in terms of the known boundary conditions $i_{m,x^-}^{(1)}$, $i_{m,y^-}^{(1)}$, and $i_{m,z^-}^{(1)}$, yielding

$$i_m^{(1)} = \frac{\dfrac{2\ell_m}{\Delta x}i_{m,x^-}^{(1)} + \dfrac{2m_m}{\Delta y}i_{m,y^-}^{(1)} + \dfrac{2n_m}{\Delta z}i_{m,z^-}^{(1)} + \kappa i_b^{(1)} + \displaystyle\sum_{m' \neq m} w_{m'}\Phi_{m',m}i_{m'}^{(1)}}{\dfrac{2\ell_m}{\Delta x} + \dfrac{2m_m}{\Delta y} + \dfrac{2n_m}{\Delta z} + \beta - \dfrac{\sigma}{4\pi}w_m\Phi_{m,m}} \qquad (7.116)$$

Inspection of Equation 7.116 reveals another reason the solution must be interative; the intensities under the summation are currently unknown and so are set equal to zero during the first iteration. Then Equation 7.116 represents a first estimate of $i_m^{(1)}$ where m includes all ordinates for which all three direction cosines are positive. Equations similar to Equation 7.116 can be written for all nodes of the enclosure. When solved in the correct sequence (so that the neighboring "upstream" boundary intensities are always known), an improved estimate of $i_m^{(p)}, p = 1, 2, \ldots, P$, emerges. This process is repeated until convergence is achieved.

In carrying out an iterative solution care must be exercised to avoid negative values of the intensity, which can occur due to linear extrapolation across a volume element. Of course, negative net intensity has no physical significance and so when it occurs during the execution of a numerical scheme it invariably leads to unbounded oscillations in the solution. The algorithm must therefore check for negative net intensities and, when they occur, take some appropriate action, which usually consists of setting the offending intensity to zero.

Many schemes have been suggested for choosing the set of direction cosines and related weights. Gaussian sets are widely documented in the finite-element literature. This is still an area of active research, however, so readers are encouraged to experiment with sets of their own invention.

7.5 IMPROVEMENTS AND APPLICATIONS

The recent literature reflects an explosion of interest in improved methods for solving the equations of radiative transfer. This section begins with a review of some of the more notable articles in which versions of the differential and discrete-ordinate approximations have been compared with each other and with other approaches. These efforts have led to useful suggestions for improvements. We then take a quick look at a potentially important phenomenon in packed beds called dependent scattering. Finally, a few applications of the equations of radiative transfer are briefly explored.

7.5.1 Further Development of the Differential and Discrete-Ordinate Approximations

Several authors have published comparisons of the available methods for the solution of the equation of radiative transfer. Fiveland [10] compared results obtained using S_2, S_4, and S_6 discrete-ordinate models for radiative transfer in a two-dimensional rectangular enclosure with results obtained by Ratzel and Howell [11] using Hottel's [3] zone method and a P_3 differential approximation. Fiveland concludes that the S_4 and S_6 discrete-ordinate approximations typically outperform the P_3 differential approximation when predicting transfer rates. The results given in the article also indicate that the P_3 approximation is slightly more accurate that the S_2 approximation. Fiveland also demonstrates that, at least for the two-dimensional situation he studied, the S_4 approximation is significantly more accurate than the S_2 approximation, while very little improvement in accuracy is achieved going from the S_4 approximation to the S_6 approximation.

Truelove [12] offers an improvement of the discrete-ordinate approximation based on careful selection of the quadrature set to match the half-range first moment (Equation 7.106 integrated over 2π-space rather than 4π-space). He reports the weights and corresponding direction cosines for the improved quadrature set and uses them to demonstrate that generally better agreement with an exact solution is obtained using this new quadrature set than with the set used by Fiveland [10]. In fact, Truelove reports that his S_4 approximation outperforms Fiveland's S_6 approximation.

Kumar, Majumdar, and Tien [13] use readily available library subroutines to solve the system of ordinary differential equations that result when the discrete-ordinate approximation is used to solve the equation of radiative transfer in a planar system. The authors investigate a wide range of quadrature sets. They conclude that, for a highly forward-scattering medium, Gauss quadrature yields the most stable and accurate results when compared with other published quadrature schemes.

Modest [14] has proposed an improvement in the differential approximation that addresses the poor performance of the P_1 approximation in optically thin situations, especially when used in two- and three-dimensional geometries. The improved model reduces to the correct form in the optically thin and optically thick limits. Good agreement is obtained when computed results using the improved approximation are compared with results obtained using Monte Carlo simulations.

Liu, Swithenbank, and Garbett [15] propose an improvement to the boundary conditions used in the P_1 approximation. The authors point out that the P_1 approximation when using the Marshak boundary condition [11] overpredicts surface heat transfer. Attributing this shortcoming to what they term the arbitrariness of the Marshak boundary condition, they propose a new condition that contains an arbitrary coefficient whose value can be adjusted to improve the agreement of computed results with exact results. The authors recommend optimum values of the adjustable coefficient based on their study of one- and two-dimensional enclosures containing an absorbing, emitting medium. The hypothesis is somewhat weakened by the fact that the optimum value of the adjustable coefficient depends on the dimensionality of the enclosure and on the quantity being calculated (heat flux or temperature distribution). Still, sensitivity of computed results to the boundary condition model used is an interesting avenue of investigation that probably needs more attention.

7.5.2 Dependent Scattering

Several investigations have been carried out based on solution of the equation of radiative transfer to predict radiative transport through and at the interface with composite materials such as a packed bed. An important issue in packed-bed radiative transfer modeling is the possible influence of *dependent scattering*. In a given situation, dependent scattering is due to one or both of two separate mechanisms. If individual scattering sites are in each others' radiative near field, the electromagnetic field surrounding the particle is not a free field, but is distorted by the presence of nearby particles. Also, even if scattering itself from one particle is not influenced by the proximity of another particle, the possibility still exists that scattered radiation will undergo constructive and destructive interference as it interacts with radiation at the same wavelength that has been scattered by other particles. It should be recognized that these are two distinctly different mechanisms.

A 1982 article by Brewster and Tien [16] describes results in which predictions of radiative transfer through packed beds of latex particles are compared with measurements. The equations of radiative transfer are solved using the discrete-ordinate approximation. The authors conclude that the most important factor for determining the importance of dependent scattering in the analysis is the interparticle spacing measured in wavelengths. In general, excellent agreement is obtained between theoretical and experimental results assuming independent scattering. Brewster and Tien suggest that independent scattering can be assumed whenever

$$ x \geq \frac{0.3\pi f_v^{1/3}}{0.905 - f_v^{1/3}} \tag{7.117} $$

where $x = \pi d/\lambda$ is the particle size parameter and f_v is the particle volume fraction.

7.5.3 Some Applications

In a recent effort Kumar and White [17] studied dependent scattering in woven fibrous insulations by considering interference in the far field among rays scattered from individual fibers. Nicolau, Raynaud, and Sacadura [18] used the discrete-ordinate approximation with experimental data to extract the optical thickness, albedo, and phase function for fiberglass and silica fibers–cellulose insulations, while Dombrovsky [19] used the P_1 differential approximation in a similar study.

In a very recent effort Lopes et al. [20] used the discrete-ordinate approach to predict the directional, spectral emissivity of a packed bed of bronze oxide spherical particles whose diameters ranged from 100 to 400 μm. This investigation draws upon a rich heritage of earlier efforts aimed at describing radiative transfer through inhomogeneous media, and so is an excellent bibliographic source for investigators wishing to pursue research in this area. Lopes et al. use a scaling factor attributed to Singh and Kaviany [21], applied to the extinction coefficient, to allow Mie theory to be used to model dependent scattering. Comparison of the predicted and measured directional, spectral emissivity in the study showed reasonably good agreement between theory and experiment for trends and values out to an angle of about 70 deg from the normal; however, for larger angles the predictions fail to match the observed fall-off of emissivity with angle.

If recent activity is a valid indicator of the future trends, we can expect to see this explosion of interest in applications of the equations of transfer to continue well into the next decade.

Team Projects

TP 7.1 Consider the situation discussed in Section 7.1.3. For the wall temperatures and temperature distribution in Example Problem 7.1, compute the total net heat flux distribution in the space between the two walls if the band-averaged spectral emissivity of the walls and the band-averaged spectral absorption coefficient of the intervening medium vary as in Table 7.4.

TP 7.2 Use the P_1 approximation described in Subsection 7.4.1 to solve a modification of the problem described in Team Project TP 7.1. In the modified problem we replace the infinite bounding walls with square plates 1.0 m on a side. The resulting cube defined by the square plates is completed by black walls maintained at a temperature of absolute zero. Compute the net heat flux at the center of the two heated walls.

Table 7.4 Variation with wavelength of the band-averaged spectral emissivity of the walls and the band-averaged spectral absorption coefficient in the intervening medium for Team Projects TP 7.1, TP 7.2, and TP 7.3

Wavelength Interval (μm)	Band-Averaged Spectral Emissivity	Band-Averaged Spectral Absorption Coefficient (m^{-1})
0 to 0.4	0.8	0.005
0.4 to 0.7	0.2	0.005
0.7 to 2.5	0.5	0.015
2.5 to 5.0	0.3	0.150
5.0 to infinity	0.1	0.015

TP 7.3   Use the S_4 approximation described in Section 7.4.2 to solve the problem described in Team Project TP 7.2.

Discussion Points

DP 7.1   Explain why the signs are opposite on the left-hand side of Equations 7.1 and 7.2, but are the same on the right-hand side.

DP 7.2   Can you think of a criterion involving the distance L and the absorption coefficient κ for selecting the minimum number of zones into which the space between the bounding planes should be divided in Example Problem 7.1?

DP 7.3   In Example Problem 7.1 the net heat flux from the left-hand wall is greater than the net heat flux arriving at the right-hand wall, indicating that the effect of the participating medium was to absorb some of the radiation emitted by the two walls. Speculate on the conditions for which a linear temperature distribution in the participating medium would produce the same net heat flux leaving one wall as the net heat flux arriving at the other wall, that is, for which there would be no net absorption of heat by the medium. What might the volumetric heat release distribution in the medium look like in this case?

DP 7.4   Comment on the similarity in shape of the curves in Figure 7.3 representing the net heat flux distributions from Example Problems 7.1 and 7.2. What do you think the curve would look like for perfectly reflecting diffuse walls ($\varepsilon = 0$)? Speculate on the shape of the volumetric heat release profile in this case.

DP 7.5   Elaborate on the process described in the last paragraph of Section 7.1. Provide a logic flow diagram for the required computer code.

DP 7.6   Provide a word-processed extension of the development in Section 7.2 for a purely scattering plane layer to the case of nongray walls and a nongray participating medium. Include the derivation of all equations needed.

DP 7.7   Why do you believe a minus $(-)$ sign rather than a plus $(+)$ sign appears as a factor on the right-hand side of Equation 7.94?

DP7.8   What is the physical interpretation of the "local mean intensity" defined by Equation 7.95? [*Hint:* Try dividing it by the speed of light c.]

DP 7.9   Explain the statement leading into Equations 7.97 through 7.99 concerning orthogonality. (Can you derive these three equations?)

DP 7.10   What became of $i_b(T)$ going from Equation 7.92 to Equations 7.97 through 7.99?

DP 7.11   What is the physical significance of the first moment of intensity defined by Equation 7.106?

Problems

P 7.1   Consider the two-flux approximation embodied in Equations 7.5a and 7.6a. Subject to this approximation, what is the net heat flux to the walls if the absorption coefficient of the gray medium is 0.2 m^{-1}, its temperature is 1200 K, and the temperature of the black walls at $x = 0$ and $x = 400$ mm is 500 K?

P 7.2   Rework Problem P7.1 without invoking the two-flux approximation. Compare the result obtained using the two-flux approximation with the result obtained using the "exact" analysis presented in Section 7.1.1 (Equations 7.3 through 7.12).

P 7.3   Show that Equations 7.11 and 7.11a converge to the same limit as $\tau_L \to 0$, i.e., in the optically thin limit.

P 7.4   Derive Equation 7.14 from Equation 7.17.

P 7.5   Derive Equation 7.15 from Equation 7.11.

P 7.6   Show that

$$\frac{dE_n}{d\tau} = -E_{n-1}(\tau), \qquad n \geq 2 \tag{7.118}$$

P 7.7   Show that

$$\int E_n(\tau)d\tau = -E_{n+1}(\tau) \tag{7.119}$$

P 7.8   Show that

$$\int_0^\tau E_2(\tau - \tau')\,d\tau' - \int_{\tau_L}^\tau E_2(\tau' - \tau)\,d\tau' = E_3(\tau) - E_3(\tau - \tau_L) \tag{7.120}$$

P 7.9   Use Equation 7.17 with a spreadsheet to develop plots of $\dot{q}(\tau)$ versus τ for a range of values of τ_L, T_0/T and T_0/T_L. Word process a brief discussion of these results.

P 7.10   Create a spreadsheet that computes estimates of $E_n(\tau)$ for $n = 1, 2,$ and 3 for τ ranging from 0.00 to 3.00 in increments of 0.01. Check your results with those tabulated in Reference 2 or elsewhere in the literature.

P 7.11   Create a spreadsheet that computes $E_3(\Delta\tau, n)$ defined by Equation 7.22. Prepare plots of this function versus n for $\Delta\tau = 0.01, 0.1, 0.5,$ and 1.0.

P 7.12   Show that $2(0)^2 E_3(\Delta\tau, 0) = 1$.

P 7.13   Rework Example Problem 7.1 using 20 zones. Based on your result, can we conclude that the solution is converged with respect to the number of zones in Example Problem 7.1?

P 7.14 Set up Example Problem 7.2 in a spreadsheet and study the sensitivity of the net heat flux distribution between the two walls to the value of $\varepsilon = \varepsilon_0 = \varepsilon_L$. Let the wall emissivity take on values of 0.1, 0.3, 0.6 (Example Problem 7.2), 0.8, and 1.0 (Example Problem 7.1). Provide a brief discussion of these results.

P 7.15 Derive Equations 7.27 through 7.32 starting with Equations 7.23 through 7.26.

P 7.16 Derive Equations 7.40 through 7.47 starting with Equations 7.38 and 7.39.

P 7.17 Derive Equations 7.56 through 7.61 starting with Equations 7.52 through 7.55.

P 7.18 Derive Equation 7.71 from Equation 7.69.

P 7.19 Develop a table and accompanying plot comparing the exact value of dimensionless net heat flux, given by Equation 7.77 (nondimensionalized by $\sigma T_0^4 - \sigma T_L^4$) and the function in the second column of Table 7.3, with the approximate value, based on Equation 7.77 with $f(\tau_L)$ replaced by the function in the third column of Table 7.3. Plot the error versus the emissivity of the walls ($\varepsilon = \varepsilon_0 = \varepsilon_L$) when $\tau_L = 0.1, 0.5$, and 1.0.

P 7.20 Work out the details of the process described at the end of Section 7.3. You should end up with expressions for two unknown boundary conditions $q''(0)$ and $q''(\tau_L)$. The results can be left in terms of the unknown source distribution $\bar{i}(\tau)$.

P 7.21 Show that the solution of Equations 7.85 through 7.87 gives

$$\frac{q''(\tau)}{\sigma T_0^4 - \sigma T_L^4} = \frac{1 + E_3(\tau) + E_3(\tau_L - \tau) - \frac{3}{2}[E_4(\tau) + E_4(\tau_L - \tau)]}{1 + \dfrac{3\tau_L}{4}} \qquad (7.121)$$

P 7.22 Derive Equation 7.96.

P 7.23 Derive Equations 7.97 through 7.99.

P 7.24 Derive Equation 7.100.

P 7.25 Rework Example Problem 7.2 using the P_1 approximation developed in Section 7.4.1. Prepare a table and plot that compare the results obtained using the two approaches.

P 7.26 Derive Equation 7.112.

P 7.27 Derive a version of Equation 7.116 for the node in the corner at $x = x_L$, $y = y_L, z = z_L$.

REFERENCES

1. Chandrasekhar, S., *Radiative Transfer*, Dover Publications, Inc., New York, 1960.
2. Kourganoff, V., *Basic Methods in Transfer Problems*, Dover Publications, Inc., New York, 1963.

3. Hottel, H. C., and A. F. Sarofim, *Radiative Transfer,* McGraw-Hill Book Company, New York, 1967.
4. Heaslet, M. A., and R. F. Warming, "Radiative transport and wall temperature slip in an absorbing planar medium," *International Journal of Heat and Mass Transfer,* Vol. 8, 1965, pp. 979–994.
5. Siegel, R., and J. R. Howell, *Thermal Radiation Heat Transfer,* 3rd ed., Hemisphere Publishing Corporation, Washington, 1992, p. 775.
6. Mengüç, M. P., and R. Viskanta, "Radiative transfer in three-dimensional rectangular enclosures containing inhomogeneous, anisotropically scattering media." *Journal of Quantitative Spectroscopy and Radiative Transfer,* Vol. 33, No. 6, 1985, pp. 533–549.
7. Carlson, B. G., and K. D. Lathrop, "Transport theory: the discrete ordinates method," in *Computing Methods in Reactor Physics,* H. Greenspan, C. N. Kelber, and D. Okrent (eds.), Gordon and Breach Science Publishers, New York, 1968, pp. 171–266.
8. Love, T. J., *Radiative Heat Transfer,* Charles E. Merrill Publishing Company, Columbus, Ohio, 1968.
9. Modest, M. F., *Radiative Heat Transfer,* McGraw-Hill, Inc., New York, 1993.
10. Fiveland, W. A., "Discrete-ordinates solutions of the radiative transport equation for rectangular enclosures," *ASME Transactions, Journal of Heat Transfer,* Vol. 106, No. 4, November 1984, pp. 699–706.
11. Ratzel, A., and J. Howell, "Two-dimensional radiation in absorbing–emitting–scattering media using the P-N approximation," ASME Paper No. 82-HT-19, 1982.
12. Truelove, J. S., "Discrete-ordinate solutions of the radiative transport equation," *ASME Transactions, Journal of Heat Transfer,* Vol. 109, No. 4, November 1987, pp. 1048–1051.
13. Kumar, S., A. Majumdar, and C. L. Tien, "The differential–discrete-ordinate method for solutions of the equation of radiative transfer," *ASME Transactions, Journal of Heat Transfer,* Vol. 112, No. 2, May 1990, pp. 424–429.
14. Modest, M. F., "The improved differential approximation for radiative transfer in multidimensional media," *ASME Transactions, Journal of Heat Transfer,* Vol. 112, No. 3, August 1990, pp. 819–821.
15. Liu, F., J. Swithenbank, and E. S. Garbett, "The boundary condition of the P_N-approximation used to solve the radiative transfer equation," *International Journal of Heat and Mass Transfer,* Vol. 35, No. 8, August 1992, pp. 2043–2052.
16. Brewster, M. Q., and C. L. Tien, "Radiative transfer in packed fluidized beds: dependent versus independent scattering," *ASME Transactions, Journal of Heat Transfer,* Vol. 104, No. 4, November 1982, pp. 573–579.
17. Kumar, S., and S. M. White, "Dependent scattering properties of woven fibrous insulations for normal incidence," *ASME Transactions, Journal of Heat Transfer,* Vol. 117, No. 1, February 1995, pp. 160–166.
18. Nicolau, V. P., M. Raynaud, and J. F. Sacadura, "Spectral radiative properties identification of fiber insulating materials," *International Journal of Heat and Mass Transfer,* Vol. 37, Supplement 1, 1994, pp. 311–324.
19. Dombrovsky, L. A., "Quartz-fiber thermal insulation: infrared radiative properties and calculation of radiative–conductive heat transfer," *ASME Transactions, Journal of Heat Transfer,* Vol. 118, No. 2, May 1996, pp. 408–414.
20. Lopes, R., L. M. Moura, D. Baillis, and J.-F. Sacadura, "Directional spectral emittance of a packed bed: correlation between theoretical prediction and experimental data," *ASME Transactions, Journal of Heat Transfer,* Vol. 123, No. 2, April 2001, pp. 240–248.
21. Singh, B. P., and M. Kaviany, "Modelling radiative heat transfer in packed beds," *International Journal of Heat and Mass Transfer,* Vol. 35, No. 6, 1992, pp. 1397–1405.

8

THE NET EXCHANGE FORMULATION FOR DIFFUSE, GRAY ENCLOSURES

Part I of this book presented the interaction of thermal radiation with material sub-stances. Certain useful engineering parameters were defined and their dependence on direction, wavelength, and temperature was studied for both theoretical and real surfaces. Now we are ready to begin the study of radiation heat transfer among surfaces. The most widely used approach to radiation heat transfer analysis is the net exchange formulation, in which the walls of an enclosure are modeled as diffuse, gray surfaces. In spite of the limitations on accuracy imposed by these assumptions, the net exchange method, because of its simplicity, remains the dominant tool for the analysis of surface-to-surface radiation heat transfer.

8.1 INTRODUCTION

The approach to modeling radiation heat exchange among surfaces developed in this chapter is at least fifty years old, having already been described in the first genera-tion of heat transfer textbooks published in the 1950s [1–4]. A decade later when the first textbooks dedicated exclusively to radiation heat transfer were published [5–8] preference for this approach, referred to here as the *net exchange method,* was well established. The most recent generation of radiation heat transfer textbooks [9–11], published in the early 1990s, offers no significant improvements on this robust and well-established approach. The treatment of the net exchange formulation given be-low draws liberally on the lucid writing on the topic by E. M. Sparrow and his stu-dents and colleagues in the 1960s. In particular, the same widely adopted terminol-ogy and nomenclature has been used.

8.2   THE ENCLOSURE

The point of departure for all radiation heat transfer analyses is definition of a suitable *enclosure*. The enclosure must always be defined such that radiation leaving any surface of the enclosure will be incident on that surface or on another surface, either real or imaginary, of the enclosure. An enclosure, then, is a collection of connected real and imaginary surfaces arranged such that this requirement is satisfied. The enclosure plays the same role in radiation heat transfer analysis as the *system* plays in thermodynamics.

8.3   THE NET EXCHANGE FORMULATION MODEL

The *net exchange formulation* requires that the actual physical situation under consideration be replaced by a model that is an acceptable approximation of reality. The model is then solved, usually exactly, for the unknown heat fluxes and/or temperatures. Of course, these heat fluxes and temperatures, while exact for the model, are only approximations of reality.

The net exchange formulation model incorporates the following assumptions:

1. All surfaces of the enclosure are diffuse emitters, absorbers, and reflectors of thermal radiation.
2. All surfaces of the enclosure are gray; that is, $\alpha = \varepsilon$ at all locations on each surface.
3. All surfaces of the enclosure are opaque; that is, $\tau = 0$ so that $\rho = 1 - \alpha = 1 - \varepsilon$ at all locations on each surface. Openings in the enclosure walls are then treated as imaginary black surfaces having either specified equivalent blackbody temperatures or a specified net heat flux.
4. Either the temperature distribution or the net heat flux distribution is specified on each surface of the enclosure.
5. The temperature or net heat flux distribution on each real or imaginary surface must be continuous. This requirement can always be satisfied by subdividing a surface having a discontinuous distribution of temperature or heat flux along the lines of discontinuity.

8.4   THE RADIOSITY AND THE IRRADIANCE

Two new quantities, the *radiosity* and the *irradiance,* must be defined before the net exchange formulation can be developed. For enclosures that conform to assumptions 1, 2, and 3 above, the radiosity B is defined as the radiation heat flux (W/m²) leaving a surface. It is defined without regard to direction or wavelength such that when multiplied by the area of a surface the result is the radiative power, in watts, leaving the surface. It includes both emitted and reflected power; that is,

$$B(\mathbf{r}) = \varepsilon(\mathbf{r}) \, \sigma T^4(\mathbf{r}) + \rho(\mathbf{r}) \, H(\mathbf{r}) \qquad (8.1)$$

where $\mathbf{r}$ is the position vector, shown in Figure 8.1, of an area element $dA(\mathbf{r})$ whose radiosity is $B(\mathbf{r})$ and whose irradiance is $H(\mathbf{r})$. The quantity $\varepsilon(\mathbf{r})$ is the hemispherical, total emissivity (which for a diffuse, gray surface is equal to the directional, monochromatic emissivity) of the surface element, and $\rho(\mathbf{r})$ is its bihemispherical, total reflectivity (which for a diffuse, gray surface is equal to the directional–hemispherical, monochromatic reflectivity). The irradiance H is defined as the total radiation heat flux (W/m^2) incident to the surface element from the hemispherical space above it.

8.5 THE INTEGRAL FORMULATION

Consider the enclosure depicted in Figure 8.1. The enclosure consists of n opaque, diffuse, gray surfaces and is referred to as a diffuse, gray enclosure. For N of these surfaces the temperature distribution is specified, and for the remaining $n - N$ surfaces the net heat flux distribution is specified; that is,

for $1 \leq i \leq N$, the surface temperature distribution $T(\mathbf{r})$ is known, and

for $N + 1 \leq i \leq n$, the surface net heat flux distribution $q(\mathbf{r})$ is known.

The surface net heat flux is defined in terms of the radiosity and the irradiance as

$$q(\mathbf{r}) \equiv B(\mathbf{r}) - H(\mathbf{r}) \qquad (8.2)$$

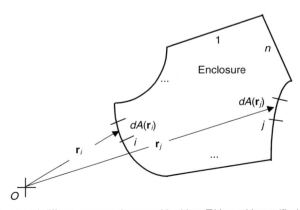

Figure 8.1 A general diffuse, gray enclosure with either $T(r)$ or $q(r)$ specified on each surface

The problem, then, is to find the unknown surface net heat flux distributions on surfaces $1 \leq i \leq N$ and the unknown surface temperature distributions on surfaces $N + 1 \leq i \leq n$. The tools available for finding the unknown temperature and net heat flux distributions are the definition of the radiosity, which may be rewritten

$$B(\mathbf{r}_i) = \varepsilon(\mathbf{r}_i)\, \sigma T^4(\mathbf{r}_i) + [1 - \varepsilon(\mathbf{r}_i)]H(\mathbf{r}_i), \qquad 1 \leq i \leq N \tag{8.3}$$

the definition of surface net heat flux, which may be rewritten

$$B(\mathbf{r}_i) = q(\mathbf{r}_i) + H(\mathbf{r}_i), \qquad N + 1 \leq i \leq n \tag{8.4}$$

and

$$H(\mathbf{r}_i)\, dA(\mathbf{r}_i) = \sum_{j=1}^{n} \int_{A_j}^{0} B(\mathbf{r}_j)\, dA(\mathbf{r}_j)\, dF_{dA(\mathbf{r}_j) \rightarrow dA(\mathbf{r}_i)}, \qquad 1 \leq i \leq n \tag{8.5}$$

where the purely geometrical factor

$$dF_{dA(\mathbf{r}_j) \rightarrow dA(\mathbf{r}_i)} = dF_{dA_j - dA_i} \tag{8.6}$$

is the fraction of radiation leaving $dA(\mathbf{r}_j)$, which arrives at $dA(\mathbf{r}_i)$ by "direct" radiation, that is, without intermediate reflections. This dimensionless factor is called the *differential–differential configuration* (or *angle, shape, view, geometry*) *factor*. Note that the terms *radiosity, irradiance,* and *configuration factor* have meaning only in the context of diffuse enclosures. Before proceeding with the solution of Equations 8.3 through 8.5 it is useful to first investigate the configuration factor more closely.

8.6   THE DIFFERENTIAL–DIFFERENTIAL CONFIGURATION (ANGLE, SHAPE, VIEW, GEOMETRY) FACTOR

What fraction dF of the diffusely distributed radiation leaving $dA(\mathbf{r}_i)$ (emitted plus reflected) arrives at $dA(\mathbf{r}_j)$ by *direct* radiation (i.e., not including intermediate reflections from third surfaces)? The answer to this question is the *differential–differential configuration* (or *angle, shape, view, geometry*) *factor,*

$$dF_{dA_i - dA_j} \equiv \frac{\text{radiation leaving } dA_i \text{ and arriving at } dA_j}{\text{radiation leaving } dA_i} \tag{8.7}$$

Although the differential–differential configuration factor is defined in terms of the radiation, emitted plus reflected, leaving one surface and arriving at another surface, its mathematical form can be determined most easily for the special case of a black surface element. In this case, only emitted radiation leaves the surface element and Equation 8.7 can be written

$$dF_{dA_i-dA_j} = \frac{i'_b\,[T(\mathbf{r}_i)]\,dA(\mathbf{r}_i)\,\cos\theta_i\,d\omega_j}{\displaystyle\int_{2\pi} i'_b\,[T(\mathbf{r}_i)]\,dA(\mathbf{r}_i)\,\cos\theta_i\,d\omega_j} \tag{8.8}$$

The notation in Equation 8.8 may be interpreted in terms of Figure 8.2. Equation 2.1 can be rewritten using the notation of Figure 8.2 as

$$d\omega_j = \frac{dA(\mathbf{r}_j)\,\cos\theta_j}{|\mathbf{r}_{ij}|^2} \tag{8.9}$$

so that Equation 8.8 becomes

$$dF_{dA_i-dA_i} = \frac{\cos\theta_i\,dA(\mathbf{r}_j)\,A\,\cos\theta_j/|\mathbf{r}_{ij}|^2}{\pi} \tag{8.10}$$

or

$$dF_{dA_i-dA_j} = \left(\frac{\cos\theta_i\,\cos\theta_j}{\pi|\mathbf{r}_{ij}|^2}\right)dA(\mathbf{r}_j) \tag{8.11}$$

It is clear from Figure 8.2 and Equation 8.11 that the differential–differential configuration factor depends only on the geometrical relationship between $dA(\mathbf{r}_i)$ and $dA(\mathbf{r}_j)$. Equation 8.11 can be rewritten in a more compact notation,

$$dF_{dA_i-dA_j} = K(\mathbf{r}_i, \mathbf{r}_j)dA(\mathbf{r}_j) \tag{8.12}$$

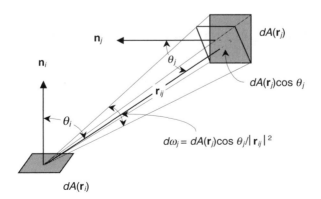

Figure 8.2 Geometry for computing the differential–differential configuration factor

where

$$K(\mathbf{r}_i, \mathbf{r}_j) \equiv \left(\frac{\cos\theta_i \cos\theta_j}{\pi |\mathbf{r}_{ij}|^2} \right) \qquad (8.13)$$

is called the *kernal,* for reasons that will be clear later.

8.7 RECIPROCITY FOR THE DIFFERENTIAL–DIFFERENTIAL CONFIGURATION FACTOR

The subscripts i and j in Equation 8.11 can be freely interchanged so that

$$dF_{dA_j - dA_j} = \left(\frac{\cos\theta_j \cos\theta_i}{\pi |\mathbf{r}_{ji}|^2} \right) dA(\mathbf{r}_i) \qquad (8.14)$$

But it is patently true that

$$\left(\frac{\cos\theta_j \cos\theta_i}{\pi |\mathbf{r}_{ji}|^2} \right) = \left(\frac{\cos\theta_i \cos\theta_j}{\pi |\mathbf{r}_{ij}|^2} \right) \qquad (8.15)$$

Therefore, combining Equations 8.11, 8.14, and 8.15 we have

$$\frac{dF_{dAj - dA_i}}{dA(\mathbf{r}_i)} = \frac{dF_{dA_i - dA_j}}{dA(\mathbf{r}_j)} \qquad (8.16)$$

or

$$dF_{dA_j - dA_i} \, dA(\mathbf{r}_j) = dF_{dA_i - dA_j} \, dA(\mathbf{r}_i) \qquad (8.17)$$

The result represented by Equation 8.17 is referred to as *reciprocity* for the differential–differential configuration factor. Additional reciprocity relations are developed in Chapter 9.

8.8 THE INTEGRAL NET EXCHANGE FORMULATION (CONTINUED)

Equation 8.5 can be simplified by invoking the reciprocity relationship, Equation 8.17, yielding

$$H(\mathbf{r}_i) = \sum_{j=1}^{n} \int_{A_j} B(\mathbf{r}_j) dF_{dA(\mathbf{r}_i) \to dA(\mathbf{r}_j)}, \qquad 1 \le i \le n \qquad (8.18)$$

or

$$H(\mathbf{r}_i) = \sum_{j=1}^{n} \int_{A_j} B(\mathbf{r}_j) K(\mathbf{r}_i, \mathbf{r}_j) dA(\mathbf{r}_j), \qquad 1 \le i \le n \qquad (8.19)$$

Equation 8.19 can now be used to eliminate the irradiance $H(\mathbf{r}_i)$ from Equations 8.3 and 8.4, yielding

$$B(\mathbf{r}_i) = \varepsilon(\mathbf{r}_i)\, \sigma T^4(\mathbf{r}_i) + [1 - \varepsilon(\mathbf{r}_i)]$$
$$\times \sum_{j=1}^{n} \int_{A_j} B(\mathbf{r}_j)\, K(\mathbf{r}_i,\mathbf{r}_j)\, dA(\mathbf{r}_j), \qquad 1 \le i \le N \tag{8.20}$$

and

$$B(\mathbf{r}_i) = q(\mathbf{r}_i) + \sum_{j=1}^{n} \int_{A_j} B(\mathbf{r}_j) K(\mathbf{r}_i,\mathbf{r}_j)\, dA(\mathbf{r}_j), \qquad N + 1 \le i \le n \tag{8.21}$$

Equations 8.20 and 8.21 represent a system of n *integral* equations in n unknown surface radiosity distributions $B(\mathbf{r}_i)$ in terms of N known surface temperature distributions and $n - N$ known surface net heat flux distributions. Once these equations have been solved for the unknown radiosity distributions, the unknown surface temperature distributions can be obtained using a rearranged version of Equation 8.20,

$$\sigma T^4(\mathbf{r}_i) = \frac{B(\mathbf{r}_i) - [1 - \varepsilon(\mathbf{r}_i)] \sum_{j=1}^{n} \int_{A_j} B(\mathbf{r}_i)\, K(\mathbf{r}_i,\mathbf{r}_j)\, dA(\mathbf{r}_j)}{\varepsilon(\mathbf{r}_i)}, \tag{8.22}$$
$$N + 1 \le i \le n$$

and the unknown surface net heat flux distributions can be obtained from a rearranged version of Equation 8.21,

$$q(\mathbf{r}_i) = B(\mathbf{r}_i) - \sum_{j=1}^{n} \int_{A_j} B(\mathbf{r}_j) K(\mathbf{r}_i,\, \mathbf{r}_j)\, dA(\mathbf{r}_j), \qquad 1 \le i \le N \tag{8.23}$$

8.9 INTEGRAL EQUATIONS VERSUS DIFFERENTIAL EQUATIONS

Students of thermal–fluid sciences are familiar with the derivation and solution of differential equations, but most probably will not have encountered integral equations before undertaking the study of radiation heat transfer. We recall that differential equations describe *diffusion* processes; that is, processes in which local transport is driven by local potential gradients. Integral equations, on the other hand, describe situations in which the value of a state variable is directly influenced by values of that, and perhaps other, state variables everywhere in the system. In the case of heat conduction, which is a diffusion process, the temperature at a point in a medium depends only on the value of the temperature at neighboring points a differential distance from the point in question. In the case of heat radiation among the surfaces of an enclosure, however, the temperature at a point on a surface is directly influenced by the temperatures at all other points on all surfaces of the enclosure.

8.10 SOLUTION OF INTEGRAL EQUATIONS

We are interested in integral equations of the general form (see Equations 8.20 and 8.21)

$$u(x) = f(x) + \lambda \int_a^b u(t) \, K(x,t) \, dt \tag{8.24}$$

which is a *Friedholm equation of the second kind.* We recognize that this is also the general form of the integral equations derived in Chapter 7 to describe radiation through a participating medium. Several alternatives are available for solving Friedholm equations.

8.11 SOLUTION BY THE METHOD OF SUCCESSIVE SUBSTITUTIONS

In the *method of successive substitutions* the equation to be solved is successively substituted into itself m times. That is, the dependent variable $u(t)$ under the integral in Equation 8.24 is replaced with the entire expression for the dependent variable $u(x)$, Equation 8.24. This substitution and subsequent evaluation of the resulting integrals is repeated successively until the emerging infinite series is recognized or at least until a sufficient number of terms has been obtained to assure convergence. After the first substitution we have

$$u_1(x) = f(x) + \lambda \int_a^b \left[f(t) + \lambda \int_a^b u(t_1) \, K(t, t_1) \, dt_1 \, K(x,t) \right] dt \tag{8.25}$$

and after the mth substitution we are left with

$$u_m(x) = f(x) + \lambda \int_a^b f(t)K(x, t) \, dt + \lambda^2 \int_a^b \int_a^b f(t_1)K(x,t)K(t,t_1) \, dt_1 dt + \cdots$$
$$+ \lambda^m \left[\int_a^b \right]^m f(t_{m-1})K(x,t)K(t,t_1) \cdots K(t_{m-2}, t_{m-1}) \, dt_{m-1} \cdots dt_1 \, dt \tag{8.26}$$

The series represented by Equation 8.26 converges if

$$|\lambda| \le \left| \frac{1}{M(b-a)} \right| \tag{8.27}$$

where M is the maximum value of the kernal K in the rectangle R defined by $a \le t \le b$ and $a \le x \le b$.

Example Problem 8.1

Use the method of successive substitutions to solve the *Voltera equation,*

$$u(x) = x + \int_0^x (t - x) u(t) \, dt \tag{8.28}$$

SOLUTION

We recognize Equation 8.28 as a Friedholm equation of the second kind with $\lambda = 1$ and $K(x, t) = t - x$. In this case $M = |x|$ and so we are assured of convergence when $x \leq 1$. The method of successive substitutions then leads to

$$u(x) = x + \int_0^x (t - x) t \, dt + \int_0^x \int_0^t (t - x)(t_1 - t) \, t_1 dt \, dt_1$$

$$+ \int_0^x \int_0^t \int_0^{t_1} (t - x)(t_1 - t)(t_2 - t_1) \, t_2 \, dt \, dt_1 \, dt_2$$

$$= x + \frac{x^3}{3} - \frac{x^3}{2} + \frac{x^5}{15} - \frac{x^5}{12} - \frac{x^5}{10} + \frac{x^5}{8} + \cdots \tag{8.29}$$

$$= x - \frac{x^3}{3!} + \frac{x^5}{5!} - \cdots = \sin x$$

This alternating sign series actually converges for all x, but as x gets larger more terms are needed to obtain an acceptable approximation for $\sin x$.

8.12 SOLUTION BY THE METHOD OF SUCCESSIVE APPROXIMATIONS

In the *method of successive approximations* an arbitrary initial guess $u_0(x)$ is substituted into the original integral equation to generate an upgraded approximation, $u_1(x)$. This procedure is repeated successively until an acceptable degree of approximation has been achieved. The convergence criterion is the same as in the method of successive substitutions, i.e., Equation 8.27. Because the method is approximate it can be terminated when

$$\Delta_p = \sqrt{\frac{1}{b - a} \int_a^b (u_p - u_{p-1})^2 \, dx} \approx 0 \tag{8.30}$$

where p represents the pth approximation.

Example Problem 8.2

Use the method of successive approximations to solve the Voltera equation, Equation 8.28.

SOLUTION

In this case it is convenient to let $u_0(x) = 0$. This is often a convenient first approximation, as is $u_0(x) = 1$. However, the choice is arbitrary; if the convergence criterion is met the method will converge regardless of the choice of $u_0(x)$. Then with $u_0(x) = 0$,

$$u_1(x) = x + \int_0^x (t - x)(0)\, dt = x \tag{8.31}$$

$$u_2(x) = x + \int_0^x (t - x) t\, dt$$

$$= x + \int_0^x t^2 dt - x \int_0^x t\, dt \tag{8.32}$$

$$= x + \frac{x^3}{3} - \frac{x^3}{2}$$

$$u_3(x) = x + \int_0^x (t - x) t\, dt + \int_0^x (t - x)\frac{t^3}{3} dt - \int_0^x (t - x)\frac{t^3}{2} dt$$

$$= x + \frac{x^3}{3} - \frac{x^3}{2} + \frac{x^5}{15} - \frac{x^5}{12} - \frac{x^5}{10} + \frac{x^5}{8} \tag{8.33}$$

$$= x - \frac{x^3}{6} + \frac{x^5}{120}$$

Then by inductive reasoning,

$$u_\infty(x) = u(x) - \frac{x^3}{3!} + \frac{x^5}{5!} - \cdots = \sin x \tag{8.34}$$

8.13   SOLUTION BY THE METHOD OF LAPLACE TRANSFORMS

Let $F(s)$ and $K(s)$ be the Laplace transforms of $f(t)$ and $K(x,t)$, respectively, where the kernal has the special form $K(x,t) = K(x - t)$ and $\lambda = 1$; that is, the equation to be solved is

$$u(x) = f(x) + \int_0^x K(t - x)\, u(t)\, dt \tag{8.35}$$

Then

$$U(s) = \frac{F(s)}{1 - K(s)} = F(s) + \frac{K(s)}{1 - K(s)} F(s) \tag{8.36}$$

so that

$$u(x) = f(x) + \int_0^x h(t - x)\, f(t)\, dt \tag{8.37}$$

where $h(x - t)$ is the function whose Laplace transform is given by

$$H(s) = \frac{K(s)}{1 - K(s)} \tag{8.38}$$

8.14 SOLUTION BY AN APPROXIMATE ANALYTICAL METHOD

The three methods outlined in Sections 8.11 through 8.13 are, or at least can be, exact analytical methods. An approximate analytical method exists that is often adequate for radiation heat transfer problems. Considering once again the Friedholm equation, Equation 8.24, we expand the unknown function $u(t)$ in a Taylor series about x:

$$u(t) = u(x) + (t - x)\frac{d}{dx}u(x) + \frac{(t - x)^2}{2!}\left[\frac{d^2}{dx^2}u(x)\right] + \cdots \tag{8.39}$$

We then introduce the expansion for $u(t)$ in the integrand of Equation 8.24, obtaining

$$\begin{aligned}
u(x) = f(x) &+ \lambda \int_a^b u(x)\, K(x,t)\, dt \\
&+ \lambda \int_a^b (t - x)\frac{du}{dx} K(x,t)\, dt \\
&+ \lambda \int_a^b \frac{(t - x)^2}{2!}\frac{d^2u}{dx^2} K(x,t)\, dt + \cdots
\end{aligned} \tag{8.40}$$

If the terms of order higher than two are neglected (herein lies the approximation), there results

$$
\left[\frac{1}{2} \int_a^b (t - x)^2 \, K(x,t) \, dt \right] \frac{d^2 u}{dx^2} + \left[\int_a^b (t - x) \, K(x, t) \, dt \right] \frac{du}{dx}
$$
$$
+ \left[\int_a^b K(x,t) \, dt - \frac{1}{\lambda} \right] u + \frac{f(x)}{\lambda} = 0 \tag{8.41}
$$

The bracketed quantities in Equation 8.41 are functions of x only after the indicated integrations are performed. It is convenient to define

$$
g_2(x) \equiv \frac{1}{2} \int_a^b (t - x)^2 K(x,t) \, dt \tag{8.42}
$$

$$
g_1(x) \equiv \int_a^b (t - x) K(x,t) \, dt \tag{8.43}
$$

and

$$
g_0(x) \equiv \int_a^b K(x,t) \, dt - \frac{1}{\lambda} \tag{8.44}
$$

With the introduction of Equations 8.42 through 8.44 into Equation 8.41, there results the ordinary, inhomogeneous, second-order differential equation with variable coefficients

$$
g_2(x) \, u''(x) + g_1(x) \, u'(x) + g_0(x) \, u(x) + \frac{f(x)}{\lambda} = 0 \tag{8.45}
$$

The solution to the integral equation, Equation 8.24, has been approximated by the solution of a differential equation.

8.15   THE FINITE NET EXCHANGE FORMULATION

An alternative to the integral net exchange formulation described in Sections 8.5 through 8.8 is the *finite net exchange formulation*. This alternative can be thought of as a numerical approach to solving the integral equations obtained with the integral net exchange formulation, Equations 8.4 through 8.6. In this approach the enclosure is subdivided into a relatively large number of relatively small surface elements and the integrals are approximated as weighted sums over these elements. In the limit of an infinite number of vanishingly small surface elements, the finite net exchange formulation converges to the integral formulation, and so is exact within the limitations of the restrictions imposed in Section 8.2.

The finite net exchange formulation can also be thought of as a less accurate approximate method if the enclosure is subdivided into only a modest number of relatively large surface elements. In this latter interpretation the model is less representative of reality because the requirements placed on the surface elements are even more restrictive than in the integral formulation. However, as in the case of the integral formulation, the equations representing the model are solved exactly. The gain of time and the reduction of effort in obtaining the approximate radiative behavior of a large and complex enclosure often adequately compensate for the sacrifice of accuracy.

In addition to assumptions 1 through 3 in Section 8.2 requiring that all surface elements of the enclosure be diffuse, gray, and opaque, the finite formulation is more restrictive concerning the distributions of surface temperature and surface net heat flux:

4'. Either the temperature or the net heat flux must be specified on each surface of the enclosure (contrast this with the fourth assumption in Section 8.2, which includes the word "distribution").

5'. The temperature or net heat flux must be *uniform* over each surface element, an assumption implicit in assumption 4' above.

To these assumptions we add two new restrictions:

6. The radiosity and irradiance must be uniform on each surface element.
7. The emissivity must be uniform on each surface element.

As a practical matter it is almost always possible to refine the surface element subdivision until requirements 5', 6, and 7 are met. However, depending on the desired accuracy of the model, it may be necessary to provide an excessively fine mesh near the intersections of two surfaces that meet at acute angles because radiation tends to "gather" in corners, with the result that gradients are steep there.

Now consider the hypothetical enclosure depicted in Figure 8.3, consisting of n finite-sized surface elements for which restrictions 1 through 3 in Section 8.2 and restrictions 4', 5', 6, and 7 above may be assumed to apply with acceptable accuracy. Assuming that the surface temperatures are specified on surfaces $1, 2, \ldots, N$ and that the surface net heat fluxes are specified on surfaces $N + 1, N + 2, \ldots, n$, Equations 8.3 through 8.5 can now be written

$$B_i = \varepsilon_i \, \sigma T_i^4 + (1 - \varepsilon_i) \, H_i, \qquad 1 \le i \le N \qquad (8.46)$$

$$B_i = q_i + H_i, \qquad N + 1 \le i \le n \qquad (8.47)$$

and

$$H_i \, A_i = \sum_{j=1}^{n} B_j \, A_j \, F_{A_j \to A_i}, \qquad 1 \le i \le n \qquad (8.48)$$

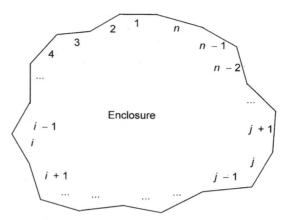

Figure 8.3 Hypothetical diffuse, gray enclosure with either T or q specified on each surface element

The quantity $F_{Aj \to Ai} = F_{ji}$ is the *finite–finite configuration factor,* and represents the fraction of diffusely distributed radiation leaving surface element A_j and arriving at surface element A_i, once again by "direct" radiation.

8.16   RELATIONSHIPS BETWEEN DIFFERENTIAL AND FINITE CONFIGURATION FACTORS

The most difficult and engineering intensive aspect of the finite–finite net exchange formulation is evaluation of the configuration factors, F_{ij}. Chapter 9 is devoted to methods for determining finite–finite configuration factors. However, before we can continue our development of the finite–finite net exchange formulation we need to develop a reciprocity relation similar to Equation 8.17. Consider the two finite surfaces A_i and A_j shown in Figure 8.4. Ultimately we seek the configuration factor from finite area A_i to finite area A_j, F_{ij}, but first we must establish expressions for the configuration factors between differential and finite area elements.

We seek an expression for the fraction of diffusely distributed radiation leaving the differential area $dA(\mathbf{r}_i)$ that arrives at the finite area element A_j by direct radiation. Note that both the power radiated into 2π-space above $dA(\mathbf{r}_i)$ and the power arriving at A_j are differential quantities, but that the ratio of the latter to the former is a finite quantity. It is clear that the fraction sought, which we call the *differential–finite configuration factor,* is obtained by integrating the differential–differential configuration factor, Equation 8.11, over the receiving surface; that is,

$$F_{dA_i \to dA_j} = \int_{A_j} dF_{dA_i \to dA_j} = \int_{A_j} K(\mathbf{r}_i, \mathbf{r}_j) \, dA(\mathbf{r}_j) \tag{8.49}$$

or

$$F_{dA_i \rightarrow dA_j} = \int_{A_j} \frac{\cos[\theta(\mathbf{r}_i, \mathbf{r}_j)] \cos[\theta(\mathbf{r}_j, \mathbf{r}_i)]}{\pi |\mathbf{r}_{ij}|^2} \, dA(\mathbf{r}_j) \qquad (8.50)$$

Next, let $dF_{A_j \rightarrow dA(\mathbf{r}_j)}$ represent the *finite–differential configuration factor,* defined as the fraction of the diffusely distributed radiation leaving A_j, which arrives at $dA(\mathbf{r}_i)$ by direct radiation. Note that this is a differential quantity because the numerator of the fraction is a differential quantity while the denominator is finite. Then, if A_j has a uniform radiosity B_j, the radiation arriving at $dA(\mathbf{r}_i)$ due to A_j is

$$B_j A_j dF_{A_j \rightarrow dA(r_i)} \qquad (8.51)$$

But another way of writing this same quantity is

$$\int_{A_j} B_j \, dF_{dA(\mathbf{r}_j) \rightarrow dA(\mathbf{r}_i)} \, dA(\mathbf{r}_j) \qquad (8.52)$$

Equating these two expressions and rearranging yields

$$dF_{A_j \rightarrow dA(\mathbf{r}_i)} = \frac{1}{A_j} \int_{A_j} dF_{dA(\mathbf{r}_j) \rightarrow dA(\mathbf{r}_i)} \, dA(\mathbf{r}_j) \qquad (8.53)$$

From Equation 8.53 we conclude that the finite–differential configuration factor from A_j to $dA(\mathbf{r}_i)$ is the mean of the differential–differential configuration factor from $dA(\mathbf{r}_j)$ to $dA(\mathbf{r}_i)$ on A_j.

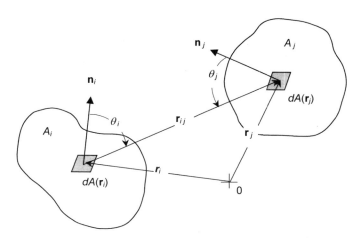

Figure 8.4 Geometry for evaluation of configuration factors

With the introduction of the reciprocity relation for differential–differential configuration factors, Equation 8.17, Equation 8.53 becomes

$$dF_{A_j \to dA(\mathbf{r}_i)} = \frac{1}{A_j} \int_{A_j} dF_{dA(\mathbf{r}_i) \to dA(\mathbf{r}_j)} \, dA(\mathbf{r}_i)$$

$$= \frac{dA(\mathbf{r}_i)}{A_j} \int_{A_j} dF_{dA(\mathbf{r}_i) \to dA(\mathbf{r}_j)}$$

(8.54)

or

$$dF_{A_j \to dA(\mathbf{r}_i)} = \frac{dA(\mathbf{r}_i)}{A_j} F_{dA(\mathbf{r}_i) \to A_j}$$

(8.55)

Equation 8.55 can be rearranged to yield the reciprocity relation

$$A_j \, dF_{A_j \to dA(\mathbf{r}_i)} = dA(\mathbf{r}_i) F_{dA(\mathbf{r}_i) \to A_j}$$

(8.56)

The indices i and j in Equation 8.56 can be interchanged to obtain a second reciprocity relation between differential–finite and finite–differential configuration factors,

$$A_i \, dF_{A_i \to dA(\mathbf{r}_j)} = dA(\mathbf{r}_j) F_{da(\mathbf{r}_j) \to A_i}$$

(8.57)

Finally, Equation 8.57 can be rearranged to obtain

$$dF_{A_i \to dA(\mathbf{r}_j)} = \frac{dA(\mathbf{r}_j)}{A_i} F_{dA(\mathbf{r}_j) \to A_i}$$

(8.58)

Integrating Equation 8.58 over surface A_j yields

$$F_{A_i \to A_j} = F_{ij} = \int_{A_j} dF_{A_i \to dA(\mathbf{r}_j)} = \frac{1}{A_i} \int_{A_j} F_{dA(\mathbf{r}_j) \to A_i} \, dA(\mathbf{r}_j)$$

$$= \frac{1}{A_i} \int_{A_j} \int_{A_i} dF_{dA(\mathbf{r}_j) \to dA(\mathbf{r}_i)} \, dA(\mathbf{r}_j)$$

(8.59)

or

$$F_{ij} = \frac{1}{A_i} \int_{A_j} \int_{A_i} K(\mathbf{r}_i, \mathbf{r}_j) \, dA(\mathbf{r}_i) \, dA(\mathbf{r}_j)$$

$$= \frac{1}{A_i} \int_{A_j} \int_{A_i} \frac{\cos[\theta(\mathbf{r}_i,\mathbf{r}_j)] \cos[\theta(\mathbf{r}_j,\mathbf{r}_i)]}{\pi |\mathbf{r}_{ij}|^2} \, dA(\mathbf{r}_i) \, dA(\mathbf{r}_j)$$

(8.60)

Now if both sides of Equation 8.60 are multiplied by A_i, there results

$$A_i F_{ij} = \int_{A_j} \int_{A_i} \frac{\cos[\theta(\mathbf{r}_i, \mathbf{r}_j)]\, \cos[\theta(\mathbf{r}_j, \mathbf{r}_i)]}{\pi |\mathbf{r}_{ij}|^2}\, dA(\mathbf{r}_i)\, dA(\mathbf{r}_j) \tag{8.61}$$

Once again, it is clear that, due to symmetry, the indices i and j can be interchanged without changing the value of the double (quadruple, actually) integral. It follows, therefore, that

$$A_i F_{ij} = A_j F_{ji} \tag{8.62}$$

which is the reciprocity relation for finite–finite configuration factors.

8.17   THE FINITE NET EXCHANGE FORMULATION (CONTINUED)

With the introduction of Equation 8.62, Equations 8.46 through 8.48 can be consolidated to obtain

$$B_i = \varepsilon_i\, \sigma T_i^4 + (1 - \varepsilon_i) \sum_{j=1}^{n} B_j F_{ij}, \qquad 1 \le i \le N \tag{8.63}$$

and

$$B_i = q_i + \sum_{j=1}^{n} B_j F_{ij}, \qquad N + 1 \le i \le n \tag{8.64}$$

8.18   SOLUTION OF THE FINITE NET EXCHANGE FORMULATION EQUATIONS

Equations 8.63 and 8.64 represent n equations in n unknowns (N unknown surface net heat fluxes and $n - N$ unknown surface temperatures). It is interesting to contrast these equations with Equations 8.20 and 8.21, which represent n unknown *distributions*. Our approach to solving these equations will be to cast them in matrix form and then use standard matrix methods to obtain solutions.

We begin by introducing the Kronecker delta function,

$$\delta_{ij} = \begin{cases} 1 & i=j \\ 0 & i \ne j \end{cases} \tag{8.65}$$

Equation 8.65 allows us to write

$$B_i = \sum_{j=1}^{n} B_j \delta_{ij}, \qquad i = 1, 2, \dots, n \tag{8.66}$$

With the introduction of Equation 8.66, Equations 8.63 and 8.64 may be rearranged to yield

$$\sigma T_i^4 = \frac{1}{\varepsilon_i} \sum_{j=1}^{n} [\delta_{ij} - (1 - \varepsilon_i)F_{ij}]B_j, \qquad 1 \leq i \leq N \qquad (8.67)$$

and

$$q_i = \sum_{j=1}^{n} (\delta_{ij} - F_{ij})B_j, \qquad N + 1 \leq i \leq n \qquad (8.68)$$

It is now convenient to define

$$\Omega_i \equiv \begin{cases} \sigma T_i^4, & 1 \leq i \leq N \\ q_i, & N + 1 \leq i \leq n \end{cases} \qquad (8.69)$$

and

$$\chi_{ij} = \begin{cases} \dfrac{1}{\varepsilon_i}[\delta_{ij} - (1 - \varepsilon_i)F_{ij}], & 1 \leq i \leq N \\ (\delta_{ij} - F_{ij}), & N + 1 \leq i \leq n \end{cases} \qquad (8.70)$$

Then, in general, we can write

$$\Omega_i = \chi_{ij}B_j, \qquad 1 \leq i \leq n \qquad (8.71)$$

where summation (over j) is implied by the repeated subscript.

We can now solve for the unknown radiosities by inverting the χ_{ij} matrix; that is,

$$B_i \equiv [\chi_{ij}]^{-1}\Omega_j, \qquad 1 \leq i \leq n \qquad (8.72)$$

where $[\chi_{ij}]^{-1}$ is the inverse of χ_{ij}. Once the radiosities of the n surfaces have been computed using Equation 8.72, the unknown surface net heat fluxes are obtained using

$$q_i = \sum_{j=1}^{n} (\delta_{ij} - F_{ij})B_j, \qquad 1 \leq i \leq N \qquad (8.73)$$

and the unknown surface temperatures are obtained using

$$T_i = \left\{ \frac{1}{\varepsilon_i \sigma} \sum_{j=1}^{n} [\delta_{ij} - (1 - \varepsilon_i)F_{ij}]B_j \right\}^{1/4}, \qquad N + 1 \leq i \leq n \qquad (8.74)$$

Team Project

TP 8.1   In anticipation of Team Project TP 9.1, in which we will solve a large surface-to-surface radiation heat transfer problem, in the current Team Project we will build a CAD model of an enclosure. The enclosure is a thermal vacuum chamber 1.5 m tall and 1.0 m in diameter. The top of the chamber is a spherical dome having a radius of 0.5 m, and the bottom of the chamber is a disk also having a radius of 0.5 m. Use a CAD program such as AutoCad to create the chamber as a stack-up of equal-area bands (this will take some thought for the dome and floor), with each band divided azimuthally into area elements. To the extent possible, each resulting pseudo-rectangular area element should be about 5 cm on a side. The drawing should be "exploded" so that each area element is a separate entity within the CAD file. This is because in Team Project TP 9.1 the individual surface elements will have to be exported one at a time into another program. Print out several views of the suitably subdivided chamber.

Discussion Points

DP 8.1   In thermodynamics a *system* is usually defined as a quantity of matter or a region of space set aside for study. Using this definition, how is an *enclosure* as defined in this chapter like a thermodynamic system and how is it different?

DP 8.2   Based on your study of Chapters 4 and 5 in Part I of this book, what are some of the physical realities that would tend to invalidate each of the five numbered assumptions in Section 8.3?

DP 8.3   In what sense can the *kernal* in Equation 8.19 be thought of as a *Green's function?*

DP 8.4   The Voltera equation, Equation 8.28, has the sine function as its solution. What would be the integral equation whose solution is the cosine function?

DP 8.5   It is stated in Section 8.15 that radiation tends to "gather" in corners. What is meant by this assertion?

DP 8.6   The solution to systems of differential equations can be approximated using finite difference methods. This leads to systems of algebraic equations that, in the implicit formulation, are solved by inverting a matrix. In this chapter we have seen that the solution to systems of integral equations can also be well approximated by solutions to a system of algebraic equations, obtained by inverting a matrix. Close inspection of the properties of the matrices in these two situations provides important insights into the fundamental difference between differential and integral equations. What is the principal difference between the matrices? What can we conclude from this difference?

DP 8.7   Think about how you would go directly from Equations 8.3, 8.4, and 8.5 to Equations 8.46, 8.47, and 8.48 *without* reformulating the problem (as was done in Section 8.15). Are your approach and the approach of Section 8.15 equivalent?

Problems

P 8.1   Two infinitely long segments of circular arc Θ face each other as shown in Figure 8.5. Find the differential–differential configuration factor from a strip $d\theta_1$ located at θ_1 on segment 1 to a strip of width $d\theta_2$ located at θ_2 on segment 2.

P 8.2   Both of the two infinitely long circular segments in Figure 8.5 are diffuse and gray, with an emissivity ϵ. Segment 1 is insulated ($q_1 = 0$) and segment 2 has a specified uniform temperature T_2. The segments are located in a vacuum chamber whose walls are black and near absolute zero temperature. Use the integral net exchange formulation to derive the integral equations needed to find the temperature distribution on segment 1 and the net heat flux distribution on segment 2. Begin by nondimensionalizing the problem in terms of $q(\theta_2)/\sigma T_2^4$ as a function of θ_2 and $T(\theta_1)/T_2$ as a function of θ_1 (note that the problem is symmetrical with respect to the horizontal plane passing through the origin).

P 8.3   Solve the equations obtained in Problem P 8.2, presenting graphical results for all combinations of

$$\Theta = \frac{\pi}{4}, \frac{\pi}{2}, \text{ and } \pi$$

and

$$\varepsilon_1 = \varepsilon_2 = \varepsilon = 0.1, 0.5, \text{ and } 1.0$$

Use an analytical approach to solve the equations; do not solve them numerically. A brief written explanation of the results should also be provided.

P 8.4   Rework Problem P 8.3 using the finite net exchange formulation. Subdivide both segments into m parallel strips and then study the sensitivity to m of the solution obtained. What value of m is adequate in your opinion?

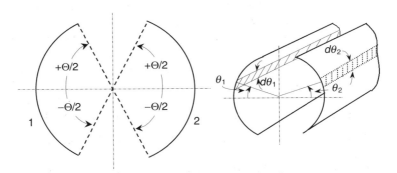

Figure 8.5   Geometry for Problems P 8.1 through P 8.4

P 8.5   Solve the Voltera equation, Equation 8.28, using the method of Laplace transforms (Section 8.13).

P 8.6   Solve the Voltera equation, Equation 8.28, using the approximate analytical method of Section 8.14.

REFERENCES

1. McAdams, W. H., *Heat Transmission,* McGraw-Hill Book Company, Inc., New York, 1954.
2. Kreith, F., *Radiation Heat Transfer for Spacecraft and Solar Power Plant Design,* International Textbook Company, Scranton, PA, 1958.
3. Jakob, M., *Heat Transfer,* Vol. II, John Wiley & Sons, Inc., New York, 1957.
4. Eckert, E. R. G., and Drake, R. M., *Heat and Mass Transfer,* McGraw-Hill Book Company, Inc., New York, 1959.
5. Sparrow, E. M., and R. D. Cess, *Radiation Heat Transfer,* Brooks/Cole Publishing Company, Belmont, CA, 1966.
6. Wiebelt, J. A., *Engineering Radiation Heat Transfer,* Holt, Rinehart and Winston, Inc., New York, 1966.
7. Hottel, H. C., and A. F. Sarofim, *Radiative Transfer,* McGraw-Hill Book Company, Inc., New York, 1967.
8. Love, T. J., *Radiative Heat Transfer,* Charles E. Merrill Publishing Company, Columbus, OH, 1968.
9. Siegel, R., and J. R. Howell, *Thermal Radiation Heat Transfer,* Hemisphere Publishing Corporation, Washington, 1992.
10. Brewster, M. Q., *Thermal Radiative Transfer and Properties,* John Wiley & Sons, Inc., New York, 1992.
11. Modest, M. F., *Radiative Heat Transfer,* McGraw-Hill, Inc., New York, 1993.

9

EVALUATION OF
CONFIGURATION FACTORS

In Chapter 8 we developed the finite net exchange formulation, a model for the radiation heat transfer analysis of enclosures consisting of diffuse, gray surfaces and subject to certain other restrictions. The finite net exchange formulation requires the use of dimensionless factors, called finite–finite configuration factors, whose values define the geometrical relationship between any pair of surface elements making up the enclosure. In this chapter we learn first some of the traditional techniques and then a more modern statistical method for evaluation of these configuration factors.

9.1 INTRODUCTION

An entire book could easily be written on the evaluation of configuration factors, and an entire one-semester course could be based on such a book. However, this hardly seems justifiable. The importance of the net exchange formulation, and therefore of the configuration factor, is rapidly being diminished by the emergence of the much more powerful statistically based methods, which are the subject of Part III of this book. The presentation of these subjects in Chapter 8 and in the current chapter, and the presentation of exchange factors in Chapter 10, is as much a concession to their historical significance as to their intrinsic value. While it is likely that upwards of 90 percent of all radiation heat transfer analyses carried out up to the date of publication of this book were formulated using the net exchange method or one of its close relatives, it seems equally likely that a similar fraction of all serious radiation heat transfer analyses performed in the next decade will be based on some variation of the Monte-Carlo ray-trace method presented in Part III of this book.

The configuration factor that is the topic of this chapter is referred to by many names in the literature and by practitioners; e.g., "view factor," "shape factor," "angle factor," "geometry factor," and so forth. Here when we speak of the "configuration factor" without a qualifying adjective we are referring to the "finite–finite" configuration factor between two finite-sized area elements.

Available methods for evaluating configuration factors may be classified as *direct* and *indirect*. In the direct methods, configuration factors are computed directly from the definition as the fraction of diffusely distributed radiation leaving one surface that arrives at another surface by direct radiation (i.e., excluding reflections). The indirect method involves what is usually called *configuration factor algebra,* presented in Sections 9.6 and 9.7. Several methods for obtaining the values of configuration factors, such as Hottel's crossed-string method [1] and various optical [2] and even mechanical [1] methods described in earlier texts, are not treated here. While of pedagogical interest, these methods are no longer used by serious practitioners. Finally, a method for computing configuration factors based on the Monte-Carlo ray-trace (MCRT) method is described in Section 9.9; and a Windows-based tool implementing this approach, program FELIX, is introduced.

9.2 EVALUATION OF CONFIGURATION FACTORS BASED ON THE DEFINITION (THE DIRECT METHOD)

The most effective way to illustrate the direct approach is through a series of examples.

Example Problem 9.1

We begin by considering a common geometry of great theoretical and practical importance, the *integrating* (or *Obrecht*) *sphere.* The integrating sphere has the distinction of being one of only a very few radiation heat transfer problems that entertain an exact analytical solution. This is because the relevant configuration factor has a constant kernal.

SOLUTION

Consider the diffuse, gray spherical enclosure shown in Figure 9.1. Let $T(\mathbf{r})$ be the specified wall temperature distribution, where $\mathbf{r}$ represents the local position vector. In this case, Equation 8.20 becomes

$$B(\mathbf{r}_1) = \varepsilon\sigma T^4(\mathbf{r}_1) + (1 - \varepsilon)\int_{A_2} B(\mathbf{r}_2)\, dF_{dA(\mathbf{r}_1)\rightarrow dA(\mathbf{r}_2)} \tag{9.1}$$

where

$$dF_{dA(\mathbf{r}_1)\rightarrow dA(\mathbf{r}_2)} = K(\mathbf{r}_1, \mathbf{r}_2)\, dA(\mathbf{r}_2) = \left(\frac{\cos\theta_1\cos\theta_2}{\pi L^2}\right) dA(\mathbf{r}_2) \tag{9.2}$$

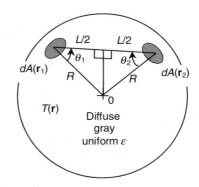

Figure 9.1 The "integrating" sphere

But $\cos\theta_1 = \cos\theta_2 = L/2R$. Thus,

$$dF_{dA(\mathbf{r}_1)\to dA(\mathbf{r}_2)} = \frac{dA(\mathbf{r}_2)}{4\pi R^2} = \frac{dA(\mathbf{r}_2)}{A_2} \tag{9.3}$$

where $A_2 = 4\pi R^2$ is the surface area of the spherical enclosure. Therefore, the kernal $K(\mathbf{r}_1, \mathbf{r}_2) = 1/A_2$ is a *constant*. In this special case, then, Equation 9.1 can be written

$$B(\mathbf{r}_1) = \varepsilon\sigma T^4(\mathbf{r}_1) + (1 - \varepsilon)\,\frac{1}{A_2}\int_{A_2} B(\mathbf{r}_2)\,dA_2 \tag{9.4}$$

We note that the expression

$$\frac{1}{A^2}\int_{A_2} B(\mathbf{r}_2)\,dA_2 \tag{9.5}$$

in Equation 9.4 is nothing more than the average radiosity over the surface of the spherical enclosure, and as such is a constant. In fact, referring to Equation 8.19 we recognize that this quantity is the irradiance H on the spherical surface. We conclude that the irradiance on the walls of a diffuse, gray spherical enclosure is uniform, regardless of the variation of temperature, and even of the emissivity, on the walls!

The irradiance H can be obtained by multiplying Equation 9.4 by dA_1 and then integrating over A_1,

$$\int_{A_1} B(\mathbf{r}_1)\,dA_1 = \varepsilon\int_{A_1} \sigma T^4(\mathbf{r}_1)\,dA_1 + (1 - \varepsilon)\int_{A_1} H\,dA_1 \tag{9.6}$$

But we recognize the left-hand side of Equation 9.6 to be HA, where now $A = A_1 = A_2$. Thus,

$$HA = \varepsilon\int_A \sigma T^4(\mathbf{r})\,dA + (1 - \varepsilon)HA \tag{9.7}$$

or, finally,

$$H = \frac{1}{A} \int_A \sigma T^4 (\mathbf{r}) \, dA \tag{9.8}$$

This intriguing and important result makes it clear why diffuse, gray spherical enclosures are referred to as "integrating" spheres. Integrating spheres are used as elements in many optical and thermal radiative experimental situations, for example, in a common apparatus for measuring surface properties.

Example Problem 9.2

Consider the case of an enclosure that is arbitrarily long in one of its dimensions, for example, the box channel shown in Figure 9.2. We seek to compute first the differential–differential configuration factor and then the differential–finite and finite–finite configuration factors between surfaces of this enclosure.

SOLUTION

Figure 9.2 is similar to Figure 8.2. The only real difference is the use of the θ, ϕ spherical coordinate system in the latter and the α, β ortho-angular coordinate system in the former. The angles α and β are both measured with respect to the x axis,

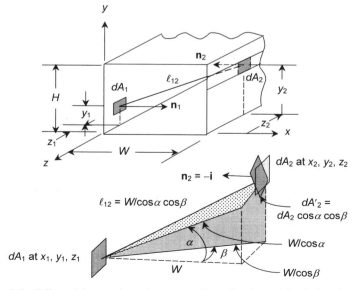

Figure 9.2 Differential area elements on opposite walls of an arbitrarily long box channel

the former in the x, y plane and the latter in the x, z plane. Then, from the definition of the differential–differential configuration factor, Equation 8.7, we have

$$dF_{dA_1 \rightarrow dA_2} \equiv \frac{\text{diffusely distributed radiation leaving } dA_1 \text{ and arriving at } dA_2}{\text{diffusely distributed radiation leaving } dA_1} \tag{9.9}$$

or

$$dF_{dA_1 \rightarrow dA_2} = \frac{i_1 \, dA_1 \, \cos\alpha \, \cos\beta \, d\omega_2}{\displaystyle\int_{2\pi} i_1 \, dA_1 \, \cos\alpha \, \cos\beta \, d\omega_2} \tag{9.10}$$

where the factor $\cos\alpha \, \cos\beta$ replaces the factor $\cos\theta$ in Equation 9.8 (see Problem P9.1).

We now seek an expression for the differential solid angle, $d\omega_2$. Applying the basic definition, Equation 2.1, to the geometry of Figure 9.2, we obtain

$$d\omega_2 = \frac{dA_2'}{\ell_{12}^2} = \frac{(\ell_{12} \, d\alpha)(\ell_{12} \, d\beta)}{\ell_{12}^2} = d\alpha \, d\beta \tag{9.11}$$

thus verifying the intuitively pleasing idea that the solid angle is simply the product of two plane angles.

With the introduction of Equation 9.11 into Equation 9.10, we obtain

$$dF_{dA_1 \rightarrow dA_2} = \frac{\cos\alpha \, \cos\beta \, d\alpha \, d\beta}{\displaystyle\int_{-\pi/2}^{\pi/2} \cos\alpha \, d\alpha \int_{-\pi/2}^{\pi/2} \cos\beta \, d\beta} = \left(\frac{\cos\alpha \, d\alpha}{2}\right)\left(\frac{\cos\beta \, d\beta}{2}\right) \tag{9.12}$$

or

$$dF_{dA_1 \rightarrow dA_2} = [\tfrac{1}{2}d(\sin\alpha)] \, [\tfrac{1}{2}d(\sin\beta)] = (\tfrac{1}{2}d\Theta)(\tfrac{1}{2}d\Phi) \tag{9.13}$$

where $\Theta = \sin\alpha$ and $\Phi = \sin\beta$ are dummy variables. In other words, when expressed in terms of ortho-angular coordinates the differential–differential configuration factor is the product of two *independent* factors. Even though the development leading to this result is based on a particular geometry, the specifics of that geometry have not yet been invoked. Therefore, the result represented by Equation 9.13 is completely general.

How difficult is it to express a problem in terms of ortho-angular coordinates? In the case at hand, illustrated in Figure 9.2, α and β are the natural coordinates. For example, suppose the area element dA_2 is allowed to slide in its plane parallel to the z axis so that it sweeps out a strip of height $dy_2 = \ell_{12} d\alpha/\cos\alpha$ and infinite length extending from $z = -\infty$ to $z = \infty$. In this case, the limits on the ortho-angular coordinate β are $-\pi/2 < \beta < \pi/2$, and Equation 9.13 becomes

$$dF_{dA_1 \rightarrow dA_2} = \tfrac{1}{2}d(\sin\alpha) \tag{9.14}$$

Note that this expression holds even if the area element dA_1 is also a strip of height dy_1 also parallel to the z axis. This is a very important result that is often useful in practical situations involving long ducts generated by translating a closed curve along its normal (see, for example, Problem P 8.1).

The differential–finite configuration $F_{dA_1-A_2}$ is then

$$
\begin{aligned}
F_{dA_1 \to A_2} &= \frac{1}{2} \int_{-y_1/\sqrt{y_1^2+W^2}}^{(H-y_1)/\sqrt{(H-y_1)^2+W^2}} d\Theta \\
&= \frac{1}{2}\left[\frac{H-y_1}{\sqrt{(H-y_1)^2+W^2}} + \frac{y_1}{\sqrt{y_1^2+W^2}} \right]
\end{aligned}
\tag{9.15}
$$

and the finite–finite configuration factor F_{1-2} is

$$
F_{1-2} = \frac{A_2}{A_1} F_{2-1} = \frac{1}{A_1}\int_{A_1} F_{dA_1-A_2} \, dA_1 = \frac{W}{H}\left[\sqrt{(H/W)^2+1} - 1\right] \tag{9.16}
$$

The result represented by Equation 9.16 has far-reaching implications. First, when computing the differential–finite distribution factor it is necessary only to determine the limits on two integrals, that is

$$
dF_{dA_1 \to A_2} = \left(\frac{1}{2}\int_{\Theta_1}^{\Theta_2} d\Theta \right)\left(\frac{1}{2}\int_{\Phi_1}^{\Phi_2} d\Phi \right) = \frac{1}{4}(\Theta_2 - \Theta_1)(\Phi_2 - \Phi_1) \tag{9.17}
$$

The quantities $\Theta_1 = \sin \alpha_1$ and $\Theta_2 = \sin \alpha_2$ (and $\Phi_1 = \sin \beta_1$ and $\Phi_2 = \sin \beta_2$) are relatively easy to determine in many geometries. However, typically one of the ortho-angular variables must be expressed parametrically in terms of the other, in which case the integrals are not separable.

9.3 EVALUATION OF CONFIGURATION FACTORS USING CONTOUR INTEGRATION

Evaluation of Equation 8.60 to obtain the finite–finite configuration factor is generally much more difficult than in the two examples in the previous section. The problem is the quadruple integration lurking behind the rather innocent-looking notation. In all but the simplest geometry this leads to nearly insurmountable mathematical difficulties. Fortunately, an elegant method exists for greatly simplifying the mathematical intricacies by reducing the order of the integration from four to two. The following development closely follows that of Sparrow [3] and Sparrow and Cess [4].

Consider the flux of a vector quantity **V** through a smooth three-dimensional surface S bounded by curve C, shown in Figure 9.3. Stokes' theorem states that

$$
\iint_S \nabla \times \mathbf{V} \cdot \mathbf{n} \, dS = \oint_C \mathbf{V} \cdot d\mathbf{s} \tag{9.18}
$$

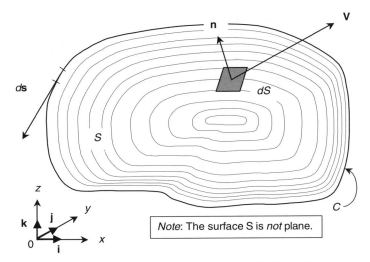

Figure 9.3   The flux of a vector quantity V through a smooth three-dimensional surface S bounded by curve C

In Cartesian coordinates we have

$$\mathbf{V}(x,y,z) = P(x,y,z)\,\mathbf{i} + Q(x,y,z)\,\mathbf{j} + R(x,y,z)\,\mathbf{k} \tag{9.19}$$

where $\mathbf{i}$, $\mathbf{j}$, and $\mathbf{k}$, are the unit vectors in the x, y, and z directions, respectively. Then

$$\nabla \times \mathbf{V} = \begin{vmatrix} \mathbf{i} & \mathbf{j} & \mathbf{k} \\ \dfrac{\partial}{\partial x} & \dfrac{\partial}{\partial y} & \dfrac{\partial}{\partial z} \\ P & Q & R \end{vmatrix} \tag{9.20}$$

or

$$\nabla \times \mathbf{V} = \left(\frac{\partial R}{\partial y} - \frac{\partial Q}{\partial z}\right)\mathbf{i} - \left(\frac{\partial R}{\partial x} - \frac{\partial P}{\partial z}\right)\mathbf{j} + \left(\frac{\partial Q}{\partial x} - \frac{\partial P}{\partial y}\right)\mathbf{k} \tag{9.21}$$

Now if α, β, and γ are the angles that $\mathbf{n}$ forms with, $\mathbf{i}$, $\mathbf{j}$, and $\mathbf{k}$, respectively, then

$$\begin{aligned} \mathbf{n} &= \mathbf{i}\cos\alpha + \mathbf{j}\cos\beta + \mathbf{k}\cos\gamma \\ &= \ell\,\mathbf{i} + m\,\mathbf{j} + n\,\mathbf{k} \end{aligned} \tag{9.22}$$

where ℓ, m, and n are the "direction cosines" whose angles α, β, and γ are illustrated in Figure 9.4. Thus, forming the inner product of Equations 9.21 and 9.22 there results

$$\nabla \times \mathbf{V} \cdot \mathbf{n} = \left(\frac{\partial R}{\partial y} - \frac{\partial Q}{\partial z}\right)\cos\alpha - \left(\frac{\partial R}{\partial x} - \frac{\partial P}{\partial z}\right)\cos\beta + \left(\frac{\partial Q}{\partial x} - \frac{\partial P}{\partial y}\right)\cos\gamma \tag{9.23}$$

Also,

$$\mathbf{V}\cdot d\mathbf{s} = (P\,\mathbf{i} + Q\,\mathbf{j} + R\,\mathbf{k})\cdot(\mathbf{i}\,dx + \mathbf{j}\,dy + \mathbf{k}\,dz)$$
$$= P\,dx + Q\,dy + R\,dz \tag{9.24}$$

Therefore, Stokes' theorem expressed in Cartesian coordinates is

$$\iint_S \left[\left(\frac{\partial R}{\partial y} - \frac{\partial Q}{\partial z} \right) \cos\alpha - \left(\frac{\partial R}{\partial x} - \frac{\partial P}{\partial z} \right) \cos\beta + \left(\frac{\partial Q}{\partial x} - \frac{\partial P}{\partial y} \right) \cos\gamma \right] dS$$
$$= \oint_C (P\,dx + Q\,dy + R\,dz) \tag{9.25}$$

Now consider the differential–differential configuration factor $dF_{dA_i \to dA_j}$ illustrated in Figure 8.2. Equation 8.11 can be rewritten

$$dF_{dA_i \to dA_j} = \left(\frac{\cos\theta_i\,\cos\theta_j}{\pi r^2} \right) dA_j \tag{9.26}$$

where

$$r \equiv |\mathbf{r}_{ij}| = \sqrt{(\Delta x)^2 + (\Delta y)^2 + (\Delta z)^2} \tag{9.27}$$

with

$$\Delta x = x_j - x_i \tag{9.28a}$$

$$\Delta y = y_j - y_i \tag{9.28b}$$

and

$$\Delta z = z_j - z_i \tag{9.28c}$$

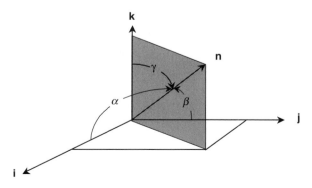

Figure 9.4   Illustration of the direction cosines

Then, referring once again to Figure 8.2,

$$
\begin{aligned}
\mathbf{r}_{ij} \cdot \mathbf{n}_i &= r \cos\theta_i \\
&= \Delta x \cos\alpha_i + \Delta y \cos\beta_i + \Delta z \cos\gamma_i \\
&= \ell_i \, \Delta x + m_i \, \Delta y + n_i \, \Delta z
\end{aligned}
\tag{9.29}
$$

and

$$
\begin{aligned}
\mathbf{r}_{ij} \cdot \mathbf{n}_j &= -r \cos\theta_j \\
&= -\ell_j \, \Delta x - m_j \, \Delta y - n_j \, \Delta z
\end{aligned}
\tag{9.30}
$$

Therefore,

$$
dF_{dA_i \to dA_j} = \frac{(\ell_i \, \Delta x + m_i \, \Delta y + n_i \, \Delta z)(-\ell_j \, \Delta x - m_j \, \Delta y - n_j \, \Delta z)}{\pi r^4} dA_j
\tag{9.31}
$$

Now recall the first equality in Equation 8.49,

$$
F_{dA_i \to A_j} = \int_{A_j} dF_{dA_i \to dA_j}
\tag{8.49}
$$

and define

$$
f \equiv \frac{1}{\pi r^4}(\ell_i \, \Delta x + m_i \, \Delta y + n_i \, \Delta z)
\tag{9.32}
$$

Then,

$$
F_{dA_i \to A_j} = \int_{A_j} (-\ell_j \, \Delta x \, f - m_j \, \Delta y \, f - n_j \, \Delta z \, f) \, dA_j
\tag{9.33}
$$

where x_i, y_i, and z_i are constants under the integral. By comparing Equation 9.33 with the Cartesian form of Stokes' theorem, Equation 9.25, we can recognize the following identities:

$$
\frac{\partial R}{\partial y_j} - \frac{\partial Q}{\partial z_j} = -\Delta x \, f = (x_i - x_j)f
\tag{9.34}
$$

$$
\frac{\partial P}{\partial z_j} - \frac{\partial R}{\partial x_j} = -\Delta y \, f = (y_i - y_j)f
\tag{9.35}
$$

and

$$
\frac{\partial Q}{\partial x_j} - \frac{\partial P}{\partial y_j} = -\Delta z \, f = (z_i - z_j)f
\tag{9.36}
$$

Now we can verify by direct substitution that solutions to this set of partial differential equations are

$$P = \frac{-m_i\,\Delta z + n_i\,\Delta y}{2\pi r^2} \tag{9.37}$$

$$Q = \frac{\ell_i\,\Delta z - n_i\,\Delta x}{2\pi r^2} \tag{9.38}$$

and

$$R = \frac{-\ell_j\,\Delta y + m_i\,\Delta x}{2\pi r^2} \tag{9.39}$$

Therefore,

$$F_{dA_i \to A_j} = \frac{1}{2\pi} \oint_{C_j} \left[\left(\frac{-m_i\,\Delta z + n_i\,\Delta y}{r^2} \right) dx_j + \left(\frac{\ell_i\,\Delta z - n_i\,\Delta x}{r^2} \right) dy_j \right.$$
$$\left. + \left(\frac{-\ell_i\,\Delta y + m_i\,\Delta x}{r^2} \right) dz_j \right] \tag{9.40}$$

or

$$F_{dA_i \to A_j} = \ell_i \oint_{C_j} \left(\frac{\Delta z\,dy_j - \Delta y\,dz_j}{2\pi r^2} \right) + m_i \oint_{C_j} \left(\frac{\Delta x\,dz_j - \Delta z\,dx_j}{2\pi r^2} \right)$$
$$+ n_i \oint_{C_j} \left(\frac{\Delta y\,dx_j - \Delta x\,dy_j}{2\pi r^2} \right) \tag{9.41}$$

We now have the option of orienting the coordinate system so that perhaps as many as two of the three direction cosines will be zero. Note that even though the surface may be three-dimensional, we need only evaluate a line integral in a plane to find the differential–finite configuration factor. The serious student will want to ponder this surprising result.

Example Problem 9.3

Compute the differential–finite configuration factor $F_{dA_1 \to A_2}$ for the geometry shown in Figure 9.5.

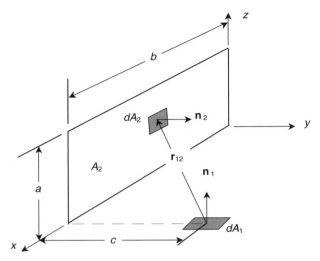

Figure 9.5 Geometry for Example Problem 9.3; configuration factor from a differential area element to a rectangular flat plate

SOLUTION

From the geometry in Figure 9.5 we note that $\mathbf{n}_1 = \mathbf{k}$ and $\mathbf{n}_2 = \mathbf{j}$. Thus, $\ell_1 = m_1 = 0$ and $n_1 = 1$. Therefore,

$$
F_{dA_i \to A_2} = \oint_{C_2} \frac{(y_2 - c)dx_2 - (x_2 - b)dy_2}{2\pi r^2}
$$

$$
= -\int_0^b \frac{c\,dx_2}{2\pi r^2} - \int_b^0 \frac{c\,dx_2}{2\pi r^2}
$$

(9.42)

since $y_2 = 0$ and $dy_2 = 0$ (note that "r" is different in the two integrals). Then,

$$
F_{dA_1 \to A_2} = \int_b^0 \frac{c\,dx_2}{2\pi\,[(b - x_2)^2 + a^2 + c^2]} + \int_0^b \frac{c\,dx_2}{2\pi\,[(b - x_2)^2 + c^2]}
$$

(9.43)

or, upon integration,

$$
F_{dA_1 \to A_2} = \frac{c}{2\pi}\left[\frac{1}{c}\tan^{-1}\left(\frac{b}{c}\right) - \frac{1}{\sqrt{c^2 + a^2}}\tan^{-1}\left(\frac{b}{\sqrt{c^2 + a^2}}\right)\right]
$$

(9.44)

Note in the integration implied by Equations 9.42 and 9.43 that the direction of integration is important because it determines the sign of the value obtained for the integral. The convention to be observed here is that the contour is followed such that the surface facing dA_1 lies to the left.

9.4 THE SUPERPOSITION PRINCIPLE

Note that Equation 9.41 may be thought of as the sum of three "basic" configuration factors with the magnitudes of the direction cosines as weighting factors. That is,

$$F_{dA_i \to A_j} = |\ell_i|(F_{dA_i \to A_j})_x + |m_i|(F_{dA_i \to A_j})_y + |n_i|(F_{dA_i \to A_j})_z \qquad (9.45)$$

where the values of the basic configuration factors $(F_{dA_i \to A_j})_x$, $(F_{dA_i \to A_j})_y$, and $(F_{dA_i \to A_j})_z$ presumably may be obtained, for example, as in Example Problem 9.3. The magnitudes of the direction cosines appear in Equation 9.45 because each of the three terms on the right-hand side of Equations 9.41 and 9.45 represents a component of a configuration factor, and so must necessarily be positive. Now, any direction cosine and any of the three contour integrals in Equation 9.41 can be negative; however, it turns out that when the direction cosine is negative, the integral it multiplies will also be negative, so that their product is always positive.

Example Problem 9.4

Consider the geometry on the left in Figure 9.6. It is similar to the geometry in Figure 9.5 except that the differential area element dA_1 is rotated about the x axis through some angle α. Compute the configuration factor for this geometry.

SOLUTION

In this case the direction cosines in Equation 9.45 are $\ell_1 = 0$, $m_1 = \mathbf{n}_1 \cdot \mathbf{j} = \cos(\pi/2 + \alpha) = -\sin \alpha$, and $n_1 = \mathbf{n}_1 \cdot \mathbf{k} = \cos \alpha$. Bearing in mind that the integral multiplying the direction cosine m_1 in Equation 9.41 will also be negative, we can write

$$F_{dA_1 \to A_2} = \sin \alpha \, (F_{dA_1 \to A_2})_y + \cos \alpha \, (F_{dA_1 \to A_2})_z \qquad (9.46)$$

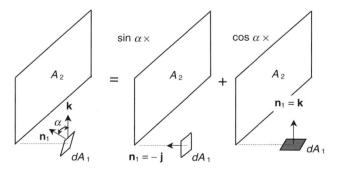

Figure 9.6 Illustration of the superposition principle

where $(F_{dA_1 \to A_2})_z$ is given by Equation 9.44 and $(F_{dA_1 \to A_2})_y$ may be obtained in a process similar to the one leading to $(F_{dA_1 \to A_2})_z$. Note that the integral corresponding to $(F_{dA_1 \to A_2})_y$ will be negative because m_1 is negative, and so the magnitude of the integral must be retained.

9.5 FORMULATION FOR FINITE–FINITE CONFIGURATION FACTORS

Recall Equation 8.59,

$$
F_{A_i \to A_j} = F_{ij} = \int_{A_j} dF_{A_i \to dA(\mathbf{r}_j)} = \frac{1}{A_i} \int_{A_j} F_{dA(\mathbf{r}_j) \to A_i} \, dA(\mathbf{r}_j)
$$

$$
= \frac{1}{A_i} \int_{A_j} \int_{A_i} dF_{dA(\mathbf{r}_j) \to dA(\mathbf{r}_i)} \, dA(\mathbf{r}_j)
$$

(8.59)

Introducing Equation 9.40 into Equation 8.59 yields

$$
F_{A_i \to A_j} = \frac{1}{2\pi A_i} \left[\oint_{C_j} \int_{A_i} \left(\frac{-m_i \, \Delta z + n_i \, \Delta y}{r^2} \right) dA_i \, dx_j \right.
$$

$$
+ \oint_{C_j} \int_{A_i} \left(\frac{\ell_i \, \Delta z - n_i \, \Delta x}{r^2} \right) dA_i \, dy_j
$$

(9.47)

$$
\left. + \oint_{C_j} \int_{A_i} \left(\frac{-\ell_i \, \Delta y + m_i \, \Delta x}{r^2} \right) dA_i \, dz_j \right]
$$

We may now apply Stokes' theorem separately to each of the three inner integrals of Equation 9.47. For the first integral we get

$$
\frac{\partial R}{\partial y_i} - \frac{\partial Q}{\partial z_i} = 0
$$

(9.48)

$$
\frac{\partial P}{\partial z_i} - \frac{\partial R}{\partial x_i} = -\frac{\Delta z}{r^2}
$$

(9.49)

and

$$
\frac{\partial Q}{\partial x_i} - \frac{\partial P}{\partial y_i} = \frac{\Delta y}{r^2}
$$

(9.50)

It may be readily verified that $P = \ln(r)$ and $Q = R = 0$ are solutions to this system of partial differential equations. Thus, the first area integral in Equation 9.47 becomes

$$
\int_{A_i} \left(\frac{-m_i \, \Delta z + n_i \, \Delta y}{r^2} \right) dA_i = \oint_{C_i} \ln(r) \, dx_i
$$

(9.51)

Similar treatment of the other area integrals in Equation 9.47 leads to

$$F_{A_i \to A_j} = \frac{1}{2\pi A_i} \oint_{C_i} \oint_{C_j} [\ln(r)\, dx_i\, dx_j + \ln(r)\, dy_i\, dy_j + \ln(r)\, dz_i\, dz_j] \quad (9.52)$$

The double integration in Equation 9.52 might still lead to formidable analytical difficulties, but, comparing Equation 9.52 with Equation 8.60, which involves quadruple integration over (generally) three-dimensional surfaces, it is clear that the order of these difficulties has been greatly reduced. Still, experience has shown that the analytical calculation of configuration factors is generally tedious. It is noted that computer software based on contour integration is commercially available for evaluating configuration factors.

9.6 CONFIGURATION FACTOR ALGEBRA

Many useful and practical (as well as many useless and impractical) configuration factors have been worked out and published in the open literature. Most radiation heat transfer texts include more or less extensive tables of configuration factors in the appendices. Using *configuration factor algebra,* tabulated configuration factors can often be manipulated to find configuration factors that are not tabulated.

The basic tools of configuration factor algebra are

1. Conservation of energy:

$$\sum_{j=1}^{n} F_{ij} = 1.0, \qquad 1 \le i \le n \qquad (9.53)$$

2. Reciprocity:

$$A_i F_{ij} = A_j F_{ji}, \qquad 1 \le i \le n, \quad 1 \le j \le n \qquad (9.54)$$

3. Consolidation of surfaces:

$$\left(\sum_{i=1}^{m} A_i \right) F_{(\Sigma A_i) \to A_j} = \sum_{i=1}^{m} A_i F_{A_i \to A_j} \qquad 1 \le j \le n, \quad m \le n \qquad (9.55)$$

where

$$\sum_{i=1}^{m} A_i = \sum A_i \qquad (9.56)$$

is the area of a composite surface consisting of m individual contiguous surface elements.

4. The whole is equal to the sum of the parts:

$$F_{A_i \to \Sigma A_j} = \sum_{j=1}^{m} F_{A_i \to A_j}, \qquad 1 \le i \le n, \quad m \le n \qquad (9.57)$$

(It is demonstrated in Example Problem 9.5 that this result follows from the principles of reciprocity and consolidation of surfaces.)

5. Symmetry: For a symmetric enclosure it is often possible to recognize by inspection that certain configuration factors are equal to each other. For example, in the case of a short cylindrical enclosure (such as the inside of a beer can) the configuration factor from the curved wall to the top is equal to the configuration factor from the curved wall to the bottom.

Other identities may be obtained by combining of two or more of these.

The most effective way to explain the use of configuration factor algebra is through examples.

Example Problem 9.5

Use the principles of reciprocity and consolidation of surfaces to prove that

$$F_{A_j \to \Sigma A_i} = \sum_{i=1}^{m} F_{A_j \to A_i}, \qquad 1 \le j \le n, \quad m \le n \qquad (9.57a)$$

(Note that Equation 9.57a is identical to Equation 9.57 except that the arbitrary indices have been reversed.)

SOLUTION

Invoking the principle of reciprocity, Equation 9.54, we have

$$A_j F_{A_j \to \Sigma A_i} = \left(\sum_{i=1}^{m} A_i \right) F_{(\Sigma A_i) \to A_j}, \qquad 1 \le j \le n, \quad m \le n \qquad (9.58)$$

which, with the introduction of the principle of consolidation of surfaces, may be written

$$A_j F_{A_j \to \Sigma A_i} = \sum_{i=1}^{m} (A_i F_{A_i \to A_j}) = A_1 F_{A_1 \to A_j} + A_2 F_{A_2 \to A_j}$$

$$+ \cdots + A_m F_{A_m \to A_{ji}}, \qquad 1 \le j \le n, \quad m \le n \qquad (9.59)$$

Applying reciprocity once again, we have

$$A_j F_{A_j \to \Sigma A_i} = \sum_{i=1}^{m} (A_j F_{A_j \to A_i}), \qquad 1 \le j \le n \qquad (9.60)$$

Finally, noting that A_j is a constant under the summation on the right-hand side of Equation 9.60 and dividing through by A_j, there results

$$F_{A_j \to \Sigma A_i} = \sum_{i=1}^{m} F_{A_j \to A_i}, \qquad 1 \le j \le n, \quad m \le n \qquad (9.57a)$$

(Note that the principle of conservation of energy, Equation 9.53, is a special case of this result with $m = n$.)

Example Problem 9.6

Find F_{2-4} ($= F_{dA_2 \to dA_4}$), where A_2 and A_4 are subdivisions of the two rectangular flat plates, shown in Figure 9.7, which share a common edge and meet at right angles.

SOLUTION

We note that F_{2-3} and $F_{2-3,4}$ are both of the same general form and are obtainable using contour integration. In fact, this configuration factor is tabulated in most elementary heat transfer texts. Then from the principle that the whole is equal to the sum of the parts, Equation 9.56, we can write

$$F_{2-3,4} = F_{2-3} + F_{2-4} \qquad (9.61)$$

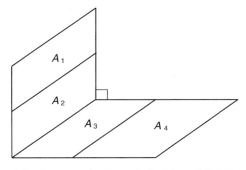

Figure 9.7   Geometry for Example Problems 9.6, 9.7, and 9.8

or

$$F_{2-4} = F_{2-3,4} - F_{2-3} \tag{9.62}$$

which is the desired result.

Example Problem 9.7

Find F_{4-2} for the geometry of Figure 9.7.

SOLUTION

There are two possible (equivalent) solutions. The most direct approach is to invoke the principle of reciprocity,

$$A_4 F_{4-2} = A_2 F_{2-4} \tag{9.63}$$

or

$$F_{4-2} = \frac{A_2 F_{2-4}}{A_4} \tag{9.64}$$

where F_{2-4} was obtained in the previous example. An alternative is to use the principle of consolidation of surfaces to write

$$A_4 F_{4-2} + A_3 F_{3-2} = (A_3 + A_4) F_{3,4-2} \tag{9.65}$$

or

$$F_{4-2} = \frac{A_3 + A_4}{A_4} F_{3,4-2} - \frac{A_3}{A_4} F_{3-2} \tag{9.66}$$

where $F_{3,4-2}$ and F_{3-2} are known configuration factors. It is left as an exercise for the student to demonstrate that Equations 9.64 and 9.66 are equivalent.

Example Problem 9.8

Find F_{1-4} for the geometry of Figure 9.7.

SOLUTION

We first note that the configuration factors $F_{1,2-3,4}$, $F_{1,2-3}$, $F_{2-3,4}$, and F_{2-3} are all considered to be known. The principle that the whole is the sum of the parts, Equation 9.56, yields

$$F_{1-3,4} = F_{1-3} + F_{1-4} \tag{9.67}$$

or

$$F_{1-4} = F_{1-3,4} - F_{1-3} \tag{9.68}$$

Equation 9.68 tells us that we now need expressions for $F_{1-3,4}$ and F_{1-3}. Application of the principle of consolidation of surfaces yields

$$A_{1,2}F_{1,2-3,4} = A_1 F_{1-3,4} + A_2 F_{2-3,4} \tag{9.69}$$

or, solving for $F_{1-3,4}$,

$$F_{1-3,4} = \frac{1}{A_1}(A_{1,2}F_{1,2-3.4} - A_2 F_{2-3,4}) \tag{9.70}$$

where $F_{1,2-3,4}$ and $F_{2-3,4}$ are both known. Similarly, application of the principle of consolidation of surfaces involving F_{1-3} yields

$$A_1 F_{1-3} + A_2 F_{2-3} = A_{1,2}F_{1-2,3} \tag{9.71}$$

or

$$F_{1-3} = \frac{1}{A_1}(A_{1,2}F_{1-2,3} - A_2 F_{2-3}) \tag{9.72}$$

Finally, after substituting Equations 9.70 and 9.72 into Equation 9.68 and rearranging, we have

$$F_{1-4} = \frac{1}{A_1}[A_{1,2}(F_{1,2-3,4} - F_{1,2-3}) + A_2(F_{2-3} - F_{2-3,4})] \tag{9.73}$$

9.7 GENERAL PROCEDURE FOR PERFORMING CONFIGURATION FACTOR ALGEBRA

At this point it is useful to note that we have consistently proceeded in the same manner in the solution in Example Problems 9.6, 9.7, and 9.8; that is, there is a methodology to be followed. The steps are as follows:

1.   Always begin by writing an expression, based on one of the five tools given in Section 9.5, *which involves the unknown configuration factor.* Which principle (conservation of energy, reciprocity, consolidation of surfaces, the whole is equal to the sum of the parts, symmetry) do we begin with? A certain amount of intuition and experience is required here. As is often the case in other disciplines, the more experience we have solving configuration factor algebra problems the easier they become.

2.   Solve the expression from step 1 for the unknown configuration factor. This is the *root equation.* The right-hand side will now generally involve a combination of known and unknown configuration factors.

3.   Repeat steps 1 and 2 for each of the remaining unknown configuration factors on the right-hand side of the root equation, always being careful not to use the same relation more than once (it is permitted and normal to use the same *principle* more than once, however).

4.   Finally, introduce the expressions for the unknown configuration factors found in step 3 into the root equation and simplify the resulting expression.

Example Problem 9.9

For the geometry of Figure 9.8 show that

$$A_1 F_{1-4} = A_2 F_{2-3} \tag{9.74}$$

SOLUTION

First, it is emphasized that $a \neq b$, so that the principle of symmetry cannot be invoked. We begin by noting that if Equation 9.74 is true, it must also be true that

$$A_4 F_{4-1} = A_3 F_{3-2} \tag{9.75}$$

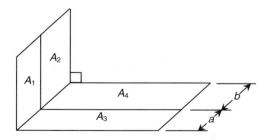

Figure 9.8   Geometry for Example Problems 9.9 and 9.10

That is, Equations 9.74 and 9.75 are completely equivalent relationships. Then, subtracting Equation 9.75 from Equation 9.74, we have

$$A_1F_{1-4} - A_4F_{4-1} = A_2F_{2-3} - A_3F_{3-2} \tag{9.76}$$

But, according to the principle of reciprocity, Equation 9.54, both sides of Equation 9.76 are identically zero. Then, since Equation 9.76 is true, the two interrelated expressions, Equations 9.74 and 9.75, must also be true.

Example Problem 9.10

Find F_{1-4} for the geometry of Figure 9.8.

SOLUTION

From application of the principle of consolidation of surfaces, Equation 9.55, we can write

$$A_{1,2}F_{1,2-3,4} = A_1F_{1-3,4} + A_2F_{2-3,4} \tag{9.77}$$

where $F_{1,2-3,4}$ is considered known. Then, from the principle that the whole is equal to the sum of the parts, Equation 9.56,

$$F_{1-3,4} = F_{1-3} + F_{1-4} \tag{9.78}$$

and

$$F_{2-3,4} = F_{2-3} + F_{2-4} \tag{9.79}$$

where F_{1-3} and F_{2-4} are considered known. Now the result from Example Problem 9.9, Equation 9.74, can be rearranged to obtain

$$F_{2-3} = \frac{A_1}{A_2}F_{1-4} \tag{9.80}$$

in which case Equation 9.79 becomes

$$F_{2-3,4} = \frac{A_1}{A_2}F_{1-4} + F_{2-4} \tag{9.81}$$

Finally, Equations 9.78 and 9.81 can be introduced into Equation 9.77, yielding

$$A_{1,2}F_{1,2-3,4} = A_1F_{1-3} + A_1F_{1-4} + A_2\frac{A_1}{A_2}F_{1-4} + A_2F_{2-4} \tag{9.82}$$

or, solving for F_{1-4},

$$F_{1-4} = \frac{1}{2A_1}(A_{1,2}F_{1,2-3,4} - A_1F_{1-3} - A_2F_{2-4})$$ (9.83)

Example Problem 9.11

Find F_{1-2} for the geometry of Figure 9.9. Surface elements A_1 and A_2 are parallel bands on the inside walls of a conical cavity formed by the intersection of the right circular cone with four parallel cutting planes normal to the axis of the cone. Surfaces A_a, A_b, A_c and A_d are *imaginary* disks formed by the intersection of the four cutting planes with the right circular cone.

SOLUTION

The configuration factor between any two parallel, coaxial disks, $F_{\text{disk-disk}}$ is easily obtainable by the contour integration method (please see Problem 9.9) and so is considered known for our purposes; that is, we know F_{a-b}, F_{a-c}, F_{a-d}, F_{b-c}, and so forth. Then our challenge is to relate the desired configuration factor F_{1-2} to the various disk-to-disk configuration factors. Once again we begin by writing an expression, in this case based on a version of the principle that the whole is equal to the sum of the parts, which includes the unknown configuration factor,

$$F_{1-2} = F_{1-c} - F_{1-d}$$ (9.84)

Equation 9.84 states that the radiation leaving ring 1 and passing through disk c must be intercepted by either ring 2 or disk d. Note that Equation 9.84 is not based entirely

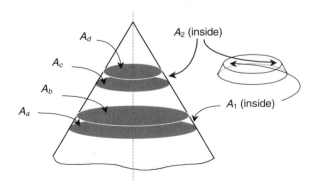

Figure 9.9 Geometry for Example Problem 9.11

on the principle that the whole is equal to the sum of the parts but also includes the idea that

$$F_{1-c} = F_{1-2,d} \tag{9.85}$$

that is, all radiation leaving surface element A_1 and arriving at the imaginary surface A_c must also arrive at the composite surface $A_{2,d} = A_2 + A_d$. This "principle" is difficult to generalize but is clearly valid and, at least in this case, useful.

The two configuration factors F_{1-c} and F_{1-d} can be obtained from a combination of the reciprocity principle and the unnamed principle embodied in Equations 9.84 and 9.85; that is,

$$A_1 F_{1-c} = A_c F_{c-1} = A_c(F_{c-a} - F_{c-b}) \tag{9.86}$$

and

$$A_1 F_{1-d} = A_d F_{d-1} = A_d(F_{d-a} - F_{d-b}) \tag{9.87}$$

Then solving Equations 9.86 and 9.87 for F_{1-c} and F_{1-d}, respectively, and introducing the results into Equation 9.84 yields

$$F_{1-2} = \frac{A_c}{A_1}(F_{c-a} - F_{c-b}) + \frac{A_d}{A_1}(F_{d-a} - F_{d-b}) \tag{9.88}$$

in which all of the configuration factors on the right-hand side are of the known disk-to-disk type.

9.8 PRIMITIVES

An interesting and even entertaining aspect of configuration factor algebra is the *primitive*. Strictly speaking, a primitive is a configuration factor that can be written down by inspection in one or two steps. Consider the following illustrative examples.

Example Problem 9.12

Surface 1 is a sphere and surface 2 is an infinite plane at any distance $H > R$ from the sphere, where R is the radius of the sphere, as illustrated in Figure 9.10. Find F_{1-2}.

SOLUTION

Imagine a second infinite plane parallel to the first with the sphere located midway between them. It is clear from the principle of symmetry that half of the diffusely dis-

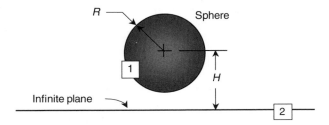

Figure 9.10 Geometry for Example Problem 9.12

tributed radiation leaving the spherical surface arrives at each of the two plane surfaces. Thus, $F_{1-2} = \frac{1}{2}$. A curious result from the principle of reciprocity is that $F_{2-1} = 0$!

Example Problem 9.13

Surface 1 is a plane square surface whose dimensions are $L \times L$, and surface 2 is a sphere of diameter D centered a distance H above this square such that $L = 2H > D$, as shown in Figure 9.11. Find F_{1-2}.

SOLUTION

The clue to solving this tricky old classic may be found in the curious constraints placed on L, H, and D. Upon reflection we realize that under these constraints the sphere lies at the center of an imaginary cubical enclosure having surface 1 as one of its six walls. Then from the principles of symmetry and conservation of energy $F_{2-1} = \frac{1}{6}$, and from the principle of reciprocity

$$F_{1-2} = \frac{A_2}{A_1}F_{2-1} = \frac{1}{6}\left(\frac{4\pi R^2}{L^2}\right) = \frac{\pi}{6}\left(\frac{R}{H}\right)^2 \qquad (9.89)$$

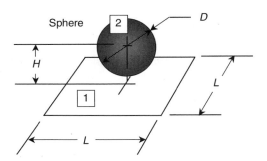

Figure 9.11 Geometry for Example Problem 9.13

A generation of students has been perplexed by the fact that the entire spherical surface area $4\pi R^2$ is used in the reciprocity relation, *this in spite of the fact that parts of the sphere cannot be "seen" from anywhere on the square!* (Note that this was not the case in Example Problem 9.12.) In other words, the configuration factor F_{1-2} would not be changed if a certain irregularly shaped portion of the top of the sphere were cut away, even though this portion of the sphere is "counted" in the reciprocity relation.

Example Problem 9.14

Consider two infinitely long concentric cylinders, and let the inside surface of the outer cylinder be surface 1 and the outside surface of the inner cylinder be surface 2, as illustrated in Figure 9.12. Find F_{1-1}.

SOLUTION

Application of the principles of conservation of energy and reciprocity lead immediately to

$$F_{1-1} = 1 - F_{1-2} = 1 - \left(\frac{A_2}{A_1}\right)F_{2-1} = 1 - \left(\frac{A_2}{A_1}\right) = 1 - \left(\frac{R_2}{R_1}\right)^2 \quad (9.90)$$

where R_1 and R_2 are the radii of the two cylinders.

Example Problem 9.15

Consider a long triangular cross-section groove of opening angle α, and let the two opposite sides of the groove be surfaces 1 and 2, as shown in Figure 9.13. Find F_{1-2}.

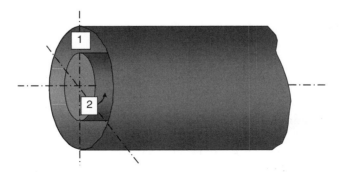

Figure 9.12   Geometry for Example Problem 9.14 (both cylinders are infinitely long)

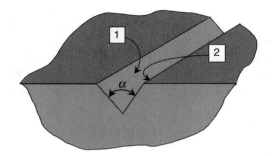

Figure 9.13   Geometry for Example Problem 9.15 (infinitely long groove)

SOLUTION

We begin by imagining a virtual surface 3 stretched over the opening of the groove. (Virtual surfaces are often useful entities in configuration factor algebra. Indeed, we have already made use of them in Example Problems 9.11, 9.12, and 9.13.) Then from the principle of conservation of energy,

$$F_{1-2} + F_{1-3} = 1.0 \tag{9.91}$$

and

$$F_{3-1} + F_{3-2} = 1.0 \tag{9.92}$$

and from the symmetry principle we have

$$F_{3-1} = F_{3-2} = \tfrac{1}{2} \tag{9.93}$$

Also, invoking the reciprocity principle and the geometry yields

$$F_{1-3} = \frac{A_3}{A_1} F_{3-1} = \frac{1}{2} \frac{A_3}{A_1} = \frac{1}{2} \left(2 \sin \frac{\alpha}{2} \right) \tag{9.94}$$

Finally, substituting Equation 9.94 into Equation 9.91 and rearranging, we obtain

$$F_{1-2} = 1.0 - \sin \frac{\alpha}{2} \tag{9.95}$$

Example Problem 9.16

Consider a right conical cavity of height H and base radius R, as shown in Figure 9.14. Let surface 1 be the conical surface of the cavity and surface 2 be the base area of the cone. Find F_{1-1}.

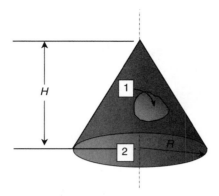

Figure 9.14   Geometry for Example Problem 9.16

SOLUTION

From the principle of conservation of energy we can write

$$F_{1-1} + F_{1-2} = 1.0 \tag{9.96}$$

and from the reciprocity principle we find that

$$F_{1-2} = \frac{A_2}{A_1} F_{2-1} = \frac{A_2}{A_1} = \frac{R}{\sqrt{R^2 + H^2}} \tag{9.97}$$

since $F_{2-1} = 1.0$. Solving Equation 9.96 for F_{1-1} and introducing Equation 9.97 into the result yields

$$F_{1-1} = 1.0 - \frac{R}{\sqrt{R^2 + H^2}} \tag{9.98}$$

9.9   A NUMERICAL APPROACH, THE MONTE CARLO RAY-TRACE METHOD

Part III of this book, which develops the Monte Carlo ray-trace (MCRT) method for radiation heat transfer modeling, introduces a Windows-based computing environment called FELIX. The student version of FELIX is included on the CD-ROM packaged with this book. Briefly, FELIX is a PC-based tool for solving radiation heat transfer problems involving directionally emitting and absorbing and bidirectionally reflecting surfaces. It uses an approach based on *distribution factors,* defined as the fraction of radiation emitted by one surface element that is absorbed by a second surface element due to direct radiation and all possible reflections. For black enclosures the distribution factor reduces to the configuration factor; there-

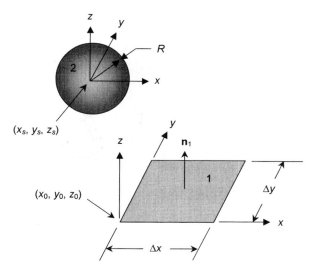

Figure 9.15   A flat plate viewing a sphere

fore, FELIX can also be used to estimate* configuration factors, even when partial blockage occurs.

Consider the black rectangular plate and black sphere shown in Figure 9.15. Our goal is to obtain an estimate of the configuration factor F_{1-2} from the plate, surface 1, to the sphere, surface 2. We begin by emitting a large number of "energy bundles," each carrying the same amount of energy, from a sequence of uniformly distributed locations on the upper side of the plate. Each energy bundle would be emitted in a random direction into 2π-space. We could choose the emission sites by, for example, visiting sequentially each point on a uniform grid laid out on the surface for this purpose. Alternatively—and more in keeping with the spirit of the

---

* On a recent offshore sail with midshipmen from the U.S. Naval Academy the author was showing the youngsters how to use a sextant to determine longitude from a noon sun shot. The process involves taking a time series of observations of the solar altitude around local solar noon and then plotting solar altitude versus time. The peak of this curve occurs at local solar noon, and knowledge of local solar noon in GMT on a given day is sufficient, with the aid of tables, to determine longitude. During the exercise it was convenient to use the time signal received by our on-board Global Position System (GPS) receiver to obtain the exact time in GMT of each observation. Inevitably, the midshipmen wanted to know why anyone who had access to GPS would bother using a sextant to obtain longitude. Of course, the real answer is that a vessel sailing offshore would normally be equipped with an accurate chronometer; but our answer, inspired by Linda Greenlaw writing in *The Hungry Ocean* [5], was that a conscientious mariner would want to verify the accuracy of GPS from time to time. The point of this anecdote is that the reader might well ask at this point, "Why would anyone having access to a powerful engine like FELIX use it to compute configuration factors?" This is, in fact, a valid question that we hope will motivate the reader to undertake the study of Part III of this book.

MCRT method—for each emission event we could draw two random numbers R_x and R_y from a random number generator and then compute the location of the emission site using

$$x = x_0 + R_x \Delta x \tag{9.99}$$

and

$$x = x_0 + R_x \Delta x \tag{9.100}$$

where x_0 and y_0 are the coordinates of the "lower left-hand" corner of the plate and Δx and Δy are the dimensions of its sides. Of course, if surface 1 was not rectangular, this approach would have to be appropriately modified.

The random direction associated with each emission event would be computed using

$$\theta = \sin^{-1}(\sqrt{R_\theta}) \tag{9.101}$$

and

$$\phi = 2\pi R_\phi \tag{9.102}$$

where R_θ and R_ϕ are the next two available random numbers. The zenith angle θ is as usual measured with respect to direction of the unit normal $\mathbf{n}_1$, and the azimuth angle ϕ is measured with respect to an arbitrary axis, say a line in the plane of surface 1 that passes through the point of emission and lies parallel to the x axis. The origin of Equation 9.102 is obvious; discussion of the origin of Equation 9.101 is deferred until Chapter 11.

For each emission event a line ("ray") is imagined whose equation is

$$\frac{x_2 - x_1}{\ell_1} = \frac{y_2 - y_1}{m_1} = \frac{z_2 - z_1}{n_1} = T \tag{9.103}$$

where the point x_1, y_1, z_1 is the known emission point on surface 1 and the point x_2, y_2, z_2 is a *candidate point* that may or may not lie on the sphere. The quantity T is the distance between the two points. To determine if the ray intersects the sphere, we solve the three equations in four unknowns represented by Equations 9.103 simultaneously with the equation for the sphere

$$(x_2 - x_s)^2 + (y_2 - y_s)^2 + (z_2 - z_s)^2 = R^2 \tag{9.104}$$

where the radius R and center of the sphere x_s, y_s, z_s are known. This is best accomplished by using Equations 9.103 to express x_2, y_2, and z_2 in terms of x_1, y_1, z_1 and T.

These expressions are then substituted into Equation 9.104 and the result solved for T. If T is real and positive, we know the emitted ray intersects the sphere and, since the sphere is black, that the ray is absorbed.

After emitting a suitably large number of rays, we estimate the configuration factor as

$$F_{1 \to 2} \cong \frac{N_{\text{hits}}}{N_{\text{shots}}} \tag{9.105}$$

where N_{shots} is the number of energy bundles emitted and N_{hits} is the number of "hits" on the sphere. In general, the accuracy of the estimate increases with the number of energy bundles emitted.

Figure 9.16 shows the convergence of Equation 9.105 as a function of N_{shots} for the special case shown in Figure 9.11 with $L = 2D = 2H$. The horizontal line in the figure is drawn at a level representing the exact value computed using Equation 9.89. The estimate of the configuration factor in this case is better than 5 percent when only 10,000 energy bundles are traced. Considering the fact that the equations governing radiation heat transfer in an enclosure are integral equations, the effect of configuration factor errors of ± 5 percent would be heat transfer errors of less than this amount. Figure 9.17 shows the distribution density (hits per unit area) of "hits" on the sphere

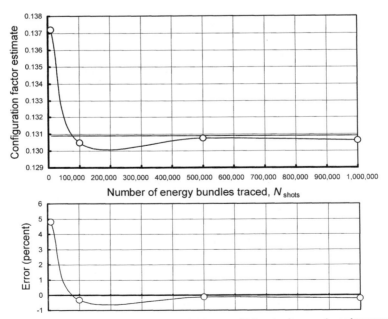

Figure 9.16 Configuration factor convergence and error with increasing number of energy bundles emitted during an MCRT estimate for the geometry in Figure 9.11 with $L = 2D = 2H$

Figure 9.17   Density of energy bundles absorbed on the sphere (energy bundles per unit area) for the case of $N_{shots} = 10^6$

for the case of $N_{shots} = 100,000$. Figure 9.17 makes it clear that the irradiance distribution on surface 2 due to the radiosity of surface 1 is not uniform, calling into question the meaning and usefulness of F_{1-2} in this case. However, if both surfaces were part of an enclosure the irradiance distribution on surface 2 would be more uniform and the basic assumptions of the net exchange formulation would be more nearly valid.

This statistical approach to estimating configuration factors has two major advantages. The most obvious of these is that it easily handles blockage. For example, suppose that the sphere in Figure 9.11 was positioned between two flat plates. Then the approach described above could easily be modified to enable the configuration factor from one plate to the other to be estimated. That is, rays that did not intersect the sphere but did intersect the second plate could be counted and their total number divided by the total number of energy bundles emitted to obtain an estimate of the configuration factor with blockage.

The second major advantage is that the accuracy of the configuration factor estimate is controllable. Recall that the net exchange method results in a model whose accuracy depends on, among other things, the fineness of the surface mesh into which the enclosure has been divided. If the spatial resolution is too low (too few surface elements whose sizes are too large), the model will not well represent reality. In such cases it is possible that increasing the accuracy of the estimate of the configuration factors will actually decrease the accuracy of the computed result. An example of this phenomenon may be seen in Figure 9.16, in which the nearest approach of the estimated value of the configuration factor to the exact value occurs at $N_{shots} = 500,000$ rather that at one million, as might be expected. This is because FELIX has modeled the sphere as a collection of 3200 flat triangular facets. Evidently in this particular example the deviation of this shape from that of a true sphere begins to become important when N_{shots} exceeds about 500,000. A configuration factor whose accuracy is continuously controllable provides a means for systematically studying this and similar phenomena.

Team Projects

TP 9.1   A numerical approach to the contour integral method for evaluating configuration factors can be implemented on a computer much more easily than can the more direct formulation based on a quadruple integral. The first step is to convert Equation 9.52 into a double summation. Then two "nested" summation loops are set up in which the boundaries of the two surfaces involved are "marched" around in the appropriate direction. At each increment around one of the contours the natural logarithm of the distance r between the two contour points is computed, weighted by the appropriate increment step size in the x, y, and z directions. The resulting values of weighted $\ln(r)$ are then added to an accumulating sum at each step. Write a computer program that implements this idea to a high degree of accuracy (i.e., small step sizes). Note that you may use a high-level programming language such as FORTRAN or C++ or you may use a spreadsheet such as EXCEL. Test the program by using it to estimate configuration factors for which you know the exact value. *Caution:* There are several traps to be avoided. For example, what do you do if two surfaces share a common edge? How could you modify your program to handle the presence of an intervening structure that partially blocks the view of one surface by another? How could the code be modified to compute F_{1-2} in Example Problem 9.13? Finally, remember that a significant simplification can usually be achieved by thoughtful orientation of one of the surfaces with respect to the Cartesian coordinate system. Submit a brief report on your team's effort.

TP 9.2   Use FELIX to estimate the elements of the configuration factor matrix for the thermal-vacuum chamber described in TP 8.1. Note that many opportunities exist for taking advantage of symmetry and reciprocity in this geometry, so that relatively few unique configuration factors need be computed using FELIX. Be sure to test each row of the matrix to assure that conservation of energy, Equation 9.53, is satisfied. Also, note that the configuration factors between any two equal-area surface elements on the hemispherical dome are the same and are known exactly (what is the value?). This can be used to determine the accuracy of the FELIX-based estimates as a function of the number of energy bundles traced. After you have obtained the full configuration factor matrix, use it with the net exchange method to estimate the net heat flux distribution on the interior walls of the chamber and the temperature distribution on the floor of the chamber if the hemispherical top is maintained at a temperature of 300 K while the cylindrical walls are maintained at 77 K and the floor is insulated.

Discussion Points

DP 9.1   At the end of Example Problem 9.1 it is stated that integrating spheres are often used as components in an experimental apparatus for measuring surface optical properties. We have already seen how a blackbody can be used as a source of infrared radiation for calibrating infrared detectors. How might an integrating sphere be used as a source of visible radiation for calibrating visible radiation detectors?

DP 9.2   Above Example Problem 9.3 it is stated that the serious student will want to ponder why contour integration works even if the surface in question is three dimensional. Why *does* it work equally well for plane and curved surfaces? Do you believe there should be *any* constraints on the surface topography?

DP 9.3   The serious student will want to give some critical thought to the material concerning signs under Equation 9.45. Why is the last sentence in that paragraph true?

DP 9.4   Even if you do not undertake Team Project TP 9.1, several interesting points are raised there that should be discussed. For example, what would happen in Equation 9.52 (or its numerical approximation) if two surfaces share a common edge?

DP 9.5   How could you modify the computer program described in Team Project TP 9.1 to handle the presence of an intervening structure that partially blocks the view of one surface by another?

DP 9.6   How could the code described in Team Project TP 9.1 be modified to compute F_{1-2} in Example Problem 9.13? Indeed, could it be?

DP 9.7   See if you can compose one or two "primitives." Do you believe there is a systematic approach to composing primitives? If so, what would it be?

DP 9.8   Example Problem 9.13 poses a riddle: Why *is* the entire sphere area used in the reciprocity relation even though parts of the spherical surface cannot be observed from the plane?

DP 9.9   Consider the infinitely long V-groove in Example Problem 9.15 (Figure 9.13). Suppose the groove was only semi-infinitely long, with one end terminated by a triangular wall, surface 4. How would you go about computing F_{4-1}? [*Hint:* Consider the special case where surface 4 is an equilateral triangle.]

DP 9.10   Consider the geometry in Example Problem 9.16 (Figure 9.14). Do you believe that this cavity can ever satisfy the conditions assumed for use of the finite net exchange formulation? Which assumption(s) might be suspect?

DP 9.11   Follow up on the discussion begun in the last paragraph of this chapter ("The second major advantage . . .").

Problems

P 9.1   Show that $\cos\theta = \cos\alpha\,\cos\beta$, where the angles θ, α, and β are defined in Figure 9.2. (*Hint:* Begin by proving that $\tan^2\theta = \tan^2\alpha + \tan^2\beta$.)

P 9.2   Verify Equation 9.17; that is, supply the missing steps.

P 9.3   Verify that Equations 9.37, 9.38, and 9.39 are solutions to Equations 9.34, 9.35, and 9.39.

P 9.4   Verify that $P = \ln(r)$ and $Q = R = 0$ are solutions of Equations 9.48, 9.49, and 9.50.

P 9.5   Verify Equation 9.52; that is, supply the missing steps between Equation 9.51 and Equation 9.52.

P 9.6   Confirm that Equations 9.64 and 9.66 are equivalent.

P 9.7   Use contour integration to verify Equation 9.93.

P 9.8   Use contour integration to show that F_{1-2} for the geometry of Figure 9.18 is

$$
\begin{aligned}
F_{1-2} = \frac{1}{\pi W} &\left(W \tan^{-1}\left(\frac{1}{W}\right) + H \tan^{-1}\left(\frac{1}{H}\right) \sqrt{H^2 + W^2} \tan^{-1}\left(\frac{1}{\sqrt{H^2 + W^2}}\right) \right. \\
&+ \frac{1}{4}\ln\left\{ \frac{(1 + W^2)(1 + H^2)}{1 + W^2 + H^2} \left[\frac{W^2(1 + W^2 + H^2)}{(1 + W^2)(W^2 + H^2)} \right]^{W^2} \right. \\
&\left.\left. \times \left[\frac{H^2(1 + H^2 + W^2)}{(1 + H^2)(H^2 + W^2)} \right]^{H^2} \right\} \right)
\end{aligned}
\tag{9.106}
$$

where $W \equiv a/b$ and $H \equiv c/b$.

P 9.9   Use contour integration to show that the configuration factor F_{1-2} between the two disks shown in Figure 9.19 is

$$
F_{1-2} = \frac{1}{2}\left\{ 1 + \left(1 + \frac{a^2}{c^2}\right)\frac{c^2}{b^2} - \sqrt{\left[1 + \left(1 + \frac{a^2}{c^2}\right)\frac{c^2}{b^2}\right]^2 - 4\left(\frac{a^2}{b^2}\right)} \right\}
\tag{9.107}
$$

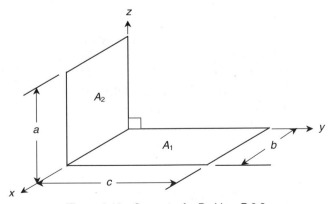

Figure 9.18   Geometry for Problem P 9.8

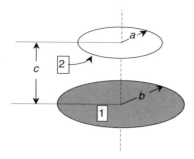

Figure 9.19 Geometry for Problem P 9.9

P9.10 For the geometry shown in Figure 9.20 use configuration factor algebra to find the configuration factor from the interior curved surface of a short circular cylinder to itself, F_{1-1}, in terms of its length L and its radius R.

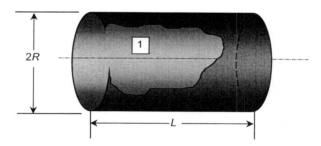

Figure 9.20 Geometry for Problem P 9.10

P9.11 Consider the long, three-walled duct whose cross section is an equilateral triangle, shown in Figure 9.21. What is the configuration factor $F_{\text{wall-wall}}$ from one wall to another wall?

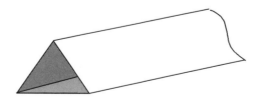

Figure 9.21 Geometry for Problem P 9.11 (duct is infinitely long)

P9.12 The walls of a certain tetrahedral enclosure consist of four identical triangles, as shown in Figure 9.22. What is the configuration factor $F_{\text{wall-wall}}$ from one wall to another wall?

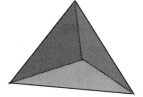

Figure 9.22 Geometry for Problems P 9.12 and P 9.13 (one face removed)

P9.13 For the geometry shown in Figure 9.22 the emissivity and either the temperature or the net heat flux is given in Table 9.1. Fill in the missing information in the table.

Table 9.1   Data for Problem P 9.13

	Wall 1	Wall 2	Wall 3	Wall 4
ε	0.4	0.1	0.5	0.2
T	400 K	800 K		
q			0 W/m^2	1000 W/m^2

P9.14 Consider a right conical enclosure formed by five equal-height parallel bands, or rings, on the curved surface and a circular disk at the bottom, as shown in Figure 9.23. Use configuration factor algebra to derive the configuration factors listed below in terms of the height H and the base radius R of the enclosure.

(a) F_{1-1} (b) F_{1-2} (c) F_{1-3} (d) F_{1-4}

(e) F_{1-5} (f) F_{1-6} (g) F_{2-2} (h) F_{2-3}

(i) F_{2-4} (j) F_{2-5} (k) F_{2-6} (l) F_{3-3}

(m) F_{3-4} (n) F_{3-5} (o) F_{3-6} (p) F_{4-4}

(q) F_{4-5} (r) F_{4-6} (s) F_{5-5} (t) F_{5-6}

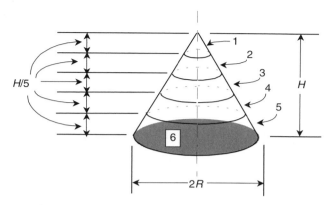

Figure 9.23 Geometry for Problems P 9.14 and P 9.15

P9.15 For the enclosure in Figure 9.23 suppose $H = 2R,$ ring surface 1 has a uniform temperature of 300 K, ring surfaces 2 through 5 are insulated ($q = 0$), and the base, surface 6, absorbs a net heat flux of 111.8 W/m^2 uniformly over its surface. If the emissivity of all the surfaces is 0.7, what are the temperatures of ring surfaces 2 through 4 and base surface 6 and what is the net heat flux from ring surface 1?

P9.16 Write a computer program using a programming language such as FORTRAN or C to directly verify the result in Example Problem 9.13. Do not use reciprocity.

P9.17 Use FELIX to verify the result in Example Problem 9.13.

REFERENCES

1. Hottel, H. C., and A. F. Sarofim, *Radiative Transfer,* McGraw-Hill Book Company, Inc., New York, 1967.
2. Love, T. J., *Radiative Heat Transfer,* Charles E. Merrill Publishing Company, Columbus, OH, 1968.
3. Sparrow, E. M., "A new and simpler formulation for radiative angle factors," *ASME Transactions, Series C, Journal of Heat Transfer,* Vol. 85, No. 2, May 1963, pp. 81–88.
4. Sparrow, E. M., and R. D. Cess, *Radiation Heat Transfer,* Brooks/Cole Publishing Company, Belmont, CA, 1966.
5. Greenlaw, L., *The Hungry Ocean,* Hyperion, New York, 1999.

10

RADIATIVE ANALYSIS OF NONDIFFUSE, NONGRAY ENCLOSURES USING THE NET EXCHANGE FORMULATION

In many situations of practical engineering interest it is not reasonable to assume diffuse emission, absorption, and reflection. However, it is often possible in these cases to assume that the reflectivity may be adequately approximated as the sum of a diffuse component and a specular component. In other realistic situations the temperature range, and, therefore, the wavelength interval, is too broad to justify the graybody assumption. In this pivotal chapter we learn how to adapt the finite net exchange formulation to the problem of nondiffuse, nongray radiative analysis.

10.1 THE "DUSTY MIRROR" MODEL

Consider the "surface" shown in Figure 10.1, which consists of the sharp edges of a stack of common single-edged razor blades aligned in a plane. Intuition correctly suggests that the surface depicted in Figure 10.1 will reflect an incident beam of visible radiation in a raylike manner, producing a very different directionality pattern, depending on whether the incident radiation is aligned with or against the "grain." Infrared radiation of sufficiently long wavelength will undergo diffraction, thereby producing a highly wavelength-dependent reflection pattern. In fact, this particular surface has been studied experimentally as an example of a *nondiffuse* reflector.

While the surface of Figure 10.1 is an extreme example, the reflectivity of most real surfaces exhibits a degree of directionality. For metals, surface finish (cold-rolled, ground, polished, turned, sand-blasted, peened, milled, annealed, etc.), sur-

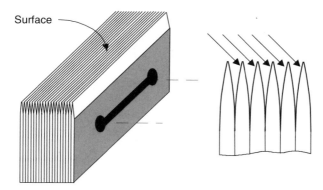

Figure 10.1 A "surface" created by aligning the edges of a stack of common single-edged razor blades

face chemistry (pure, anodized, oxidized, etc.), state of contamination (painted, oily, dirty, etc.), and temperature all influence the directionality of emission, absorption and reflection. For nonmetals, grain structure (fibrous, granular, crystalline, amorphous, etc.), surface treatment (sanded, painted, etc.), aging, and temperature influence these surface radiation properties.

One of the best examples of scholarly ferment occurred in 1962 when R. A. Seban provided a written discussion of a seminal article by Sparrow, Eckert, and Jonsson [1]. The original article presents what was at the time a new theory for modeling radiative exchange in enclosures consisting of two kinds of surfaces: surfaces that are purely diffuse reflectors and surfaces that are purely specular reflectors. In his remarkably insightful discussion published with the article Seban suggested an extension to the theory that allows treatment of enclosures consisting of surfaces that have both diffuse and specular components of reflectivity. The development in this chapter is based on Seban's extension of the theory presented by Sparrow and his co-authors.

In many cases of practical engineering interest, reflection from a surface can be modeled with acceptable accuracy as the sum of a diffuse component and a specular component; that is,

$$\rho = \rho^d + \rho^s \tag{10.1}$$

where ρ^d is the diffuse component of reflectivity and ρ^s is the specular component. The relationship between the reflectivity defined by Equation 10.1 and the reflectivity used in the net exchange formulation will be clarified later in this chapter. The model defined by Equation 10.1 and illustrated in Figure 10.2 might be referred to as the "dusty mirror" model.

In the dusty mirror model, emission and absorption could still be assumed diffuse, especially if the diffuse component of reflectivity is predominant. In any case, the analysis of an enclosure made up of directional surfaces could always be improved

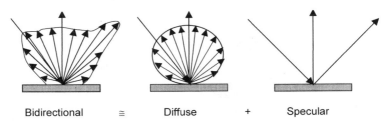

Bidirectional   ≅   Diffuse   +   Specular

Figure 10.2   The "dusty mirror" model for the reflectivity of certain real surfaces

by assuming diffuse-specular reflections even if diffuse emission and absorption are retained.

10.2   ANALYSIS OF ENCLOSURES MADE UP OF DIFFUSE–SPECULAR SURFACES

We consider enclosures for which one of the two following requirements is met:

1. Some surfaces are diffuse reflectors and the remaining surfaces are specular reflectors, or
2. All surfaces have both diffuse and specular components of reflectivity

We note that requirement 1 is a limiting special case of requirement 2. In both cases, diffuse emission and absorption are assumed. Requirements 2 and 3 in Section 9.1 and requirements 4′, 5′, 6, and 7 in Section 9.13 are still assumed to be in force, with one of the two requirements above replacing requirement 1 in Section 8.1.

10.3   THE EXCHANGE FACTOR

Lin and Sparrow [2] evidently were the first to introduce the concept of the exchange factor under that name, although other authors writing at about the same time introduced the same concept under other names [3, 4]. The *differential–differential exchange factor* from differential surface element dA_i to differential surface element dA_j on the walls of an enclosure is defined as the *sum of the fractions* of the diffuse radiation leaving dA_i, which arrives at dA_j both directly and by all possible specular reflections.

The student is cautioned that this definition is commonly misstated or stated in a misleading way in the literature, which often neglects to state clearly that the exchange factor is a *sum of fractions* rather than a fraction. The identification in other texts of the exchange factor as a fraction is misleading because it gives the false impression that it cannot exceed unity. In general, the number of possible reflective paths from dA_i to dA_j is limitless, and so the mathematical expression for the exchange factor is an infinite sum. Each term in the summation is a quantity somewhat analogous to the configuration factor. That is, each term represents a fraction of diffuse radiation leaving dA_i, which arrives at dA_j. The difference between these various terms is the manner by which the diffuse radiation gets from dA_i to dA_j. The first term of the summation is actually the configuration factor $dF_{dA_i \to dA_j}$, which can also be considered as the fraction of the diffuse radiation leaving dA_i, which arrives at dA_j after zero reflections. A typical term in the exchange factor sum is given by [5].

$$dP_{ij}^{nk} = df_{ij}^{nk} \prod_{m=1}^{n} \rho_m^s \qquad (10.2)$$

In Equation 10.2 the factor df_{ij}^{nk} is the fraction of the diffuse radiation leaving dA_i, which arrives at dA_j after n ideal ($\rho^s = 1$) specular reflections following a certain sequence k of reflections, and is thus a purely geometric quantity. Generally, more than one possible sequence may exist for diffusely emitted radiation to travel from dA_i to dA_j by n specular reflections, and so the superscript k is needed to distinguish among the various nth-order sequences. This idea is illustrated in Figure 10.3.

Each time a reflection occurs, a certain fraction of the radiation is absorbed by the reflecting surface. Therefore, to obtain the actual amount that arrives at dA_j, the quantity df_{ij}^{nk} must be multiplied by the product of the specular reflectivities of all the specularly reflecting surfaces encountered. The subscript m then denotes the mth surface from which a reflection takes place.

Having presented and explained a typical term in the expression for the exchange factor, the entire expression may now be given as [5]

$$
\begin{aligned}
dE_{dA_i \to dA_j} &= dF_{dA_i \to dA_j} + \sum_{n=1}^{\infty} \sum_{k=1}^{l(n)} dP_{ij}^{nk} \\
&= dF_{dA_i \to dA_j} + \sum_{n=1}^{\infty} \sum_{k=1}^{l(n)} \prod_{m=1}^{n} \rho_m^s \, df_{ij}^{nk}
\end{aligned}
\qquad (10.3)
$$

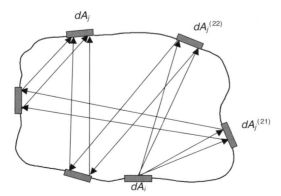

Figure 10.3 Two possible second-order specular reflections from differential surface element dA_i to differential surface element dA_j

This expression is much more complicated than the rather simple (and incomplete) expressions typically found in other sources. The idea behind the second summation (over k) in Equation 10.3, which accounts for the fact that there may be more than one nth-order sequence of reflections carrying radiation from one surface element to another, has typically been omitted in previous treatments.

The first term on the right-hand side of Equation 10.3, $dF_{dA_i \rightarrow dA_j}$, is the familiar differential–differential (diffuse) configuration factor from dA_i to dA_j. It should not be surprising that the exchange factor reduces to the configuration factor when the enclosure contains no specularly reflecting surfaces ($\rho_m^s = 0$, $m = 1, 2, ..., n$). The outside summation in Equation 10.3 is over all numbers of reflection, while the inside summation is over all possible sequences for a given number of reflections. The function $l(n)$ gives the total number of possible sequences for which radiation can travel from dA_i to dA_j by a given number n of specular reflections. Definition of $l(n)$ is highly dependent on the geometry of the enclosure.

We return now to the idea that the exchange factor, unlike the configuration factor, is not a fraction and can in fact exceed unity. This is because the same energy that has arrived at the second surface from the first may get other chances, through continuing specular reflections, to arrive there again. A convincing illustration of this fact is the exchange factor $E_{A'-A}$ from an arbitrary surface element A' inside an enclosure whose walls have uniform specular reflectivity ρ^s to the entire enclosure whose surface area is A. From the definition of the exchange factor this quantity can be obtained directly as

$$E_{A' \rightarrow A} = \int_A dE_{A' \rightarrow dA} = 1 + \rho^s + (\rho^s)^2 + ... = \frac{1}{1 - \rho_s} \qquad (10.4)$$

The exchange factor of Equation 10.4 is always greater than or equal to unity [5].

10.4 RECIPROCITY FOR THE EXCHANGE FACTOR

The exchange factor is subject to the same reciprocity relations as the configuration factor; that is,

$$dA_i \, dE_{dA_i \rightarrow dA_j} = dA_j dE_{dA_j \rightarrow dA_i} \qquad (10.5)$$

$$A_i \, dE_{A_i \rightarrow dA_j} = dA_j E_{dA_j \rightarrow A_i} \qquad (10.6)$$

and

$$A_i \, E_{A_i \rightarrow A_j} = A_j E_{A_j \rightarrow A_i} \qquad (10.7)$$

Development of Equations 10.5 through 10.7 is left as an exercise for the student.

10.5   CALCULATION OF EXCHANGE FACTORS

Exchange factors are usually calculated using some variation of what is usually referred to as the *image method* [1]. It is clear that to find the differential–differential exchange factor $dF_{dA_i \to dA_j}$ the quantities df_{ij}^{nk} must be evaluated for all n and k in terms of the location and geometry of dA_i and dA_j. The common approach is to locate the differential area element on the surface of the enclosure from which the diffuse radiation traveling from dA_i to dA_j is *first* reflected. We identify these area elements by the symbol $dA_j^{(nk)}$, where, as before, k identifies a specific sequence of n specular reflections between surface elements dA_i and dA_j.

Figure 10.3 shows ($k =$) two possible relationships between dA_i, dA_j, and $dA_j^{(nk)}$ for the case of ($n =$) two specular reflections from dA_i to dA_j. Once the area elements $dA_j^{(nk)}$ have been located, the quantities df_{ij}^{nk} may be computed as

$$df_{ij}^{nk} = dF_{dA_i \to dA_j^{(nk)}} \qquad (10.8)$$

Thus, for all possible n and k the geometry (local curvature and surface normal) and location of the $dA_j^{(nk)}$ must be determined in terms of the geometry and locations of dA_i and dA_j. This is a formidable task for all but the simplest enclosures.

10.6   THE IMAGE METHOD FOR CALCULATING EXCHANGE FACTORS

The principles elaborated in the previous section are difficult to carry out in practice. In general, they are directly applicable in only a few limited cases of regular enclosures, such as rectangular, conical, and cylindrical cavities. The person who first applied these principles to the analysis of a spherical cavity earned a Ph.D. for his efforts [5]! Still, it is instructive to study some of the classical problems to see clearly how the theory translates into practice.

Example Problem 10.1

Consider the long, rectangular channel with three diffusely reflecting walls and one specularly reflecting wall shown in Figure 10.4. Walls 1, 2, and 4 of the (real) enclosure are diffuse reflectors and wall 3 is a specular reflector. Find E_{1-4}, E_{2-4}, and E_{1-1}.

SOLUTION

Following the development of Sparrow et al. [1] we begin by completing an *image enclosure* made up of image, or virtual, surfaces, as shown in Figure 10.4. Then virtual surface 2(3) [read this: "Surface 2 as seen in surface 3."] indicates the virtual im-

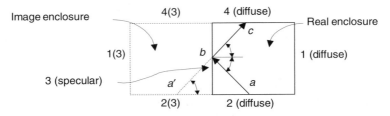

Figure 10.4 Geometry for Example Problem 10.1

age of surface 2 reflected in surface 3. Real ray **abc,** leaving diffuse surface 2 and arriving at diffuse surface 4 by way of a single specular reflection from specular surface 3, is exactly equivalent to the virtual ray **a'bc.** Thus, we can write from the finite version of Equation 10.3 (with $n = k = 1$)

$$E_{2-4} = F_{2-4} + \rho_3^s F_{2(3)-4} \tag{10.9}$$

Similarly, we can write

$$E_{1-4} = F_{1-4} + \rho_3^s F_{1(3)-4} \tag{10.10}$$

and

$$E_{1-1} = \rho_3^s F_{1(3)-1} \tag{10.11}$$

Example Problem 10.2

Now consider the case of another long, rectangular cross-section channel having two adjacent diffuse walls and two adjacent specular walls, as shown in Figure 10.5. Let

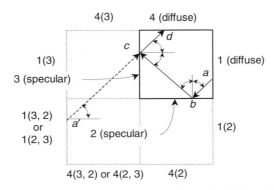

Figure 10.5 Geometry for first part of Example Problem 10.2

walls 1 and 4 be diffuse reflectors and walls 2 and 3 be purely specular reflectors. Find E_{1-4} and E_{1-1}.

SOLUTION

As in the previous example, we begin by completing an image enclosure made up of real and virtual surfaces. Inspection of the new geometry allows us to write directly from Equation 10.3, with $k = 2$ when $n = 1$ and $k = 1$ when $n = 2$,

$$E_{1-4} = F_{1-4} + \rho_3^s\, F_{1(3)-4} + \rho_2^s\, F_{1(2)-4} + \rho_2^s \rho_3^s\, F_{1(2,3)-4} \qquad (10.12)$$

The first term on the right-hand side of Equation 10.12 is the familiar diffuse configuration factor, and the second and third terms are completely analogous to the similar terms in Equations 10.9 and 10.10. The last term on the right-hand side of Equation 10.12 is based on Figure 10.5, which establishes that the ray **abcd** is exactly equivalent to the ray **a′cd.**

The second unknown exchange factor, E_{1-1}, is a little less straightforward. Careful inspection of Figure 10.6 reveals that a ray leaving surface 1 from some locations and in some directions can be reflected by two specular reflections back to surface 1, but that rays leaving surface 1 from other locations and in other directions cannot. For example, ray **abcd** (and ray **dcba**) in Figure 10.6 describe valid two-specular-reflection paths originating and terminating on surface 1, but ray **abcd** in Figure 10.5 does not. It may be concluded that virtual surface 1(2, 3) and virtual surface 1(3, 2) are partially obscured from surface 1. (Note that in Figure 10.6 we associate virtual surface 1(2, 3) with rays like **abcd** and virtual surface 1(3, 2) with rays like **dcba.**) Then application of Equation 10.3 with $k = 1$ when $n = 2$ and $k = 2$ when $n = 2$ yields

$$E_{1-1} = \rho_3^s\, F_{1(3)-1} + \rho_2^s \rho_3^s\, F^*_{1(2,3)-1} + \rho_3^s \rho_2^s\, F^*_{1(3,2)-1} \qquad (10.13)$$

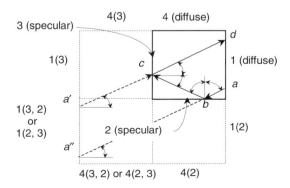

Figure 10.6   Geometry for second part of Example Problem 10.2

where $F^*_{1(2,3)-1}$ and $F^*_{1(3,2)-1}$ are *partial configuration factors* from the parts of virtual surfaces $1(2, 3)$ and $1(3, 2)$, respectively, visible from surface 1. How are the partial configuration factors related to the configuration factors $F_{1(2,3)-1}$ and $F_{1(3,2)-1}$? A clue is provided by Figure 10.6, in which radiation arriving at surface 1 along path **abcd** appears to originate from point a' on virtual surface $1(2, 3)$, while radiation arriving at surface 1 along path **dcba** appears to originate from point a'' on virtual surface $1(3, 2)$. This suggests, and, in fact, it can be shown, that

$$F^*_{1(2,3)-1} + F^*_{1(3,2)-1} = F_{1(2,3)-1} = F_{1(3,2)-1} \tag{10.14}$$

therefore,

$$E_{1-1} = \rho_3^s F_{1(3)-1} + \rho_2^s \rho_3^s F_{1(2,3)-1} \tag{10.15}$$

10.7 NET EXCHANGE FORMULATION USING EXCHANGE FACTORS

We now consider an enclosure made up of n finite surfaces, N of which have specified surface temperatures and $n - N$ of which have specified net heat fluxes. Note that we are bypassing the differential formulation in favor of the finite formulation. This is justified because, in fact, the prohibitive majority of all practical radiation problems are formulated from the finite point-of-view. Also, "near-differential" spatial resolution can always be achieved by subdividing the enclosure into a sufficiently large number of small surface elements. It is further assumed that all n surfaces have both diffuse and specular components of reflectivity. Note that this specification does not preclude the possibility that the reflectivity is purely diffuse on some surfaces and purely specular on others since these are but limiting special cases of diffuse-specular surfaces. Otherwise, all of the requirements listed in Section 10.1 are assumed to apply to within acceptable accuracy.

The principles applied in Sections 8.15 and 8.17 to formulate the finite net exchange method using configuration factors are now applied to the situation involving exchange factors. That is, the defining expression for radiosity for the surfaces whose temperatures are specified, and the defining expression for net heat flux for the surfaces whose net heat fluxes are specified, are written down:

$$B'_i = \varepsilon_i \sigma T_i^4 + \rho_i^d H'_i, \qquad 1 \le i \le N \tag{10.16}$$

and

$$q_i = q_{i,\text{out}} - q_{i,\text{in}}, \qquad N + 1 \le i \le n \tag{10.17}$$

where

$$H'_i = \sum_{j=1}^{n} B'_j E_{ij}, \qquad 1 \le i \le n \tag{10.18}$$

We note three differences between the above equations and the corresponding set describing radiant exchange within a purely diffuse enclosure, Equations 8.46 through 8.48:

1. The coefficient of the irradiance H_i' in Equation 10.16 is ρ_i^d rather than $1 - \varepsilon_i$, where ε_i is related to ρ_i^d by

$$\varepsilon_i = 1 - \rho_i = 1 - \rho_i^d - \rho_i^s \qquad (10.19)$$

   It is for this reason that the radiosity is given the new symbol B' in Equation 10.16.

2. Equation 10.17 is reticent on the relationship among the net heat flux, the radiosity and the irradiance.

3. The irradiance is defined in terms of the exchange factor (Equation 10.18) instead of in terms of the configuration factor (Equation 8.48).

The radiosity in Equation 10.16 is considered to consist of the sum of a diffusely emitted component and *the diffuse part* of the reflected component; specular reflections are completely accounted for by the new definition of the irradiance in terms of the exchange factor rather than the configuration factor. It is noted that under these changes the radiosity of a surface is a fictitious quantity whose value cannot be measured, while the irradiance remains a real, measurable quantity.

We can now see why Equation 10.17 is reticent on the relationship among surface net heat flux, radiosity, and irradiance: the radiosity is no longer simply related to the radiation leaving a surface. The heat flux arriving at surface element i, $q_{i,\text{in}}$, is still equal to the irradiance; that is,

$$q_{i,\text{in}} = H_i', \qquad N + 1 \leq i \leq n \qquad (10.20)$$

However, the radiation leaving surface element i is the sum of a diffuse component and a specular component; that is,

$$q_{i,\text{out}} = B_i' + \rho_i^s H_i', \qquad N + 1 \leq i \leq n \qquad (10.21)$$

Introducing Equations 10.20 and 10.21 into Equation 10.17 and simplifying then yields

$$q_i = B_i' - (1 - \rho_i^s) H_i', \qquad N + 1 \leq i \leq n \qquad (10.22)$$

or

$$B_i' = q_i + (1 - \rho_i^s) H_i', \qquad N + 1 \leq i \leq n \qquad (10.23)$$

Combining Equations 10.16, 10.18, and 10.23 yields the working equations for the finite net exchange method applied to enclosures having diffuse–specular surfaces:

$$B_i' = \varepsilon_i \sigma T_i^4 + \rho_i^d \sum_{j=1}^{n} B_j' E_{ij}, \qquad 1 \leq i \leq N \qquad (10.24)$$

and

$$B_i' = q_i + (1 - \rho_i^s) \sum_{j=1}^{n} B_j' E_{ij}, \qquad N + 1 \le i \le n \qquad (10.25)$$

Equations 10.24 and 10.25 represent n equations in n unknown radiosities. With the introduction of the Kronecker delta function, Equation 8.65, and Equation 8.66 (with $B = B'$), Equation 10.24 can be written

$$\sigma T_i^4 = \frac{1}{\varepsilon_i} \sum_{j=1}^{n} (\delta_{ij} - \rho_i^d E_{ij}) B_j', \qquad 1 \le i \le N \qquad (10.26)$$

and Equation 10.23 becomes

$$q_i = \sum_{j=1}^{n} [\delta_{ij} - (1 - \rho_i^s)E_{ij}]B_j', \qquad N + 1 \le i \le n \qquad (10.27)$$

Now, following the procedure introduced in Section 8.18, we define

$$\Omega_i = \begin{cases} \sigma T_i^4, & 1 \le i \le N \\ q_i, & N+1 \le i \le n \end{cases} \qquad (10.28)$$

and

$$X_{ij} = \begin{cases} \dfrac{1}{\varepsilon_i}(\delta_{ij} - \rho_i^d E_{ij}), & 1 \le i \le N \\ [\delta_{ij} - (1 - \rho_i^s)E_{ij}], & N + 1 \le i \le n \end{cases} \qquad (10.29)$$

Then, in general, we can write

$$\Omega_i = X_{ij} B_j', \qquad 1 \le i \le n \qquad (10.30)$$

where, as in Equation 8.71, the repeated j subscript implies summation over j. Then, as in Section 8.18, we can solve for the unknown radiosities by inverting the X_{ij} matrix; that is,

$$B_i' = [X_{ij}]^{-1} \Omega_j, \qquad 1 \le j \le n \qquad (10.31)$$

Finally, the unknown surface net heat fluxes are obtained from

$$q_l = \sum_{j=1}^{n} [\delta_{ij} - (1 - \rho_i^s)E_{ij}]B_j' \qquad 1 \le i \le N \qquad (10.32)$$

and the unknown surface temperatures are given by

$$T_i = \left[\frac{1}{\sigma \varepsilon_i} \sum_{j=1}^{n} (\delta_{ij} - \rho_i^d E_{ij}) B_j' \right]^{1/4}, \qquad N + 1 \le i \le n \qquad (10.33)$$

It is noted at this point that the finite net exchange formulation for diffuse–specular enclosures converges to the finite net exchange formulation for purely diffuse enclosures in the limit as the specular component of reflectivity of the surfaces of the enclosure approaches zero. Therefore, the formulation in this chapter is more general, and thus of greater value, than the formulation in Chapter 8.

It is further noted that the approximation associated with Figure 10.2, referred to here as the dusty mirror model, is the best that the net exchange formulation can do to treat bidirectional reflections. Even this approximation, which can easily lead to poor results in problems of practical interest, often requires the use of exchange factors which themselves are exceedingly difficult to obtain. Finally, the dusty mirror model still leaves us with the awkward contradiction that bidirectionally reflecting surfaces would also tend to be directionally emitting and absorbing, a situation not addressed by the net exchange formulation.

10.8 TREATMENT OF WAVELENGTH DEPENDENCE (NONGRAY BEHAVIOR)

The net exchange formulation is fully capable of treating wavelength-dependent problems. However, the treatment is cumbersome to say the least when at least one of the surfaces of the enclosure has a specified net heat flux, which is often the case.

Consider the common situation in which one or more of the surfaces of an enclosure has a wavelength-dependent emissivity $\varepsilon_\lambda(\lambda)$ that varies significantly over the wavelength interval of the radiation within the enclosure. Suppose further that temperature variations within the enclosure are sufficient to assure that those surfaces having a strongly wavelength-dependent emissivity exhibit significantly different absorptivities for radiation arriving from different parts of the enclosure.

The approach to this situation is to divide the thermal radiation spectrum up into a sufficiently large number of sufficiently narrow but finite-width bands to capture the essential features of the wavelength-dependent emissivity. The number of such subdivisions depends both on the required accuracy and the degree of variability of the emissivity with wavelength. This is illustrated in Figure 10.7. Then, to an acceptable degree of approximation, we can apply Kirchhoff's law for an opaque surface and obtain

$$\alpha(\Delta \lambda_k) = \varepsilon (\Delta \lambda_k) = 1 - \rho (\Delta \lambda_k) \qquad (10.34)$$

in each band k of width $\Delta \lambda_k$. A more compact way of writing Equation 10.34 is

$$\alpha_k = \varepsilon_k = 1 - \rho_k \qquad (10.35)$$

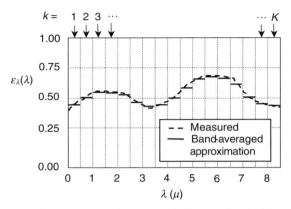

Figure 10.7 Wavelength dependence of a hypothetical spectral emissivity, as measured and the corresponding band-averaged approximation

10.9 FORMULATION FOR THE CASE OF SPECIFIED SURFACE TEMPERATURES

Now consider an enclosure made up of n opaque, diffusely absorbing and emitting, diffuse–specularly reflecting ("dusty mirror" model), nongray surface elements having specified temperatures. In this case, we can write for each surface element i in each wavelength band k,

$$B'_{ik} = \varepsilon_{ik} e_b(\Delta\lambda_k, T_i) + \rho^d_{ik} H'_{ik}, \qquad 1 \le i \le n, \quad 1 \le k \le K \qquad (10.36)$$

where

$$H'_{ik} = \sum_{j=1}^{n} B'_{jk} E_{ijk} \qquad 1 \le i \le n, \quad 1 \le k \le K \qquad (10.37)$$

Equations 10.36 and 10.37 represent $2(n + K)$ equations in a like number of band-averaged spectral radiosities and irradiances. The quantity $e_b(\Delta\lambda_k, T_i)$ in Equation 10.36 is defined

$$e_b(\Delta\lambda_k, T_i) \equiv \int_{\Delta\lambda_k} e_{b\lambda}(\lambda, T_i)d\lambda = \sigma_i^4 \, \Phi(\Delta\lambda_k \, T_i) \qquad (10.38)$$

where the function $\Phi(\Delta\lambda_k T_i) = \Phi_{ik}$ is the fraction of the power emitted by a blackbody at temperature T_i in the wavelength interval $\Delta\lambda_k$. Referring back to Equation 2.85 we see that

$$\Phi(\Delta\lambda_k T_i) = F_{\lambda T_i \to (\lambda + \Delta\lambda_k) T_i} = F_{0 \to (\lambda + \Delta\lambda_k) T_i} - F_{0 \to \lambda T_i} \qquad (10.39)$$

where the unsubscripted λ in Equation 10.39 refers to the value of λ at the beginning of the wavelength interval $\Delta\lambda_k$. With the introduction of the Kronecker delta function, Equations 10.36, 10.37, and 10.38 can be combined to yield

$$\sigma T_i^4 \, \Phi_{ik} = \sum_{j=1}^{n} \left(\frac{\delta_{ij} - \rho_{ik}^d \, E_{ijk}}{\varepsilon_{ik}} \right) B'_{jk}, \qquad 1 \le i \le n, \quad 1 \le k \le K \quad (10.40)$$

Let us now take stock before continuing. The left-hand side of Equation 10.40 is a vector of nK known components once we have established the wavelength intervals. The right-hand side consists of an nK-by-nK matrix, whose elements are all considered known, operating on a vector of nK unknown band-averaged spectral radiosities. Looking ahead, we know we are going to have to invert the matrix to solve the system of nK equations in nK unknown radiosities. It is, therefore, in our best interest to choose the wavelength intervals in some sort of optimal way so that we obtain adequate spectral sampling while maintaining mathematical tractability.

Representing the matrix in Equation 10.40 with the symbol X_{ijk}, we can then write

$$B'_{jk} = X_{ijk}^{-1} \, \sigma T_i^4 \, \Phi_{ik}, \qquad 1 \le j \le n \quad 1 \le k \le K \quad (10.41)$$

where now summation is implied over the thrice-repeated *(i)* subscript. Finally, when the band-averaged spectral radiosities have been obtained using Equation 10.41, the *total* net heat fluxes can be obtained from

$$q_i = \sum_{k=1}^{K} \sum_{j=1}^{n} [\delta_{ij} - (1 - \rho_{jk}^s) E_{ijk}] B'_{jk} \qquad 1 \le i \le n \quad (10.42)$$

10.10   FORMULATION FOR THE GENERAL CASE OF SPECIFIED TEMPERATURE ON SOME SURFACES AND SPECIFIED NET HEAT FLUX ON THE REMAINING SURFACES

The problem is significantly more difficult if one or more of the surfaces has a specified net heat flux. First, let us consider the case where all of the surfaces of the enclosure have specified net heat fluxes; that is, where the vector q_i is known for all i, $1 \le i \le n$. In this case Equation 10.42 is the only relationship available between the specified *total* net heat fluxes and the unknown *band-averaged spectral* radiosities. This creates difficulties because Equation 10.42 cannot be inverted to obtain the band-averaged spectral radiosities. If the radiosities were known they could in principle be used in Equation 10.40 to obtain the unknown temperatures (although this in itself is complicated by the transcendental form of the function of T_i on the left-hand side of Equation 10.40).

The reason Equation 10.42 cannot be inverted is, of course, that the summation over wavelength "destroys" any knowledge we might have had about the wavelength dependence of the band-averaged spectral radiosity B'_{jk}. The concept of (band-averaged) spectral *net* heat flux is contradictory and so is meaningless. As an illustration, it is easy to imagine a water-cooled black surface that, for all practical purposes, absorbs radiation in one wavelength interval (for example, in the solar spectrum) while emitting radiation in a completely different (in this case much longer) wavelength interval. In this case, it clearly makes no sense to speak of the "wavelength interval" of the excess heat carried away by the coolant.

The problem of specified temperatures on some surfaces and specified net heat fluxes on the remaining surfaces is challenging. A solution strategy is outlined in the block diagram of Figure 10.8. The first step is to estimate (i.e., guess) the values of the unknown elements of the T_i vector. Next, the elements of the three-dimensional X_{ijk} matrix and the two-dimensional Φ_{ik} matrix are evaluated based on the combination of known (specified) and estimated surface temperatures. Note that the elements of the X_{ijk} matrix may be a function of the surface temperatures if the band-averaged spectral emissivities are temperature dependent. The next step is to use Equation 10.41 to compute estimates of the two-dimensional B'_{ik} matrix based on the current values of the elements of the temperature vector. Then estimates of the values of the

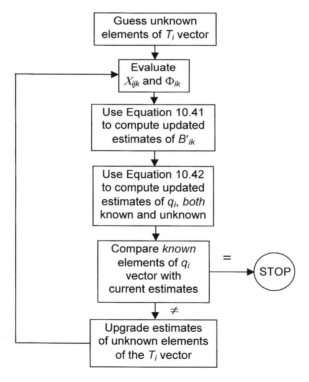

Figure 10.8 Block diagram for solving the problem of nongray, nondiffuse radiation in an enclosure having some specified net heat fluxes

net heat fluxes, *both known and unknown,* are computed using Equation 10.42. The current estimates of the net heat fluxes are then compared with the corresponding known values, for example, by computing the least-mean-square differences. Alternatively, the root-mean-square of the differences in the ensemble of heat fluxes for all surfaces could be checked for convergence.

If the global difference between the known and estimated values of net heat flux vector elements is acceptable, the solution is terminated. If the difference is not tolerable, however, it is necessary to upgrade the estimates of the unknown elements of the surface temperature vector. Many schemes are available for obtaining these upgraded estimates in a way that assures convergence, for example, the genetic algorithm described in Section 13.7.2. Numerical implementation of the scheme diagrammed in Figure 10.8 would probably make a good master's thesis topic.

10.9 AN ALTERNATIVE APPROACH FOR AXISYMMETRIC ENCLOSURES

In 1979 Mahan, Kingsolver, and Mears [6] introduced an alternative to the net-exchange method applied to enclosures whose walls have a specular component of reflectivity. The alternative, which is applicable only to axisymmetric enclosures such as parabolic reflectors used for illumination, is exact in principle, assuming that a sufficiently large number of terms are retained in the final expression; otherwise, it is an approximate method that permits a trade-off between modeling effort and accuracy of the result obtained.

Consider the axisymmetric enclosure shown in Figure 10.9 formed by the rotation through 360 deg about the x axis of a function $y = f(x)$ with $a \leq x \leq b$. If $d^2y/dx^2 \leq 0$ on the interval $[a, b]$, then the walls of the enclosure are everywhere concave. The analysis that follows applies only to enclosures of this general form. We further limit the method to enclosures for which temperature, heat flux, and surface properties vary only with x. Even subject to these limitations, the method applies to a wide range of useful devices.

We begin the analysis by subdividing the enclosure into a finite number n of conic sections whose lengths may be variable, depending on the local curvature of the directrix $y = f(x)$, with more divisions being used where the magnitude of the curvature is large. Each of these ring elements is then further subdivided into infinitesimal area elements,

$$dA = r \, d\theta \, ds \tag{10.43}$$

Let one such area element, dA_i, be a source of radiant energy, and let a beam of this energy sweep over ring element k, which has a specular component of reflectivity. Then the part of this energy that is specularly reflected from ring element k is incident to the interior of the enclosure over a zone indicated by the irregularly shaped lightly shaded region in Figure 10.9. We define the *differential partial exchange fac-*

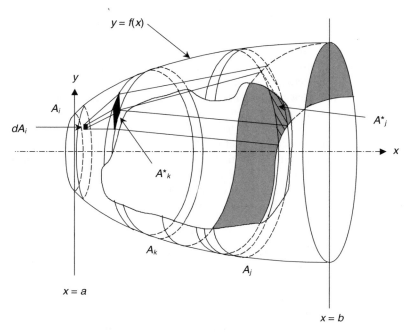

Figure 10.9 Axisymmetric enclosure showing nomenclature for developing the partial exchange factor method

tor $dP_{dA_i(dA_k)-dA_j}$ as the fraction of diffusely distributed energy leaving dA_i and arriving at dA_j after one intervening specular reflection in dA_k with $\rho_k^s = 1.0$. When this definition is applied to the geometry of Figure 10.9 we have

$$dP_{dA_i(dA_k)-dA_j} = dF_{dA_i-dA_k} \qquad (10.44)$$

with the restriction that dA_k lies in the darkly shaded area A_k^*. Then, the partial exchange factor from A_i to A_j with one intervening specular reflection in A_k is

$$P_{dA_i(A_k)-A_j} = \int_{A_k^*} dF_{dA_i-dA_k} \qquad (10.45)$$

where Equation 10.45 was obtained by integrating the left-hand side of Equation 10.44 over A_j^* and, correspondingly, the right-hand side over A_k^*. Equation 10.45 is now multiplied through by dA_i, after which reciprocity is invoked for the left-hand side, yielding

$$A_j dP_{A_j-dA_i(A_k)} = dA_i \int_{A_k^*} dF_{dA_i-dA_k} \qquad (10.46)$$

Finally, upon integration of Equation 10.46 over ring i followed by application of reciprocity to the left-hand side, there results

$$P_{A_i(A_k)-A_j} = \frac{1}{A_i} \int_{A_i} \int_{A_k^*} dF_{dA_i - dA_k} dA_i \qquad (10.47)$$

In principle, the quadruple integrals in Equation 10.47 can be evaluated once the ordinary diffuse configuration factors and appropriate limits of integration have been identified. In practice, this usually must be done numerically. Once the partial exchange factors $P_{i(k)-j}$ have been obtained there remains only the monumental bookkeeping task of converting them into exchange factors. This is accomplished by identifying all possible Markov chains of partial exchange factors that begin on ring i and end on ring j. Such a chain of order m would be expressed symbolically as

$$V_{i-j}^m = [\rho_{k_1}^s P_{i(k_1)-k_2}][\rho_{k_2}^s P_{k_1(k_2)-k_3}] \cdots [\rho_{k_{m-1}}^s P_{k_{m-2}(k_{m-1})-k_m}][\rho_{k_m}^s P_{k_{m-1}(k_m)-kj}] \quad (10.48)$$

In general, there will be more than one, perhaps many, chains of a given order linking rings i and j. The specular component of the exchange factor E_{i-j} is the sum of all such chains of all orders that link rings i and j.

Because the order of the highest-order chain and the number of chains of a given order linking rings i and j both increase with n, the number of possible chains can quickly become unmanageably large. Fortunately, the more links involved in a given chain, the less important will be its contribution to the value of E_{i-j}. For example, for an enclosure divided into ten axial rings the contribution of a typical third-order chain will be less than one percent of the contribution of a typical first-order chain [6]. The method may be considered "exact" in the sense that if $E_{i-j}^{(m)}$ is the mth-order approximation of the exchange factor E_{i-j}, then

$$\lim_{m \to n} [E_{i-j}^{(m)} - E_{i-j}] = 0 \qquad (10.49)$$

The details of the procedure outlined above and application of the method to a parabolic reflector may be found in Reference 6.

Team Projects

TP 10.1 We will build up a surface like the one shown in Figure 10.1 and then evaluate its bidirectional reflectivity for illumination at two azimuthal angles θ, 0 deg (with respect to the alignment of the edges) and 90 deg, both at a range of zenith angles ϕ. We will use collimated sunlight as the source and our rudimentary radiometer as the detector, as suggested in Figure 10.10. As a practical matter this will be a difficult measurement to obtain with high precision because the sun is a moving source and the reflecting surface is relatively small. Use at least three values of the zenith angle $\theta > 0$

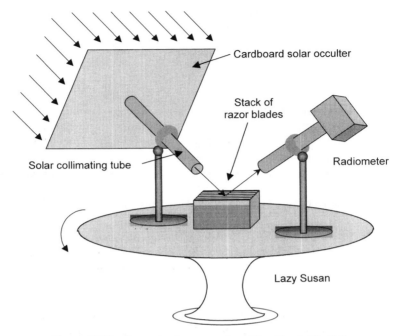

Figure 10.10 Apparatus for measuring bidirectional reflectivity

deg. Present your results as a ratio of the measurements to measurements obtained for a sheet of white typing paper at the same angles.

TP 10.2 If you visit a flooring store to buy floor tile, you will find that they have an ingenious apparatus for using one tile to show the customer how an entire floor covered with the tile would look. It consists of a four-walled box whose walls are plane mirrors. The 12-in. tile just fits on the floor of the box, and the customer views the tile through the open top of the box. The illusion produced is of an infinite room whose floor is covered with the tiles. (A similar effect is obtained when being fitted for a jacket at a clothing store.) Obtain four suitably sized rectangular mirrors and fabricate one of these devices. Cut sheets of cardboard that can be used to cover one or more of the walls. What can we learn about exchange factors using this device? What would be the exchange factor from one wall to another in the case where none of the mirrors is covered by a cardboard sheet?

Discussion Points

DP 10.1 How should the two sides of the cardboard solar occulter in Figure 10.10 be treated? Should they be white? Black? Aluminum foil? Please justify your answer.

DP 10.2   What do you think the reflection pattern might be for visible radiation incident to the surface of Figure 10.1 at a zenith angle of $\theta = 45$ deg for the two values of azimuth angle ϕ suggested in TP 10.1? Sketch reflected intensity as a function of the zenith angle θ for these two cases.

DP 10.3   Criticize the experiment proposed in TP 10.1, bearing in mind that each measurement of the bidirectional reflectivity of the surface formed by stacking razor blades is compared with the same measurement for a sheet of white typing paper. If bidirectional reflectivity data for the typing paper were available for the solar spectrum, how might these data be used to obtain bidirectional reflectivity data for the surface formed by the stack of razor blades (see Problem P 3.13)?

DP 10.4   A version of the experiment proposed in TP 10.1 has actually been carried out using proper equipment and the results reported in the literature. See if you can find the article.

DP 10.5   Two infinite plane dusty mirrors face each other at a finite distance. Both mirrors have the same values of diffuse and specular components of reflectivity, ρ^d and ρ^s. Find the expression for the exchange factor from one mirror to the other. Investigate the limit of this expression as the specular component of the reflectivity approaches unity.

DP 10.6   Compare the set of equations describing net exchange among diffuse surfaces, Equations 8.67 and 8.68, with the corresponding set of equations describing net exchange among diffuse-specular surfaces, Equations 10.26 and 10.27. How do they differ? Could you explain and justify the differences to a bright undergraduate who had completed a basic heat transfer course? How would you do this?

DP 10.7   Consider the last sentence in the paragraph under Equation 10.40. What kind of strategy involving computer resources, time (clock and CPU), and number of wavelength bands K in the trade-space might lead to an optimum solution?

DP 10.8   The final paragraph of Section 10.8 mentions "genetic algorithms" (GAs). Have you heard of these before? If so, discuss them with someone who has not, and be prepared to lead a classroom discussion of GAs. If you have not heard of them go to the literature and read something basic. Then share what you have read with a classmate who has not heard of them. Be prepared to lead a classroom discussion of GAs. (Hint: See Section 13.7.2)

DP 10.9   Explain the conclusion in Section 10.9 that "for an enclosure divided into ten axial rings the contribution of a typical third-order chain will be less than one percent of the contribution of a typical first-order chain."

DP 10.10   Obtain a copy of Reference 6 and prepare a brief presentation to the class of the details not included in Section 10.9.

Problems

P 10.1   Verify Equation 10.4.

P 10.2   Verify Equation 10.9.

P 10.3   Verify Equation 10.6.

P 10.4   Verify Equation 10.10.

P 10.5   Verify Equation 10.14.

P 10.6   Demonstrate that the finite net exchange formulation for diffuse–specular enclosures converges to the finite net exchange formulation for purely diffuse enclosures in the limit as the specular component of reflectivity of the surfaces of the enclosure approaches zero.

P 10.7   For the infinitely long channel shown in Figure 10.11 surfaces 1 and 2 are diffuse and surface 3 is specular. The duct cross section is an equilateral triangle. Sketch the virtual enclosure, label the virtual walls, and derive expressions for the exchange factors $E_{1\text{-}1}$ and $E_{1\text{-}2}$.

P 10.8   For the geometry described in Problem 10.7 and illustrated in Figure 10.11, suppose the emissivity of surfaces 1 and 2 is 0.6 and the reflectivity of surface 3 is

$$\rho_3 = \rho_3^d + \rho_3^s = 0.5 + 0.2 \qquad (10.50)$$

Further, let the wall temperatures be $T_1 = 400$ K, $T_2 = 600$ K, and $T_3 = 200$ K.
(a) Evaluate all required exchange factors.
(b) Evaluate the net heat fluxes from the three surfaces.

P 10.9   Rework Problem 10.8 for the case where surface 3 is insulated rather than at a specified temperature. Find the temperature of surface 3 and the net heat flux from surfaces 1 and 2.

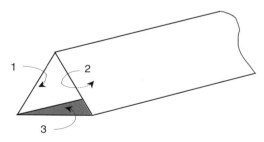

Figure 10.11   Geometry for Problems P 10.7, P 10.8, and P 10.9 (duct is infinitely long)

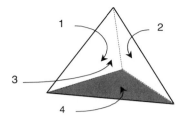

Figure 10.12 Geometry for Problems P 10.10, P 10.11, P 10.12, P 10.13, and P 10.14

P 10.10 The walls of a certain tetrahedral enclosure consist of four identical triangles, as shown in Figure 10.12. Surfaces 1, 2, and 3 are purely diffuse and surface 4 is specular. Sketch the virtual enclosure, label the virtual surfaces, and derive expressions for $E_{1\text{-}1}$ and $E_{1\text{-}2}$.

P 10.11 For the enclosure described in Problem 10.10 and depicted in Figure 10.12 the emissivity of surfaces 1, 2, and 3 is 0.3, and the reflectivity of wall 4 is

$$\rho_3 = \rho_3^{\mathrm{d}} + \rho_3^{\mathrm{s}} = 0.4 + 0.2 \qquad (10.51)$$

Further, let the surface temperatures be $T_1 = 800$ K, $T_2 = 600$ K, $T_3 = 200$ K, and $T_4 = 400$ K.
(a) Evaluate and list all required exchange factors.
(b) Evaluate the net heat fluxes from all four surfaces.

P 10.12 Rework Problem 10.11 for the case where surface 4 is insulated rather than at a specified temperature. Find the temperature of surface 4 and the net heat flux from surfaces 1, 2, and 3.

P 10.13 Suppose the emissivity of surfaces 1, 2, and 3 in Problem 10.11 is wavelength dependent such that

$$\varepsilon_\lambda (\lambda) = \begin{cases} 0.1, & 0 \le \lambda < 1 \ \mu\mathrm{m} \\ \lambda/10, & 1 \le \lambda < 9 \ \mu\mathrm{m} \\ 0.9, & \lambda \ge 9 \ \mu\mathrm{m} \end{cases} \qquad (10.52)$$

Evaluate the net heat fluxes from all four surfaces.

P 10.14 Suppose the emissivity of surfaces 1, 2, and 3 in Problem 10.12 varies with wavelength as given by Equation 10.52. Describe *in detail* how you would find the temperature of surface 4 and the net heat flux from surfaces 1, 2, and 3.

P 10.15 Use the partial exchange factor method described in Section 10.9 to verify the results for the parabolic reflector given in Reference 6.

REFERENCES

1. Sparrow, E. M., E. R. G. Eckert, and V. K. Jonsson, "An enclosure theory for radiative exchange between specularly and diffusely reflecting surfaces," *ASME Transactions, Series C, Journal of Heat Transfer,* Vol. 84, No. 4, November 1962, pp. 294–300.
2. Lin, S. H., and E. M. Sparrow, "Radiant interchange among curved specularly reflecting surfaces—application to cylindrical cavities," *ASME Transactions, Series C, Journal of Heat Transfer,* Vol. 87, No. 2, May 1965, pp. 299–307.
3. Bobco, R. P., "Radiation heat transfer in semigray enclosures with specularly and diffusely reflecting surfaces," *ASME Transactions, Series C, Journal of Heat Transfer,* Vol. 86, No. 1, February 1964, pp. 123–130.
4. Sarofim, A. F., and H. C. Hottel, "Radiative exchange among non-Lambert surfaces," *ASME Transactions, Series C, Journal of Heat Transfer,* Vol. 88, No. 1, February 1966, pp. 37–44.
5. Kowsary, F., *Radiative Characteristics of Spherical Cavities Having Partially or Completely Specular Walls,* PhD Dissertation, Department of Mechanical Engineering, Virginia Polytechnic Institute and State University, Blacksburg, VA, November 1989.
6. Mahan, J. R., J. B. Kingsolver, and D. T. Mears, "Analysis of diffuse–specular axisymmetric surfaces with application to parabolic reflectors," *ASME Transactions, Journal of Heat Transfer,* Vol. 101, No. 4, November 1979, pp. 689–694.

PART III

THE MONTE CARLO
RAY-TRACE METHOD

11

INTRODUCTION TO THE MONTE CARLO RAY-TRACE METHOD

Arguably over 90 percent of all thermal radiation problems encountered by engineers to date have been worked using some variation of the net exchange method presented in Part II of this book. In many—perhaps most—of these cases the results obtained were of acceptable accuracy, either because high precision was not required or because the limiting assumptions of the method were a good approximation of reality. In this chapter we introduce an important modern alternative to the net exchange formulation capable of providing accuracies limited only by our knowledge of the directionality and wavelength dependence of the surface properties.

11.1 COMMON SITUATIONS REQUIRING A MORE ACCURATE ANALYTICAL METHOD

The common situations requiring a more accurate analytical approach than that offered by the net exchange method may be divided into two categories: those in which the assumptions of the net exchange formulation are strictly invalid, and those requiring exceptionally high accuracy. The first category includes situations in which surfaces have been specifically engineered to be directional emitters, absorbers, or bidirectional reflectors. This is a potentially important approach to the design of "low observable" military aircraft and ground and seaborne vehicles. In this application, surfaces would be engineered, for example, through microgrooving or the application of second-surface mirror or interference filter coatings, so that emitted or reflected radiation in specific wavelength intervals is directed into one or more onboard "radiation traps" or otherwise in a direction away from hostile infrared search-and-track (IRST) weaponry. In such cases the diffuse and diffuse–specular models fail to provide an accurate description of directionality. More to the point, they do not sample the directionality effects that have been intentionally engineered into the surfaces.

The second category, that requiring exceptionally high accuracy, includes

1. Situations where better than one-percent accuracy is required in the modeling of, for example, thermal radiometer concepts, where models are aimed at predicting instrument accuracy, precision, and level of thermal contamination of the signal.

2. Situations involving the prediction of jet engine exhaust plume infrared emission and modification of plume and engine hot part signatures by the atmosphere. In such applications a one-percent uncertainty can translate into several kilometers of difference of predicted IRST lock-on capability, and thus into several precious seconds of uncertainty in predicted lock-on advantage by heat-seeking missiles.

3. Situations where infrared imaging is used to validate computational fluid dynamic (CFD) codes for predicting jet engine exhaust plume fluid dynamic and thermal development [1]. An essential step in this process is the prediction of the jet plume spectral thermal radiation field based on the predicted distributions of temperature and chemical species concentration. The predicted image for some narrow wavelength band of interest is then compared with the image obtained using a suitably filtered thermographic imaging system. The accuracy of the subsequent CFD code validation depends directly on the accuracy with which the spectral image can be predicted.

Some of the scenarios described above involve gaseous radiation. In fact, nonstatistical methods such as those presented in Chapter 7 exist for treating gaseous radiation, just as nonstatistical methods exist for dealing with radiation among surfaces. However, unlike the statistical treatment developed in this book, the traditional treatments of gaseous radiation are not easily coupled with directionally dependent surface radiation models. The treatment of gaseous radiation in the Monte Carlo ray-trace (MCRT) method will be seen to be a natural extension of the treatment of radiation among surfaces.

11.2 A BRIEF HISTORY OF THE MONTE CARLO RAY-TRACE METHOD IN RADIATION HEAT TRANSFER

The term "Monte Carlo" was first applied to the probablistic approach used to describe neutron transport through fissile materials during the atomic bomb project in the early 1940s. After WWII its use expanded to include the solution of a wide variety of diffusion problems. The so-called "random walk" method was used to solve heat conduction problems in domains whose geometry precluded exact closed-form analytical and even finite-difference approaches.

The earliest radiation heat transfer applications seem to have appeared in 1964 in a series of papers by J. R. Howell and M. Perlmutter [2–4]. These were followed in 1966 by a paper by R. C. Corlett [5] describing a Monte Carlo method for direct heat transfer calculation through a vacuum, and the 1966 version of the radiation heat transfer textbook by Sparrow and Cess [6] briefly treats the topic. In 1968, Howell [7] published a review article on the application of the Monte Carlo method to heat transfer problems, and in another 1968 article J. S. Toor and R Viskanta [8] describe a "numerical experiment" in which the Monte Carlo method is used to evaluate the importance of bidirectional reflection. In a 1969 article A. Haji-Sheikh and E. M. Sparrow [9] describe a method for estimating the error associated with the Monte Carlo solution of radiation heat transfer problems.

The twenty years since the author's first published contribution [10] to this rapidly emerging discipline have seen an astonishingly rapid growth in computer speed, storage, and availability. In that relatively brief period of time the MCRT method has evolved from an expensive and very approximate estimation tool, in which a few tens of thousands of rays were typically traced, to a cost-effective and highly accurate approach involving tens of millions of rays traced to obtain "exact" solutions to complex thermal radiation problems. One of the author's recent doctoral students, F. J. Nevárez-Ayala, has produced a robust Windows-based MCRT environment for thermal radiation modeling called FELIX [11]. The student version of FELIX packaged with this book and described in Appendix C has been used to solve many of the example problems in Part III of this book.

11.3 SECOND LAW IMPLICATIONS

At this juncture it is convenient to think of a thermal radiative analysis as leading to an "image" of the enclosure as viewed from one of its surface elements, the "observer." Indeed, an image in the literal sense of the word might be the goal of a thermal radiative analysis in which a map is sought of spectral intensity incident over the two-dimensional angle of incidence in the hemispherical space above the observer. However, in most cases involving radiation among surfaces we seek mappings of local temperature or local net heat flux over the surfaces of the enclosure. In cases involving gaseous radiation the image might be the flux in a given narrow wavelength band intercepting an imaginary "virtual screen" oriented normal to the line-of-sight of an observer. For example, this would be a viable model of an infrared search-and-track weapon system. In any case, there is a sense in which we can speak of the *information content* [12] of this image. This information content, which completely describes the image, may be stored in a pseudo-three-dimensional matrix whose "dimensions" are (1) the direction θ, ϕ that defines the angular resolution, (2) the wavelength λ that defines the spectral resolution, and (3) the spectral intensity $i_\lambda(\lambda, \theta, \phi)$ that defines the amplitude and is characterized by a *dynamic range*. By "dynamic range" it is meant the ratio of the maximum value observed to the minimum resolvable value (i.e., the "noise floor").

Thus, what a single surface element, the observer, looking into its 2π-space, "sees" can be stored in the pseudo-three-dimensional matrix represented in Figure 11.1. ("Pseudo" because one of the "dimensions" itself consists of two dimensions, θ and ϕ.) A matrix such as the one illustrated, which exists for each observer element (i.e., one for each surface element making up the walls of the enclosure), represents the "image" seen by the observer element. If the enclosure is subdivided into n surface elements, there will be n image matrices such as the one depicted in Figure 11.1. It is clear that a "problem" of practical interest potentially contains a large amount of information.

Now radiation heat transfer practitioners are usually interested only in the properly weighted integral of the image over wavelength and direction, i.e., in the net heat flux for each surface element. However, in general, the complete image must be created—its information content known and used—during the course of the analysis before it is ultimately "smeared," or degraded, by integration.

The organization, or order, represented by the matrix in Figure 11.1 has a relatively low entropy, while the integrated result, which is a single number, has the maximum possible entropy. The entropy can be decreased by dividing up the box in the figure into more "volume" elements; that is, by increasing the spatial and spectral resolution and the number of significant figures to which the spectral intensity is known at each wavelength and in each direction. Note that this requires information input, or "intelligence." Conversely, the entropy can be increased by combining adjacent elements of the matrix, i.e., by information-destroying integration. This would amount to, for example, decreasing the spatial or spectral resolution, or both.

Consider the situation depicted in Figure 11.2a, which shows a box with opaque walls partitioned into two sections, one containing two marbles and the other containing one marble. When the partition is removed, as shown in Figure 11.2b, information is lost: because we cannot see into the box we no longer know how the marbles are distributed. This idea is illustrated in Figure 11.3. However, the integrated information, How many marbles are in the box? is retained.

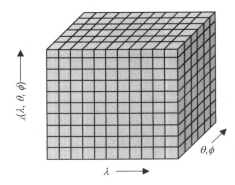

Figure 11.1 Matrix representation of the "image" of an enclosure as viewed by an "observer" surface element on an interior wall

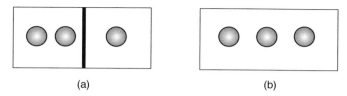

Figure 11.2 Box containing three marbles (a) with a partition in place and (b) with the partition removed

In general, we can say that all processes, or analyses, that provide exactly the same image (i.e., exactly the same information) require the same amount of knowledge. Now, when we consider the division of labor between the engineer and the computer in solving a radiation heat transfer problem, we clearly want the computer to do most of the work. Putting aside the philosophical question of what computers can "know," at least in the context of the above interpretation of knowledge as information, this means that we would like for the computer to bring as much knowledge to the process as possible. In summary, we seek a method of radiation heat transfer analysis that leads to a minimum of entropy (= a maximum of knowledge = a maximum "resolution" of the "image") with a minimum of effort on the part of the engineer.

If a lot is known about the directionality and wavelength dependence of surfaces and participating media, we would like to have an analytical tool at our disposal capable of exploiting this information. Conversely, if we have a tool at our disposal capable of providing a high-quality image, it behooves us to obtain a compatible degree of knowledge about the directionality and wavelength dependence of surface properties.

Example Problem 11.1

The spectral distribution of the hemispherical, spectral emissivity of a certain hypothetical surface coating at a temperature of $1725°C$ is as given in Figure 11.4.

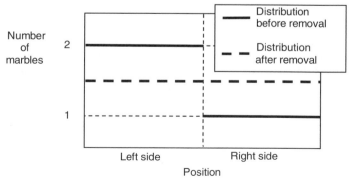

Figure 11.3 When the partition in Figure 11.2a is removed information is lost about how the marbles are distributed

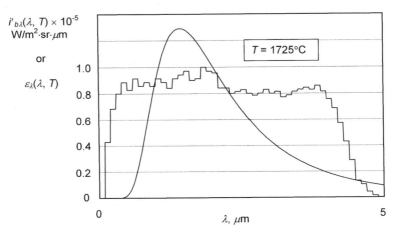

Figure 11.4 Hypothetical hemispherical, spectral emissivity data and corresponding radiation distribution function for a blackbody at 1725°C

Calculate the hemispherical, total emissivity and the loss in spectral resolution, the latter in terms of the increase in entropy, associated with this totalization process.

SOLUTION

The spectral emissivity dataset consists of 50 entries in the 0- to 5-μm wavelength range. We estimate the total emissivity using

$$\varepsilon(T) \cong \frac{\displaystyle\sum_{j=1}^{50} \varepsilon_{\lambda_i} i_{b\lambda}(\lambda_i, T)}{\displaystyle\sum_{i=1}^{50} i_{b\lambda}(\lambda_i, T)} \tag{11.1}$$

with the help of a spreadsheet. When we do this we obtain $\varepsilon \cong 0.853$. But what is meant by "the loss in spectral resolution, in terms of the increase in entropy"? The integration process implied by Equation 11.1 has destroyed information about the spectral distribution of emissivity. This loss of knowledge can be interpreted in terms of an increase in entropy using the relation

$$\Delta S = -k \ln\left(\frac{1}{n}\right) \quad \text{(J/K)} \tag{11.2}$$

where k is the Boltzmann constant and n is a number (50 in this example) representing the spectral resolution of the original dataset. In the current example Equation 11.2 yields $\Delta S = 5.40 \times 10^{-23}$ J/K.

11.4 THE RADIATION DISTRIBUTION FACTOR

If we could chose any "factor" we wanted to help us do radiation heat transfer analysis, our choice probably would not be the traditional configuration factor or exchange factor used in the net exchange method. This is because

1. Except in the most simple enclosures both factors are difficult to evaluate.
2. They carry only a limited amount of directional information (limited to diffuse emitting and absorbing, and diffuse and/or specularly reflecting).
3. Their use requires the inversion of a large and often ill-conditioned matrix.

We will presently learn about a statistical method, called the Monte Carlo ray-trace method, that is fully capable of computing both the configuration factor and the exchange factor. However, we will also learn that it is equally capable of computing a much more powerful factor, called the *radiation distribution factor,* which has the potential to contain all directional and spectral information relevant to the ultimate disposition of thermal radiation emitted from a given surface element. As described in this book, the radiation distribution factor is similar to Gebhart's *absorption factor* [13] except that the latter as defined and used in the reference accounts only for direct radiation and all possible diffuse reflections between diffuse surfaces, while the distribution factor includes directional absorption and emission and bidirectional reflections. In fact, the absorption factor defined and used by Toor and Viskanta [8] is closer to the distribution factor as defined in this book. The term "distribution factor" is preferred because it is more descriptive of how the factor is actually used; that is, to distribute radiation emitted from one surface or volume element to one of the other surface and volume elements of the enclosure.

The *total radiation distribution factor* from surface element i to surface element j, D_{ij}, is defined as the fraction of the total radiation emitted from surface element i that is absorbed by surface element j, due both to direct radiation and to all possible reflections within the enclosure. Note that this definition does not restrict emission, absorption, and reflection to any particular directional model; that is, the radiation distribution factor is a completely directional tool. Later we will define radiation distribution factors for exchange between surface elements and volume elements, and among volume elements, of radiatively active gases and other participating media that may fill an enclosure. However, we will begin by confining our discussion to radiation exchange among surface elements. The distribution factor as defined and used in this book is the "exchange fraction" referred to by Howell [7]. The development presented here has its origins in two papers by Mahan and Eskin presented in 1981 [10] and 1984 [15], and conforms to that of the "absorption factor" studied by Maltby and Burns [15].

In the case of wavelength-dependent surface properties coupled with radiation over a sufficiently broad range of wavelengths, it is convenient to define the *band-averaged spectral radiation distribution factor* D_{ijk}, corresponding to the wavelength band $\Delta\lambda_k$, as the fraction of power emitted in wavelength interval $\Delta\lambda_k$ by surface element i that is absorbed by surface element j, both directly and due to all possible reflections within the enclosure.

It is often adequate to use the simpler *total, diffuse–specular radiation distribution factor D'_{ij}*, defined as the fraction of total radiation emitted diffusely from surface element i that is absorbed by surface element j, due to direct radiation and to all possible diffuse and specular reflections within the enclosure. Use of the total, diffuse–specular radiation distribution factor leads to results comparable to those obtained using the net exchange formulation. Another possibility is to base the analysis on the *band-averaged spectral, diffuse–specular radiation distribution factor D'_{ijk}*.

Once the radiation distribution factors for an enclosure have been obtained, they are much easier to use than either configuration factors or exchange factors. This is especially true if all of the surface temperatures are specified and net heat fluxes are to be computed. In this case there is no matrix to invert and the required heat fluxes can be computed directly in terms of the distribution factors and specified surface temperatures. If some surfaces of the enclosure do have specified net heat fluxes, then (at least in the case of a total analysis) the order of the resulting matrix to be inverted to obtain the unknown surface temperatures is limited to the number of surfaces having specified net heat fluxes.

11.5   THE TOTAL, DIFFUSE–SPECULAR RADIATION DISTRIBUTION FACTOR

We begin our study of the MCRT method with consideration of the total, diffuse–specular radiation distribution factor, D'_{ij}, defined above. This factor is applicable to situations in which radiation is limited to a restricted wavelength interval in which surface properties may be considered independent of wavelength. In practice, this usually means that the temperature range of the surfaces of the enclosure is restricted to on the order of a few hundred kelvins. As has already been stated, results obtained using the total, diffuse–specular radiation distribution factor are comparable to those obtained using the diffuse–specular net exchange formulation, with the advantage that the distribution factors are generally much easier to obtain than the exchange factors, and the subsequent heat transfer analysis using the distribution factors is much simpler.

Consistent with the definition of the total, diffuse–specular radiation distribution factor we can write for the radiant power emitted by surface element i and absorbed by surface element j

$$Q_{ij} = \varepsilon_i A_i \, \sigma T_i^4 D'_{ij} \quad \text{(W)} \tag{11.3}$$

where ε_i is the hemispherical, total emissivity of surface i, A_i is its surface area (m^2), and T_i is its temperature (K). Equation 11.3 is, in fact, the definition of the distribution factor just as Fourier's law of heat conduction, Equation 1.1, is the definition of the thermal conductivity. Note that an alternative way of writing Equation 11.3 that emphasizes this interpretation is

$$D'_{ij} \equiv \frac{\partial(Q_{ij})}{\partial(\varepsilon_i A_i \sigma T_i^4)} \tag{11.3a}$$

A major advantage of the distribution factor approach introduced here over other Monte Carlo-based methods is that distribution factors are calculated separately from their subsequent use in Equation 11.3. This means that the influence of different temperature and/or net heat flux distributions on the enclosure walls can be evaluated without recomputing the distribution factors each time, that is, without redoing the Monte Carlo ray trace.

11.6 PROPERTIES OF THE TOTAL, DIFFUSE–SPECULAR RADIATION DISTRIBUTION FACTOR

It can be shown (see Problems P 11.2, P 11.3, and P 11.4) that the total, diffuse–specular radiation distribution factor has the following three useful properties:

1. Conservation of energy

$$\sum_{j=1}^{n} D'_{ij} = 1.0, \qquad 1 \leq i \leq n \tag{11.4}$$

2. Reciprocity

$$\varepsilon_i A_i D'_{ij} = \varepsilon_j A_j D'_{ji}, \qquad 1 \leq i \leq n, \quad 1 \leq j \leq n \tag{11.5}$$

3. Combination of conservation of energy and reciprocity

$$\sum_{i=1}^{n} \varepsilon_i A_i D'_{ij} = \varepsilon_j A_j, \qquad 1 \leq j \leq n \tag{11.6}$$

In Equations 11.4, 11.5, and 11.6, n is the number of surface elements making up the enclosure, ε is the hemispherical, total emissivity of a given surface element, and A is its surface area.

Equation 11.6, which is obtained by summing both sides of Equation 11.5 over i and then substituting Equation 11.4 into the result, is useful for detecting and eliminating errors made during calculation of the distribution factors for an enclosure. It can also be used to provide a statistically meaningful measure of the accuracy with which the distribution factor matrix for a given enclosure has been computed. The conservation of energy relation, Equation 11.4, and the reciprocity relation, Equation 11.5, are also useful for detecting errors or for finding unknown distribution factors from known distribution factors using distribution factor algebra. However, note that these relations cannot be used *both* for error detection *and* for finding unknown distribution factors in the same enclosure.

Finally, it is noted that distribution factors can also be defined for radiation entering an enclosure through an opening o with a specified directional distribution (e.g., collimated or diffuse). Equation 11.3 is not applicable in this situation because a heat flux, rather than a temperature, is specified for the imaginary surface element representing the opening. In this case, the appropriate relation for defining the distribution of this radiation on the surface elements making up the enclosure is

$$Q_{oj} = Q_o D_{oj}, \qquad 1 \le j \le n \qquad (11.7)$$

where Q_o is the power (W) entering the enclosure through opening o and D_{oj} is the fraction of this power absorbed by surface element j.

11.7 THE MONTE CARLO RAY-TRACE METHOD

The Monte Carlo ray-trace method is a statistical approach in which the analytical solution of a problem is bypassed in favor of a numerical simulation whose outcome may be expected to be the same as that of the analysis but which is easier to carry out. In the case of a thermal radiation problem, a given quantity of radiation energy is uniformly divided into a large number N_i of discrete energy bundles. These energy bundles are followed from their emission by surface element i, through a series of reflections on other surface elements, to their absorption by one of the surface elements j of the enclosure. The properties of the enclosure and the laws of probability are used to determine the number of energy bundles N_{ij} absorbed by a given surface element j.

A consequence of the definition of the radiation distribution factor is that it is numerically equal to the ratio of N_{ij} to N_i in the limit as N_i tends to infinity, and its value can be well estimated using large but finite values of N_i. Because the radiation distribution factor is determined using the Monte Carlo ray-trace method, it is possible to consider enclosures whose surfaces are directional emitters and absorbers and bidirectional reflectors, and whose surface properties vary with wavelength.

The general approach in the MCRT method is to emit a large number N_i of energy bundles from randomly selected locations on a given surface element i and then to trace their progress through a series of reflections until they are finally absorbed on a surface element, say j, where $j = i$ is a possibility.

The number of energy bundles traced depends on the desired accuracy and the available time and computer resources. If $1000n$ energy bundles are emitted from each surface element, where n is the total number of surface elements into which the enclosure has been divided, then on the average the distribution factors will be estimated to three significant figures. However, depending on the shape of the enclosure and the degree of directionality of the surface properties, some of the distribution factors will be estimated to a greater or lesser accuracy. For this reason it is recommended that at least 10,000n energy bundles be emitted from each surface element. It is not unusual in applications where a very high accuracy is required (and where justified by the high degree of knowledge of the directionality of the surface properties) to emit $n \times 10^6$ energy bundles per surface element. In any case, a convergence study should always be performed to assure convergence of the distribution factor matrix with the number of energy bundles traced. The uncertainty and associated confidence interval in distribution factor estimates is the subject of Chapter 15.

A word about terminology. We are careful to refer to the path followed by an energy bundle as a "ray," thus explaining the origin of the term "ray trace." The energy bundle itself should not be referred to as a ray (or as a "photon"). The reason for this formalism will become apparent when band-averaged spectral ray traces are studied in Chapter 13.

Whenever the path of an energy bundle intersects a surface of the enclosure, the fate of the energy bundle is determined by treating the surface properties as probabilities. For example, if the hemispherical, total absorptivity of the surface is α, then the probability that the incident energy bundle will be absorbed is $P_\alpha = \alpha$. In the MCRT formulation a random* number R_α, whose available discrete values are uniformly distributed between zero and unity, is drawn and its value compared with α ($= P_\alpha$). If $R_\alpha < \alpha$ then the energy bundle is deemed to be absorbed; otherwise, it is reflected. If the energy bundle is absorbed, a counter N_{ij} is incremented and a new energy bundle is emitted from a different randomly selected location on surface element i. If it has been determined that the energy bundle was reflected, a similar process is used to determine the direction of reflection. However, in the case of reflection the details depend on the model for reflectivity, i.e., diffuse, specular, or bidirectional. This process continues until the requisite number of energy bundles has been emitted from surface i, after which the logic moves on to surface $i + 1$. A logic block diagram for using the MCRT method to obtain total, diffuse–specular radiation distribution factors appears in Figure 11.5.

The steps to be followed to estimate the values of the total, diffuse–specular radiation distribution factors from a rectangular surface element to the surface elements of an enclosure (including the source element) are described below. The last digit in each of the following subsection numbers corresponds to the circled numbers in Figure 11.5.

11.7.1 Determine the Location of Emission of the Energy Bundle

Care must be taken to ensure that each energy bundle is emitted from a random location on the surface element. Otherwise, the result obtained for the distribution factors will be biased. Consider a rectangular surface element lying in the x, y-plane with its edges aligned with the x and y axes, as represented in Figure 11.6. Even though

---

* Appendix D makes the point that it is more appropriate to speak of "pseudo-random" numbers, there being no such thing as a truly random number. Nevertheless, we will continue to use the term "random number."

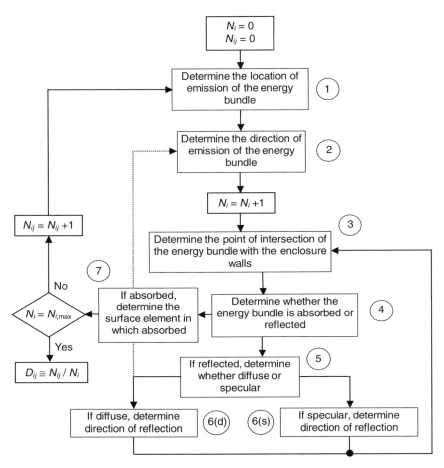

Figure 11.5 Logic block diagram for implementing the Monte Carlo ray-trace method for a given source surface in a diffuse–specular, gray enclosure

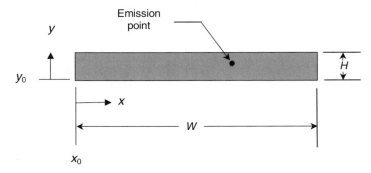

Figure 11.6 A hypothetical rectangular surface element

the process described here is particular to this specific geometry, the general principle behind the process will be clear. Exploring new ways to apply this principle to other less regular surface element shapes such as triangles is one of the genuine joys of formulating a Monte Carlo code. (A general, method applicable to triangular pixels—which turn out to be very useful for closely approximating arbitrary surface topography—is presented in Appendix C.)

We begin by drawing two random numbers, N_x and N_y, each of which is uniformly distributed between zero and unity. For now let us accept that "uniformly distributed between zero and unity" means as a practical matter that if we have a population of, say, 99,999 random numbers whose values are uniformly distributed between zero and unity exclusive, then their values are 1×10^{-5}, 2×10^{-5}, 3×10^{-5}, . . ., $99,999 \times 10^{-5}$, with no missing numbers in the sequence and no repeated numbers. (The number 99,999 is used in this example because, in fact, the random number generators we use do not permit the values zero and unity, although, clearly, we can get arbitrarily close to these two limits if our population is sufficiently large.) Further, if our access to this population is really random then we have an equal a priori probability of drawing any one of them at any given draw, i.e., one chance out of 99,999. The reader interested in learning more about pseudorandom number generators suitable for use in Monte Carlo-type applications is referred to Appendix D.

Once drawn, the two random numbers R_x and R_y are related to the coordinates of the randomly designated location of emission according to

$$x_1 = x_0 + R_x W \tag{11.8}$$

and

$$y_1 = y_0 + R_y H \tag{11.9}$$

11.7.2 Determine the Direction of Emission of the Energy Bundle

A diffuse emission is treated exactly the same as a diffuse reflection. By definition a diffusely reflected energy bundle retains no knowledge of its history before suffering the reflection; for all it "knows" it has no past and has just been born as a newly emitted energy bundle. Therefore, we will refer back to this step when we reach step 6d (as indicated by the dashed path in the block diagram of Figure 11.5).

The direction of a diffuse emission is characterized by two independent coordinates: the angle θ with respect to the local unit surface normal **n** directed into the interior of the enclosure, and the angle ϕ with respect to a local unit surface tangent $\mathbf{t}_1$. These relationships are illustrated in Figure 11.7. The angles θ and ϕ are related to two random numbers, R_θ and R_ϕ, uniformly distributed between zero and unity, by

$$\phi = 2\pi R_\phi \tag{11.10}$$

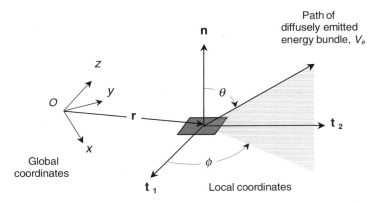

Figure 11.7   Relationship between the local and global coordinates

and

$$\theta = \sin^{-1} (\sqrt{R_\theta}) \qquad (11.11)$$

Equation 11.10 is rather obvious, but derivation of Equation 11.11 is left as an exercise for the student (see Problem P 11.5).

The direction defined by the angles given by Equations 11.10 and 11.11 are with respect to the *local* $\mathbf{n}$, $\mathbf{t}_1$, $\mathbf{t}_2$ coordinates of the surface element. To be useful this direction must be interpreted in terms of the *global* x, y, z coordinate system in which the energy bundles navigate.

The normal vector to a surface can always be found by computing the gradient of the equation for the surface, ∇S, and then dividing the result by the magnitude of the gradient; that is, if

$$S(x, y, z) = 0 \qquad (11.12)$$

is the equation for the surface, then

$$\mathbf{n} = \pm \frac{\nabla S}{|\nabla S|} \qquad (11.13)$$

The "$\pm$" decision in Equation 11.13 is made in such a way as to ensure that the normal is directed toward the interior of the enclosure. For example, consider a spherical enclosure of radius R. The equation for the surface of a sphere is

$$x^2 + y^2 + z^2 - R^2 = 0 \qquad (11.14)$$

Then,

$$\nabla S = 2x\mathbf{i} + 2y\mathbf{j} + 2z\mathbf{k} \qquad (11.15)$$

and

$$\mathbf{n} = -\frac{\nabla S}{|\nabla S|} = \frac{-x\mathbf{i} - y\mathbf{j} - z\mathbf{k}}{\sqrt{x^2 + y^2 + z^2}} = -\frac{x}{R}\mathbf{i} - \frac{y}{R}\mathbf{j} - \frac{z}{R}\mathbf{k} \qquad (11.16)$$

where the minus sign has be chosen to ensure that the surface normal is inward-directed.

If an enclosure has plane surface elements and it is possible to do so, it is often convenient to align the plane surfaces normal to the axes of the global coordinate system. Then, for example, if a plane surface element is aligned normal to the x axis, the surface normal is the unit vector in the x direction, i.e.,

$$\mathbf{n} = \mathbf{i} \qquad (11.17)$$

The orientation of the first unit tangent vector $\mathbf{t}_1$ is somewhat arbitrary. The only restriction is that the vector must be tangent to the surface at the point of interest. Also, to maintain control of the statistics of the random number sequence used, the tangent vector should be consistently oriented within a given surface built up from a collection of surface elements. This latter restriction is illustrated in Figure 11.8, which shows two pairs of consistently oriented first unit tangent vectors, labeled $\mathbf{t}_a$ and $\mathbf{t}_b$. However, a choice of $\mathbf{t}_a$ for surface element 1 and $\mathbf{t}_b$ for surface element 2, or $\mathbf{t}_a$ for surface element 2 and $\mathbf{t}_b$ for surface element 1, would not be consistent. A consistent mathematical procedure for computing the tangent vector is described below.

The equation for the first unit tangent vector in the case of a plane surface element is easy to obtain if the surface element is oriented normal to one of the global coordinate axes. In this case the first unit tangent vector will be one of the two unit vectors of the global coordinate system lying in the plane of the surface element. For ex-

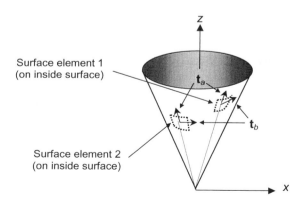

Figure 11.8 Illustration of the principle of consistently oriented first unit tangent vectors (demonstrated for a conical cavity)

ample, if the surface element is oriented normal to the x axis of the global coordinate system, then either

$$\mathbf{t}_1 = \mathbf{j} \tag{11.18}$$

or

$$\mathbf{t}_1 = \mathbf{k} \tag{11.19}$$

with the choice being arbitrary unless there are other surface elements in the same plane, in which case all tangent vectors for these surface elements should be the same for consistency, as indicated in Figure 11.8.

If a plane surface element cannot be conveniently oriented normal to one of the coordinate axes of the global coordinate system, then we must appeal to a coordinate transformation to relate $\mathbf{n}, \mathbf{t}_1, \mathbf{t}_2$ to $\mathbf{i}, \mathbf{j}, \mathbf{k}$. This is illustrated in Figure 11.9.

We define the *direction cosines l, m, n* such that

$$\mathbf{n} \cdot \mathbf{i} = \cos \alpha = \ell \tag{11.20}$$

$$\mathbf{n} \cdot \mathbf{j} = \cos \beta = m \tag{11.21}$$

and

$$\mathbf{n} \cdot \mathbf{k} = \cos \gamma = n \tag{11.22}$$

which means that

$$\mathbf{n} = \ell \mathbf{i} + m \mathbf{j} + n \mathbf{k} \tag{11.23}$$

The angles α, β, and γ in Equations 11.20, 11.21, and 11.22 are the angles that the normal vector $\mathbf{n}$ makes with the $\mathbf{i}$, $\mathbf{j}$, and $\mathbf{k}$ axes, respectively. Clearly, use of Equation 11.23 requires that the direction cosines l, m, and n be known, but of course they are because it is the user who has oriented the surface element. For example, comparison of Equations 11.16 and 11.23 reveals that $l = -x/R$, $m = -y/R$, and

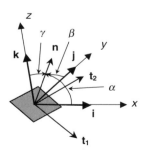

Figure 11.9 Example of a plane surface element not oriented normal to one of the axes of the global coordinate system

$n = -z/R$ for a spherical enclosure of radius R whose center lies at the origin of the coordinate system.

The first unit tangent vector for the arbitrary situation illustrated in Figure 11.9 may now be computed by dividing the cross product of the unit normal vector with *any one* of the three unit vectors **i, j,** or **k** by the magnitude of this cross product. This produces a unit vector that is orthogonal to the unit normal vector and therefore lies in the plane of the surface element. Then, for example,

$$\mathbf{t}_1 = \frac{\mathbf{n} \times \mathbf{i}}{|\mathbf{n} \times \mathbf{i}|} \tag{11.24}$$

or

$$\mathbf{t}_1 = \frac{n}{\sqrt{m^2 + n^2}} \mathbf{j} - \frac{m}{\sqrt{m^2 + n^2}} \mathbf{k} = t_{1,y} \mathbf{j} - t_{1,z} \mathbf{k} \tag{11.25}$$

where $t_{1,y}$ and $t_{1,z}$ are the y and z components of the first unit tangent vector.

It is clear that Equations 11.24 and 11.25 cannot be used in the case where $m^2 + n^2 = 0$. This will occur whenever $\mathbf{n} \cdot \mathbf{i} = 1.0$, that is, when **n** and **i** are colinear. In a numerical environment an ambiguous result (at best) will be obtained when $m^2 + n^2$ is smaller than the precision of the processor. For this reason it is necessary, before applying Equation 11.25, to compute $1.0 - \mathbf{n} \cdot \mathbf{i}$ and then compare the result with a number somewhat larger than but on the same order as the precision (smallest computable number) of the processor. If the value obtained is smaller than the precision of the processor then either **j** or **k** is used in Equation 11.24 instead of **i.** Finally, there is no a priori reason for choosing unit vector **i** over the unit vectors **j** or **k** for use in Equation 11.24; the choice is arbitrary.

When expressions for **n** and $\mathbf{t}_1$ have been found in terms of **i, j,** and **k,** then the second unit tangent vector is

$$\mathbf{t}_2 = \mathbf{n} \times \mathbf{t}_1 \tag{11.26}$$

Once **n,** $\mathbf{t}_1$, and $\mathbf{t}_2$ have been expressed in terms of the global coordinate system, the angles θ and ϕ given by Equations 11.10 and 11.11 and illustrated in Figure 11.7 may be used to compute the direction cosines of the diffusely emitted ray. This is a two-step process. First, the components of the unit vector in the direction of diffuse emission in terms of the local coordinate system are

$$\mathbf{V}_{e,n} = \mathbf{n} \cos\theta \tag{11.27}$$

$$\mathbf{V}_{e,t_1} = \mathbf{t}_1 \sin\theta \cos\phi \tag{11.28}$$

and

$$\mathbf{V}_{e,t_2} = \mathbf{t}_2 \sin\theta \sin\phi \tag{11.29}$$

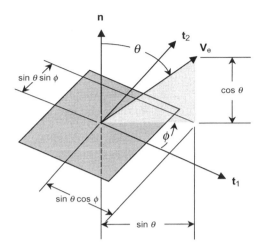

Figure 11.10   Illustration of the relationships among V_e, n, t_1, and t_2

These relationships are illustrated in Figure 11.10. Then in terms of the global coordinate system we can write

$$\mathbf{V}_{e,x} = V_{e,x}\,\mathbf{i} = (n_x \cos\theta + t_{1,x} \sin\theta \cos\phi + t_{2,x} \sin\theta \sin\phi)\,\mathbf{i} \qquad (11.30)$$

$$\mathbf{V}_{e,y} = V_{e,y}\,\mathbf{j} = (n_y \cos\theta + t_{1,y} \sin\theta \cos\phi + t_{2,y} \sin\theta \sin\phi)\,\mathbf{j} \qquad (11.31)$$

and

$$\mathbf{V}_{e,z} = V_{e,z}\,\mathbf{k} = (n_z \cos\theta + t_{1,z} \sin\theta \cos\phi + t_{2,z} \sin\theta \sin\phi)\,\mathbf{k} \qquad (11.32)$$

Navigation of the energy bundles within an enclosure is guided by successive sets of values of direction cosines. At this point it is convenient to recognize that the magnetudes of the components of the unit vector $\mathbf{V}_e$ *are* the direction cosines that define the direction in the global coordinate system of the path of the diffusely emitted energy bundle. This is because $\mathbf{V}_e$ has been designated a unit vector. Thus, we have $\ell = V_{e,x}$, $m = V_{e,y}$, and $n = V_{e,z}$.

11.7.3   Determine the Point of Intersection of the Emitted Energy Bundle with the Enclosure Walls

All possible points of intersection between the path of the energy bundle and the enclosure are found by solving the equations describing surfaces of the enclosure simultaneously with the equations of the line describing the path of the energy bundle. In performing this step it must be recognized that the equations of a line describe an infinitely long line and that, with some exceptions, such as spheroids, equations of surfaces describe surfaces of infinite extent. The problem this causes is well illustrated in Figure 11.11, which shows the possible intersections of a line, representing the path of an energy bundle, with the walls of a right conical cavity consisting of the upper branch of a cone truncated by a plane at $z = h$.

In the situation depicted in Figure 11.11 an energy bundle has been emitted from point a on the inside surface of the conical cavity in the direction of point b. However, when the equations of the line representing the path of the energy bundle,

$$\frac{x - x_a}{\ell} = \frac{y - y_a}{m} = \frac{z - z_a}{n} \tag{11.33}$$

are solved simultaneously with the equation of the plane containing the disk,

$$z = h \tag{11.34}$$

and the equation of the cone,

$$x^2 + y^2 + z^2 \tan^2 \theta \tag{11.35}$$

three solutions are obtained:

$$\left. \begin{array}{l} x = x_a \\ y = y_a \\ z = z_a \end{array} \right\} \quad \text{(point } a\text{)} \tag{11.36}$$

$$\left. \begin{array}{l} x = x_a + \dfrac{\ell}{n}(z_b - z_a) \\[2mm] y = y_a + \dfrac{m}{n}(z_b - z_a) \\[2mm] z = h \end{array} \right\} \quad \text{(point } b\text{)} \tag{11.37}$$

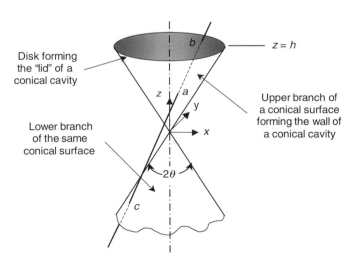

Figure 11.11 Intersections of a straight line with a conical surface

and point c, whose coordinates, in terms of the direction cosines l, m, and n and the coordinates of point a, are left as an exercise for the student (see Problem P 11.8).

From Figure 11.11 it is clear that point b is the desired point of intersection of the path of the energy bundle and the conical surface. The problem arises when it is time to instruct the computer to correctly identify point b. Several strategies are available for accomplishing this, but the most general and therefore the most powerful is a three-step process:

1. First, eliminate the point of origin (in this case point a), which will always be one of the *candidates*. This may be accomplished by comparing the candidate coordinates with the source coordinates. If they are the same to within the precision of the computer being used, the candidate point is rejected.
2. Next, eliminate the *back candidates,* that is, the ones behind the source surface (point c in Figure 11.11). This is easily accomplished by forming the dot product of the surface normal at the source point with the vector

$$\mathbf{V}_c = (x_c - x_a)\mathbf{i} + (y_c - y_a)\mathbf{j} + (z_c - z_a)\mathbf{k} \tag{11.38}$$

   directed from the source point toward the candidate point. If the value of this dot product is positive the candidate lies in front of the source surface element; otherwise it lies behind it and must be rejected.
3. Finally, of the remaining candidates that lie in front of the source surface element (called *forward candidates*), the correct receiving surface is the one nearest to the source element. For example, in the situation depicted in Figure 11.12, point b is now on the conical surface rather than on the disk surface, but

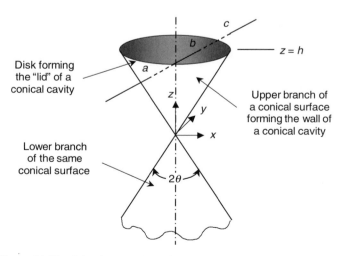

Figure 11.12 Other intersections of a straight line with a conical surface

the path of the energy bundle still intersects the plane $z = h$ (point c), only now outside of the cavity. Therefore, there are two candidate points in front of the source surface element, but the one on the conical surface, point b, is closer than the one on the plane $z = h$, point c.

Before continuing, it is appropriate to consider the consequences in a computer program of division by zero. Bitter experience has taught us that we should never use an expression in a computer code that involves division by a variable over whose value we have no control. We know that when we do this we inevitably get a "divide check" or some other equally disagreeable condition that either causes the program to "crash" or else leads to a nasty error message, or both. In other cases the processor will apply some "standard fixup," which may lead to a totally incorrect or misleading result. Looking back at Equations 11.33 and 11.37 we see that both involve division by direction cosines whose values will be zero whenever the path of an energy bundle is perpendicular to the appropriate axis. This problem can be avoided by rewriting Equation 11.33

$$\frac{x - x_a}{\ell} = \frac{y - y_a}{m} = \frac{z - z_a}{n} = T \tag{11.39}$$

where the new parameter T is the distance between the point x_a, y_a, z_a and point x, y, z. Equation 11.39 constitutes three equations in four unknowns representing the path of the energy bundle:

$$x - x_a = \ell T \tag{11.40}$$

$$y - y_a = mT \tag{11.41}$$

and

$$z - z_a = nT \tag{11.42}$$

where the unknowns are x, y, z, and T. The fourth equation needed to solve for the four unknowns is the equation of the candidate surface. For example, if the surface is the plane $z = h$ then Equations 11.37 become

$$x = x_a + \ell T$$
$$y = y_a + mT \tag{11.43}$$
$$z = h$$

where

$$T = \frac{h - z_a}{n} \tag{11.44}$$

Note that n cannot be zero in Equation 11.44 because $h - z_a \neq 0$. Equations 11.43 and 11.44 are, of course, identical to Equations 11.37.

11.7.4 Determine Whether the Energy Bundle Is Absorbed or Reflected

Recall that the enclosure in the current development is composed of walls whose reflectivity has a diffuse component and a specular component; that is, whose reflectivity is given by

$$\rho = \rho^s + \rho^d \tag{11.45}$$

where ρ^s is the specular component of reflectivity and ρ^d is the diffuse component. As already discussed in Chapter 10, Equation 11.45 seems to have been first suggested by R. A. Seban in a written review of an article by Sparrow, Eckert, and Jonsson [14]. Seban's idea is illustrated in Figure 11.13. While this model is clearly inferior to an accurate bidirectional reflectivity model, it is superior to simply assuming that all surfaces are either diffuse or specular. Remember that, because the equations governing radiative exchange in an enclosure are integral equations, errors associated with inaccurate portrayals of directionality tend to average out on a global basis (although local results may have large errors).

With the introduction of Equation 11.45 the absorptivity of the walls is given by

$$\alpha = 1 - \rho^d - \rho^s \tag{11.46}$$

We interpret the absorptivity as *the probability that an energy bundle incident to a surface will be absorbed.* This probabilistic interpretation of the surface optical properties is the essence of the Monte Carlo ray-trace method. A random number R_α, uniformly distributed between zero and unity, is drawn and its value compared to that of the absorptivity α of the surface element containing the intersection of the path of the

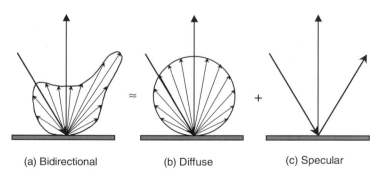

(a) Bidirectional	(b) Diffuse	(c) Specular

Figure 11.13 Approximation of (a) the bidirectional reflectivity as the sum of (b) a diffuse component and (c) a specular component

energy bundle with the enclosure wall. If $R_\alpha < \alpha$, then the energy bundle is absorbed; otherwise, it is reflected. If the energy bundle is absorbed, then the counter N_{ij} is incremented and control is returned to the point in the program where a new energy bundle is to be emitted. Otherwise, we continue to step 5.

At this juncture it is relevant to ask, How does the computer "know" the local value of absorptivity? After all, the enclosure will typically be composed of many surface elements, each generally having different surface optical properties. In general, the challenge here is to systematically search across the surface elements making up the enclosure to find the one that contains the coordinates x, y, z of the intersection of the path of the energy bundle with the walls of the enclosure. The index of this surface element is determined and then a look-up table is consulted to obtain the absorptivity. A seemingly straightforward operation, it is replete with subtleties and can be a major sink of computer resources if not approached with intelligence. The procedure, represented by step 7 in Figure 11.5, is best illustrated by example, and so its elaboration is deferred until Chapter 12.

11.7.5   If the Energy Bundle Is Reflected, Determine Whether the Reflection Is Diffuse or Specular

At this point it is necessary to introduce a new parameter, the *specularity ratio* r_s, defined as the ratio of the specular component of reflectivity to the overall reflectivity,

$$r_s = \frac{\rho^s}{\rho} = \frac{\rho^s}{\rho^s + \rho^d} \tag{11.47}$$

This specularity ratio may be interpreted as the probability that a reflection from the surface in question is specular. A random number R_r, uniformly distributed between zero and unity, is drawn and its value compared to the value of the specularity ratio to determine the type of reflection, diffuse or specular. If $R_r < r_s$, the reflection is specular; otherwise, it is diffuse. If the reflection is specular, we continue to step 6s; otherwise we continue to step 6d.

11.7.6(s)   Determine the Direction of the Specular Reflection

A reflection is said to be specular, or "mirrorlike," if its individual rays conform to two rules:

1. The incident ray, the reflected ray, and the surface normal all lie in the same plane.
2. The angle between the reflected ray and the surface normal is the same as the angle between the incident ray and the surface normal.

These two rules, which are illustrated in Figure 11.14, may be combined to yield

$$\mathbf{V}_r = \mathbf{V}_i - 2(\mathbf{V}_i \cdot \mathbf{n})\mathbf{n} = \mathbf{V}_i + 2(|\mathbf{V}_i| \cos\theta)\,\mathbf{n} \tag{11.48}$$

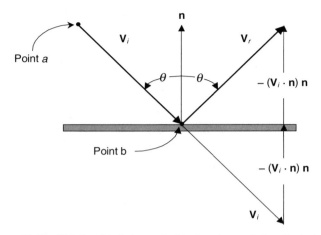

Figure 11.14   Relationship between incident and specularly reflected rays

where

$$\mathbf{V}_i = (x_b - x_a)\mathbf{i} + (y_b - y_a)\mathbf{j} + (z_b - z_a)\mathbf{k} \tag{11.49}$$

Note that $\mathbf{V}_r$ and $\mathbf{V}_i$ are *not* unit vectors.

In Equation 11.49, x_b, y_b, z_b are the coordinates of the point where the energy bundle intersects the receiving surface and x_a, y_a, z_a are the coordinates of the point on the source surface. Then, noting that

$$T_r = \sqrt{(V_{r,x})^2 + (V_{r,y})^2 + (V_{r,z})^2} \tag{11.50}$$

we write the direction cosines of the reflected ray

$$\ell_r = \frac{V_{r,x}}{T_r} \tag{11.51}$$

$$m_r = \frac{V_{r,y}}{T_r} \tag{11.52}$$

and

$$n_r = \frac{V_{r,z}}{T_r} \tag{11.53}$$

Once again, we see that the physical length T_r cannot be zero; that is, the reflected ray must go somewhere.

11.7.6(d) Determine the Direction of the Diffuse Reflection

It has already been noted that a diffuse reflection is exactly like a diffuse emission. Therefore, the logic for determining the direction of a diffuse reflection is exactly as described in step 2.

11.8 COMPUTATION OF THE ESTIMATE OF THE DISTRIBUTION FACTOR MATRIX

The elements of the distribution factor matrix, D'_{ij}, are estimated as

$$D'_{ij} \cong \frac{N_{ij}}{N_i} \tag{11.54}$$

where N_{ij} is the number of energy bundles emitted from surface element i and absorbed by surface element j, and N_i is the number of energy bundles emitted from surface element i. In the limit as N_i approaches infinity the estimate becomes exact. In practice, a modest number of energy bundles, on the order of a few hundreds of thousands per surface element, produces high-accuracy estimates of the distribution factors.

This completes the rather detailed description of the logic, represented by the block diagram of Figure 11.5, for obtaining the total, diffuse–specular radiation distribution factor. In the next chapter the logic of the current chapter for the case of a total, diffuse–specular analysis is applied in an extended example. However, first we consider how the total, diffuse–specular radiation distribution factor is used in radiation heat transfer modeling.

11.9 USE OF THE TOTAL, DIFFUSE–SPECULAR RADIATION DISTRIBUTION FACTOR FOR THE CASE OF SPECIFIED SURFACE TEMPERATURES

First we consider the special case where all of the surface elements making up the enclosure, shown in Figure 11.15, have specified temperatures. Note that the view of some surface elements by others is partially or even totally blocked, a situation not easily handled by the net exchange method because of the difficulty in obtaining the required exchange factors. Referring to Equation 11.3, we write for the radiation heat flux absorbed by surface element i,

$$q_{i,a} = \frac{Q_{i,a}}{A_i} = \frac{1}{A_i} \sum_{j=1}^{n} \varepsilon_j A_j \sigma T_j^4 D'_{ji} = \varepsilon_i \sum_{j=1}^{n} \sigma T_j^4 D'_{ij} \tag{11.55}$$

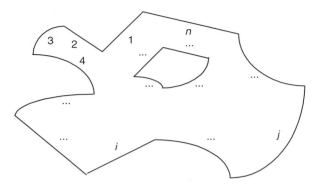

Figure 11.15 A general enclosure whose surface elements are partially obscured from each other by blockage

where reciprocity for the total, diffuse–specular distribution factor, Equation 11.5, has been used to eliminate A_i. The radiation heat flux emitted by surface element i is

$$q_{i,e} = \frac{Q_{i,e}}{A_i} = \frac{\varepsilon_i A_i \sigma T_i^4}{A_i} = \varepsilon_i \sigma T_i^4 \tag{11.56}$$

Then the *net* heat flux *from* surface element i is

$$q_i = q_{i,e} - q_{i,a} = \varepsilon_i \sigma T_i^4 - \varepsilon_i \sum_{j=1}^{n} \sigma T_j^4 D_{ij}' \tag{11.57}$$

or, with the introduction of the Kronecker delta,

$$\delta_{ij} = \begin{cases} 1, & i = j \\ 0, & i \neq j \end{cases} \tag{11.58}$$

Equation 11.57 can be written

$$q_i = \varepsilon_i \sum_{j=1}^{n} \sigma T_j^4 (\delta_{ij} - D_{ij}') \tag{11.59}$$

Equation 11.59 is an *explicit* expression for the net heat flux from all of the surfaces of the enclosure in terms of the specified surface temperatures. In contrast to the net exchange formulation, there is no matrix to invert. Best of all, the distribution factors can be estimated to almost any degree of desired accuracy regardless of the geometry. In fact, herein lies one of the capital advantages of the MCRT method: for preliminary studies relatively low-accuracy distribution factors can be obtained quickly and inexpensively; then, as the design matures, increasingly accurate distribution factors can be obtained and used at justifiably greater cost. We will also see in Chapter 15 that the MCRT method is capable of providing statistically meaningful estimates of the accuracy actually obtained. In most cases, and especially those involving curved surface elements having a specular component of reflectivity, the distribution factors are significantly easier to obtain than the corresponding exchange factors.

11.10   USE OF THE TOTAL, DIFFUSE–SPECULAR RADIATION DISTRIBUTION FACTOR FOR THE CASE OF SOME SPECIFIED SURFACE NET HEAT FLUXES

We now address the situation where surface elements $1, 2, \ldots, N$ have specified net heat fluxes and surfaces $N + 1, N + 2, \ldots, n$ have specified temperatures. To begin, let us consider the special case where $N = 1$; that is, where only the first of n surface elements has a specified net heat flux, with the remaining surfaces having specified temperatures. In this case Equation 11.59 may be written

$$q_1 = \varepsilon_1 \left[(1 - D'_{11}) \sigma T_1^4 - \sum_{j=2}^{n} \sigma T_j^4 D'_{1j} \right] \tag{11.60}$$

We immediately see that Equation 11.60 can be solved explicitly for the unknown surface temperature T_1 in terms of the known surface net heat flux q_1 and the known surface temperatures; that is,

$$\sigma T_1^4 = \frac{1}{1 - D'_{11}} \left(\frac{q_1}{\varepsilon_1} + \sum_{j=2}^{n} \sigma T_j^4 D'_{1j} \right) \tag{11.61}$$

In the more general case where several surface elements have specified net heat fluxes we can rearrange Equation 11.59 to obtain

$$q_i - \varepsilon_i \sum_{j=N+1}^{n} \sigma T_j^4 (\delta_{ij} - D'_{ij}) = \varepsilon_i \sum_{j=1}^{N} \sigma T_j^4 (\delta_{ij} - D'_{ij}), \qquad 1 \leq i \leq N \tag{11.62}$$

Equation 11.62 represents N equations in the N unknown surface temperatures in terms of the N known surface net heat fluxes and the $n - N$ known surface temperatures. It can be rewritten symbolically as

$$\Theta_i = \Psi_{ij} \Omega_j, \qquad 1 \leq i \leq N \tag{11.63}$$

where

$$\Theta_i \equiv q_i - \varepsilon_i \sum_{j=N+1}^{n} \sigma T_j^4 (\delta_{ij} - D'_{ij}), \qquad 1 \leq i \leq N \tag{11.64}$$

is a known vector,

$$\Psi_{ij} = \varepsilon_i \sum_{j=1}^{N} (\delta_{ij} - D'_{ij}), \qquad 1 \le i \le N \qquad (11.65)$$

is a known matrix, and

$$\Omega_j = \sigma T_j^4, \qquad 1 \le j \le N \qquad (11.66)$$

is an unknown vector whose elements are sought. We obtain the unknown surface temperatures by inverting the matrix defined by Equation 11.65 and then using it to operate on the vector defined by Equation 11.64, that is,

$$\Omega_i = [\Psi_{ij}]^{-1} \Theta_j, \qquad 1 \le i \le N \qquad (11.67)$$

Finally, the unknown surface heat fluxes are computed using Equation 11.59 applied over the range $N + 1 \le i \le n$.

   It is important to note here that the order of the square matrix that we need to invert in this case is only N rather than n. Recall that in the net exchange method we would have had to invert a matrix of order n ($\ge N$) regardless of the number of surfaces whose net heat fluxes were specified.

Team Projects

TP 11.1   A "quadric" is any surface whose equation is given by

$$Ax^2 + By^2 + Cz^2 + Dxy + Exz + Fyz + Gx + Hy + Iz + J = 0 \qquad (11.68)$$

For example, if $A = B = C = 1.0$, $D = E = F = G = H = I = 0$, and $J = -R^2$, Equation 11.68 becomes the equation of a sphere with radius R; and if $A = B = 1$, $C = -\tan^2\theta$, and $D = E = F = G = H = I = J = 0$, Equation 11.68 becomes the equation for a conical surface whose axis of symmetry is the z axis and whose cone half-angle is θ. Equation 11.68 is also capable of representing plane, cylindrical, paraboloidal and ellipsoidal surfaces. Write a computer subroutine that, when called with values for A, B, C, D, E, F, G, H, I, and J and values of x_a, y_a, z_a, ℓ, m, and n from Equation 11.39, returns values of x, y, z, and T. Where would such a subroutine fit into the block diagram of Figure 11.5?

TP 11.2   Set up a spreadsheet to provide plots of quadric surfaces over specified ranges of x, y, and z. Then modify the spreadsheet to show lines representing the rays of a ray trace intersecting the quadric surface. This will require that points of intersection between the line and the surface be computed. The line on the plot will then be defined by connecting the point of origin of the ray with the point(s) of intersection with the quadric.

Discussion Points

DP 11.1 Several situations are identified in Section 11.1 in which the net exchange method might not be adequate, either because the underlying assumptions are not strictly valid or because greater accuracy is needed. Can you think of additional situations in which the net exchange method might not be adequate?

DP 11.2 In Section 11.3 the concept of a figurative "image" is introduced to suggest the degree of detail required to obtain a given accuracy in a radiation heat transfer analysis. Discuss the relationship between the "sharpness" of this figurative image and the spatial and spectral resolution of the heat transfer solution obtained.

DP 11.3 What do you believe is the "point" of Section 11.3? What is the key result or conclusion of this section? Put another way, what is the most important sentence in the section?

DP 11.4 Compare the word definition of the configuration factor (given in Chapter 8) with the word definition of the total radiation distribution factor given in this chapter. Parse the two definitions word by word and carefully point out the differences.

DP 11.5 There is a sense in which the radiation distribution factor represents a *sensitivity,* or a partial derivative, as indicated in Equation 11.3a. Does this suggest an independent application of the radiation distribution factor beyond its use as a tool for computing net heat flux distributions from specified temperature distributions?

DP 11.6 The reciprocity relation for radiation distribution factors, Equation 11.5, is similar to the reciprocity relation for configuration factors, Equation 8.62. Why does the emissivity show up in the former relation but not in the latter?

DP 11.7 Equation 11.7 suggests an alternative formulation to the one, based on Equation 11.3, that is elaborated in Sections 11.8 and 11.10. Suppose the quantity Q_o in Equation 11.7 were defined as the total, diffusely distributed heat flux from surface o. How might the formalism in Sections 11.8 and 11.10 be modified to accommodate this alternative definition of the radiation distribution factor?

DP 11.8 Experience has shown that an improved estimate of the distribution factor matrix D'_{ij} for a fixed meshing of the enclosure walls does not always improve the accuracy of the heat transfer results obtained and, in fact, can degrade the accuracy of the results. In other words, tracing more rays without increasing the number of surface elements defining the enclosure does not automatically improve the accuracy of the model and may even degrade it. Why do you believe this is so?

DP 11.9 Referring to step 1 in the logic block diagram of Figure 11.5, how would you find a random emission location r, θ on a plane circular-disk surface element? Test your proposed strategy by implementing it in a spreadsheet environment and making a scattergram involving 1000 random points. Does the distribution appear to be uniform?

DP 11.10 Without actually deriving the equation, where do you believe the sine function comes from in Equation 11.11? You might study this question by implementing Equation 11.11 in a spreadsheet environment and creating a bar graph of the first 1000 values of θ returned. Compare that result with $\theta = \pi R_\theta / 2$. What do you conclude?

DP 11.11 In Section 11.7.2 it is stated that "to maintain control of the statistics of the random number sequence used, the tangent vector should be consistently oriented within a given surface built up from a collection of surface elements." What does it mean to "maintain control of the statistics of the random number sequence"?

DP 11.12 In the second sentence under Equation 11.23 it is stated that the direction cosines of the local surface normal are known "because it is the user who has oriented the surface element." What is meant by this? [*Hint*: Refer to Equations 11.12 through 11.17 and the related text.]

DP 11.13 How would you actually implement the test described in the full paragraph between Equations 11.25 and 11.26 in, say, a FORTRAN or C environment?

DP 11.14 Supply the formal mathematical steps going from Equations 11.27 through 11.29 to Equations 11.30 through 11.32.

DP 11.15 How would you actually implement the logic of the paragraph numbered 1 in Section 11.7.3 in, say, a FORTRAN or C environment?

DP 11.16 Draw some sketches that illustrate the principle behind the logic of the paragraph numbered 2 in Section 11.7.3. Use them to explain the principle to your classmates.

DP 11.17 Taking into consideration the paragraph below the paragraph numbered 3 in Section 11.7.3, what metric would you use to implement the logic described in paragraph 3? That is, which variable would you use and how would you use it?

DP 11.18 Write a short piece of computer code (a few lines should suffice) to implement step 4 described in Section 11.7.4. Be prepared to share the result with your classmates.

DP 11.19 Show formally using vector analysis that Equations 11.48 and 11.49 conform to the first rule for specular reflection.

DP 11.20 Suppose two surfaces of an enclosure had specified net heat fluxes and the remaining surfaces had specified temperatures. How would Equations

11.60 and 11.61 change? In other words, derive the equations for the temperatures of the two surfaces, say surfaces 1 and 2 out of n, having specified net heat fluxes.

DP 11.21 The point made in the final paragraph of the chapter is very important. Do you see why a square matrix of order only $N < n$ rather than n itself must be inverted (see Discussion Point DP 11.20)?

Problems

P 11.1 The entropy of a system may be defined as $S = -k \ln(P)$, where k is Boltzmann's constant $(= 1.380 \times 10^{-23}$ J/K) and P is the probability that the system will have a given configuration. Consider the box illustrated in Figure 11.2. Relative to the hypothesis that two marbles are in the left-hand side of the box and one is in the right-hand side, compute the change in entropy of the system ΔS when the partition is removed. Discuss the sign of the change.

P 11.2 Derive Equation 11.4 starting with Equation 11.3.

P 11.3 Derive Equation 11.5.

P 11.4 Derive Equation 11.6 from Equations 11.4 and 11.5.

P 11.5 Derive Equation 11.11.

P 11.6 Verify Equation 11.25.

P 11.7 Derive Equation 11.37.

P 11.8 Find point c referred to directly below Equation 11.37.

P 11.9 Find the unit normal and tangent vectors in terms of the direction cosines ℓ, m, and n for the conical enclosure illustrated in Figure 11.11.

P 11.10 Find the unit normal and tangent vectors in terms of the direction cosines ℓ, m, and n for the curved wall of a cylindrical enclosure of radius R. Orient the z axis along the axis of the cylinder.

P 11.11 Find the unit normal and tangent vectors in terms of the direction cosines ℓ, m, and n for the parabolic reflector shown in Figure 11.16.

P 11.12 Consider the wedge-shaped cavity depicted in Figure 11.17. Show that the equation for the sloped surface is

$$S(y, z) = z - \frac{y_{max} - y}{\tan \alpha} = 0 \tag{11.69}$$

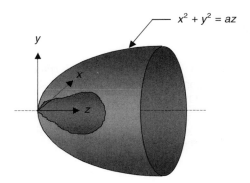

Figure 11.16 The parabolic reflector of Problem P 11.11

P 11.13 For the wedge-shaped cavity of Figure 11.17, show that the expression for the unit normal vector to the sloped surface is (see Problem P 11.12)

$$\mathbf{n} \equiv -\frac{\nabla S}{|\nabla S|} = \frac{-\left(\dfrac{1}{\tan \alpha}\mathbf{j} + \mathbf{k}\right)}{\sqrt{\left(\dfrac{1}{\tan \alpha}\right)^2 + 1}} = \frac{-1}{\sqrt{1 + \tan^2 \alpha}}\mathbf{j} - \frac{\tan \alpha}{\sqrt{1 + \tan^2 \alpha}}\mathbf{k} \quad (11.70)$$

P 11.14 For the wedge-shaped cavity of Figure 11.17, show that the expression for the first unit tangent vector to the sloped surface is (see Problem P 11.12)

$$\mathbf{t}_1 \equiv \frac{\mathbf{n} \times \mathbf{i}}{|\mathbf{n} \times \mathbf{i}|} = -\frac{\tan \alpha}{\sqrt{1 + \tan^2 \alpha}}\mathbf{j} + \frac{1}{\sqrt{1 + \tan^2 \alpha}}\mathbf{k} \quad (11.71)$$

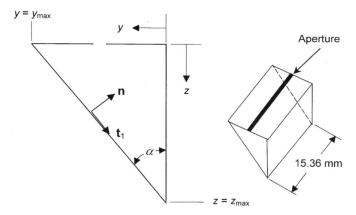

Figure 11.17 Illustration of a wedge-shaped cavity showing unit normal and tangent vectors on one of the surfaces (Problems P 11.12 through P 11.16)

P 11.15 Suppose for the geometry illustrated in Figure 11.17 with $\alpha = 40$ deg, Equations 11.10 and 11.11 yield the angles $\phi = 102.4$ deg and $\theta = 34.6$ deg for a certain diffuse reflection. What are the direction cosines of the path of the reflected energy bundle?

P 11.16 Suppose the diffuse reflection described in Problem P 11.15 originated from the center of the sloped surface (in both the x and y directions) for a cavity having a wedge angle of 40 deg and $y_{max} = 180.5$ μm. Which surface of the cavity does the reflected energy bundle strike (show on a scaled drawing)? What are the coordinates of this point?

REFERENCES

1. Nelson, E. L., J. R. Mahan, L. D. Birckelbaw, J. A. Turk, D. A. Wardwell, and C. E. Hange, *Temperature, Pressure, and Infrared Image Survey of an Axisymmetric Heated Exhaust Plume,* National Aeronautics and Space Administration Technical Memorandum 110382, February 1996.

2. Howell, J. R., and M. Perlmutter, "Monte Carlo simulation of thermal transfer through radiant media between grey walls," *ASME Transactions, Series C, Journal of Heat Transfer,* Vol. 86, February 1964, pp. 116–122.

3. Howell, J. R., and M. Perlmutter, "Monte Carlo solution of radiant heat transfer in a nongrey, nonisothermal gas with temperature dependent properties," *American Institute of Chemical Engineers Journal,* Vol. 10, July 1964, pp. 562–567.

4. Perlmutter, M., and J. R. Howell, "Radiant transfer through a grey gas between concentric cylinders using Monte Carlo," *ASME Transactions, Series C, Journal of Heat Transfer,* Vol. 86, May 1964, pp. 169–171.

5. Corlett, R. C., "Direct Monte Carlo calculation of radiative heat transfer in vacuum," *ASME Transactions, Series C, Journal of Heat Transfer,* Vol. 88, November 1966, pp. 176–182.

6. Sparrow, E. M., and R. D. Cess, *Radiation Heat Transfer,* Wadsworth Publishing Company, Inc., Brooks/Cole Publishing Company, Belmont, CA, 1966.

7. Howell, J. R., "Application of Monte Carlo to heat transfer problems," *Advances in Heat Transfer,* Vol. 5, Academic Press, New York, 1968, pp. 1–54.

8. Toor, J. S., and R. Viskanta, "A numerical experiment of radiant heat interchange by the Monte Carlo method," *International Journal of Heat and Mass Transfer,* Vol. 11, May 1968, pp. 883–897.

9. Haji-Sheikh, A., and E. M. Sparrow, "Probability distributions and error estimates for Monte Carlo solutions of radiation problems," *Progress in Heat and Mass Transfer,* Vol. 2, Pergamon Press, Oxford, 1969, pp. 1–11.

10. Mahan, J. R., and L. D. Eskin, "Application of Monte Carlo techniques to transient thermal modeling of cavity radiometers having diffuse-specular surfaces," 5th American Meteorological Society Conference on Atmospheric Radiation, Toronto, Ontario, Canada, 16–18, June 1981.

11. Nevárez-Ayala, F. J., *A Monte Carlo Ray-Trace Environment for Optical and Thermal Radiative Design,* PhD Dissertation, Department of Mechanical Engineering, Virginia Polytechnic Institute and State University, Blacksburg, VA, Spring, 2002.

12. Tribus, M., *Thermostatics and Thermodynamics,* Van Nostrand, New York, 1961.

13. Gebhart, B., *Heat Transfer,* 2nd ed., McGraw-Hill, New York, 1971, p. 158.

14. Mahan, J. R., and L. D. Eskin, "The radiation distribution factor—its calculation using Monte Carlo techniques and an example of its application," First UK National Heat Transfer Conference, Leeds, Yorkshire, England, 4–6 July 1984, EFCE Publication Series No. 39, Vol. 2, pp. 1001–1012.

15. Maltby, J. D., and P. J. Burns, "Performance, accuracy, and convergence in a three-dimensional Monte Carlo radiative heat transfer simulation," *Numerical Heat Transfer,* Part B, Vol. 19, 1991, pp. 191–209.

16. Sparrow, E. M., E. R. G. Eckert, and V. K. Jonsson, "An enclosure theory for radiative exchange between specularly and diffusely reflecting surfaces," *ASME Transactions, Series C, Journal of Heat Transfer,* Vol. 84, No. 4, November 1962, pp. 294–300.

12

THE MCRT METHOD FOR DIFFUSE–SPECULAR, GRAY ENCLOSURES: AN EXTENDED EXAMPLE

In Chapter 11 we defined the radiation distribution factor and introduced the Monte Carlo ray-trace (MCRT) method as a means for estimating its value. In this chapter a detailed numerical example is presented for the important special case of a diffuse–specular, gray enclosure. Results obtained using the MCRT method in this application are essentially equivalent to those that would be obtained using the net exchange method extended to diffuse–specular enclosures, as described in Chapter 10. However, the MCRT method accomplishes this with four distinct advantages: (1) it is easier to obtain the required "factors," (2) the resulting equations of exchange are solved directly without inverting a matrix, (3) a statistically meaningful estimate of the accuracy of the results is available, and (4) the user can trade computer time against accuracy in virtually any desired proportion by changing the number of rays traced.

12.1 DESCRIPTION OF THE PROBLEM

Consider the notional inverted-cone thermal radiation detector shown in Figure 12.1. It consists of a 1.0-cm-diameter cylindrical cavity open at the top and closed at the bottom by an inverted cone whose height is the same as the cylinder, 1.0 cm. The walls of the cylinder consist of a thin-film heat flux gauge coated with a thin absorber layer, and the conical surface is a mirror whose reflectivity is $\rho = \rho^d + \rho^s = 0.04 + 0.92$. That is, 92 percent of the incident radiation is reflected specularly from the mirror, 4 percent is reflected diffusely, and the remaining 4 percent is absorbed. The absorptive coating on the thin-film heat flux gauge has a hemispherical, total ab-

339

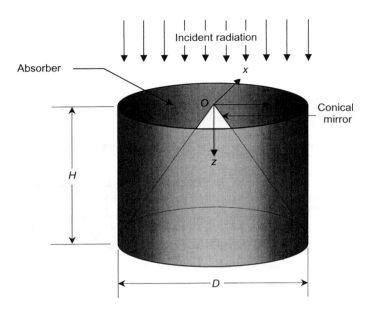

Figure 12.1 A notional cavity-type thermal radiation detector

sorptivity α of 0.9 and a specularity ratio r_s of 0.12. These values of reflectivity for the conical mirror and absorptivity and specularity ratio for the absorber layer are representative of those for currently available materials. They have been chosen, along with the shape of the cavity, to ensure that most radiation incident to the detector opening will suffer multiple incidences on the absorber layer and thus will have a high probability of being absorbed and therefore "counted."

The detector illustrated in Figure 12.1 would normally be illuminated by radiation gathered by a collimating optical system. In principle, if the conical mirror were perfectly specular, all collimated thermal radiation entering on axis through the opening, or aperture, would be reflected from the conical mirror onto the cylindrical surface. Then if the absorber layer on the conical surface was perfectly absorbing (black), all of the radiation incident to the detector aperture would be absorbed and sensed by the heat flux sensor on the cylindrical surface. However, no mirror is perfectly specular (or even perfectly reflecting for that matter) and no practical absorber is perfectly black. Some of the radiation reflected from the best mirror will be diffusely reflected—4 percent in the case at hand—and some will be absorbed. Further, not all of the radiation incident to the absorber layer will be absorbed—only 90 percent in the case at hand—and at least some of the reflected radiation will be diffusely distributed—10 percent in this case. Still, because of the *cavity effect* the overall performance of the detector will be considerably superior to that of a plane thermal radiation detector coated with the same absorptive layer. Specifically, multiple reflections between the conical mirror and the cylindrical walls will produce an *effective,* or *apparent, absorptivity* α_e [1] of the detector well above 90 percent. Diffuse reflections will degrade the performance of the detector because they will reduce the average number of reflections that an energy bundle will suffer before escaping back through the aperture.

12.2 GOALS OF THE ANALYSIS

A Monte Carlo ray-trace analysis of the notional thermal radiation detector might have the following two goals:

1. Estimate the effective absorptivity α_e of the detector. The effective absorptivity is defined as the fraction of (collimated in this case) radiation incident to the detector aperture that is ultimately absorbed *on the cylindrical walls*. Note that radiation absorbed on the mirror surface is lost and therefore not included when computing the effective absorptivity of the detector.
2. Estimate the thermal noise of the detector. That is, estimate the heat flux to (or from) the cylindrical wall due to the temperatures of the cylindrical wall and the conical mirror, and due to "out-of-field" (i.e., noncollimated) radiation entering the detector through the aperture.

The verb "estimate" used in the above statement of the goals should not be interpreted to mean that the values obtained would be relatively poor approximations of reality. To the contrary, the Monte Carlo ray-trace method offers the best available means of obtaining high accuracy. We use the word "estimate" to imply that the user has the possibility of specifying *in advance of the analysis* the accuracy of the results to be obtained. For the problem at hand an accuracy of one-half percent is a reasonable goal, limited mainly by the accuracy with which the surface properties are typically known. It is a curious and useful property of cavities that we can estimate their effective absorptivity with greater accuracy than we know the surface properties themselves (see Discussion Point DP 12.2)!

Pursuit of the first goal does not require subdivision of the cavity into surface elements. However, to study self-contamination of the detector—that is, radiation emitted from the detector walls that is ultimately absorbed on the cylindrical walls—we must subdivide the cavity into surface elements whose temperatures and heat fluxes can, at least in principle, be different from one another. That is, there is no a priori reason to think that the cavity will be isothermal or have a uniform net wall heat flux distribution. It is noted that a calculation of the thermal condition on the cavity surface would in general require that heat conduction in the cavity walls also be considered, but this is beyond the scope of this book. However, we could reasonably assume that the conical mirror is thin-walled and insulated so that the net heat flux from it is zero, in which case the temperature distribution on its surface would be determined entirely by a radiative heat balance. Further, we could reasonably assume that the cylindrical wall is maintained at a uniform temperature, for example, by actively controlled heating of the substrate upon which the thin-film heat flux gauge is mounted, so that its net heat flux distribution would be determined by a radiation balance. These are the assumptions that should be used in Team Project TP 12.1.

12.3 SUBDIVISION OF THE CAVITY WALLS INTO SURFACE ELEMENTS

Subdivision of the cavity walls into surface elements must be done respecting the desired spatial resolution and accuracy of the result to be obtained. For the moment we will defer elaboration of this point and assume that application of a suitable criterion has led us to divide the cavity walls into surface elements about 1.0 mm on a side. However, before continuing with this example it is useful to consider the following aside.

It is of overriding importance that to the extent possible the surface elements of an enclosure have approximately the same surface area. This assures that, on the average, all surface elements have an a priori equal opportunity to be visited by an energy bundle during the MCRT analysis. For example, if one surface element had 10 times the surface area as another, the former would have an a priori probability of absorbing 10 times the number of energy bundles as the latter. The term "a priori probability" has a special meaning here: it means that, if the geometrical effects of location and orientation and the effect of surface absorptivity are removed from consideration, then the probability that a given surface element will absorb an energy bundle emitted by another surface element is proportional to its surface area.

When the surface elements of an enclosure have widely varying areas, then for the ones with the smallest areas to absorb a statistically significant number of energy bundles, the largest ones will be greatly oversampled. The consequence of this is that the distribution factors to the smallest surface elements will be estimated with much less statistical certainty than will be the distribution factors to the largest surface elements. This will effectively invalidate the statistical treatment, developed in Chapter 15, which leads to a statement of the accuracy of the analysis within a given confidence interval.

Consider a ray trace from a given source surface element in which one receiver surface element, because of its small size relative to other receiver surface elements or because of its relatively low absorptivity, absorbs orders of magnitude fewer energy bundles than do the other receiver surface elements. Then the precision of the estimate of the distribution factor to that receiver surface element will be much more sensitive to the total number of energy bundles traced than in the case of the other receiver surface elements. For example, if one surface element absorbs 100 energy bundles while another absorbs 10,000 energy bundles in a ray trace involving one million emitted energy bundles, then the distribution factor estimate for the former is 0.000100, while the estimate for the latter is 0.010000. This means that a change of one absorbed energy bundle leads to a 1 percent change in the estimate of the distribution factor for the former surface but to a change of only 0.01 percent for the latter.

In some cases it is not practical for all surface elements of an enclosure to have more or less the same area. For example, an enclosure may involve some inherently small surfaces whose size, if divided into the total enclosure surface area, would lead to an excessively large number of surface elements. However, in such cases it is still possible to achieve the goal of estimating all of the distribution factors to the same precision. This is accomplished by using the reciprocity principle, Equation 11.5. For example, if surface element i has a much larger area than surface element j, then, upon completion of the Monte Carlo ray trace, the distribution factor from surface i to surface j, Equation 11.54, is replaced by

$$D'_{ij} = \frac{\varepsilon_j A_j}{\varepsilon_i A_i} D'_{ji} \tag{12.1}$$

Equation 12.1 will yield approximately the same value for the distribution factor estimate as that obtained by the original ray trace, but with a precision comparable to that for the estimate of D'_{ji}. It is important to note here that if we use Equation 12.1 to estimate some of the distribution factors for an enclosure, then we forfeit the possibility of using Equation 11.6 to estimate the global precision of the distribution factor estimates.

Finally, it should be emphasized that, although the discussion leading up to Equation 12.1 explicitly addresses the problem of widely varying surface element areas, it also implicitly addresses the similar problem of widely differing surface emissivities. In fact, the form of Equation 12.1 makes it clear that our goal should really be to size the surface elements such that they all have more or less the same value of the product of surface area and emissivity. This may be an impractical goal to achieve for a variety of reasons, in which case Equation 12.1 may be used whenever the product $\varepsilon_j A_j$ is much less than the product $\varepsilon_i A_i$.

Returning to the example at hand, we have decided to subdivide the enclosure of Figure 12.1 into surface elements approximately 1.0 mm on a side. This can be done almost exactly on the cylindrical surface, as shown in Figure 12.2. The height of the cylindrical wall is 10 mm, so we simply divide it vertically into 10 equal segments by passing a series of parallel cutting planes normal to the z axis with 1.0-mm spacing. The circumference of the cylindrical wall is $10\pi \cong 31.4$ mm, so we could divide it into either 31 or 32 circumferential segments, yielding segment (arc) widths of either $1.01+$ mm or $0.98+$ mm, respectively. It is preferable to use an even number of segments because this creates a plane of symmetry that divides the cavity into two identical subcavities. In fact, if 32 divisions are used, the cavity can be divided into four identical subcavities by using two orthogonal planes of symmetry. The creation of identical subcavities based on symmetry permits the numerical effort to be greatly reduced.

Division of the conical surface into 1.0-by-1.0-mm-square, equal-area surface elements is more problematic. It is convenient to divide the conical surface up into the same number of circumferential divisions as the cylindrical surface. We then divide it into a number of axial divisions that produces equal-area segments having a surface area as close to 1.0 mm^2 as possible. The number of axial divi-

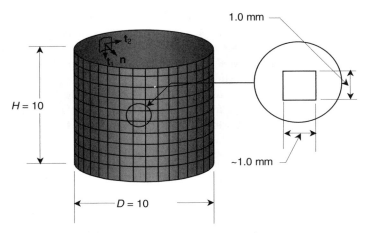

Figure 12.2 Subdivision of the (inside) cylindrical wall of the cavity in Figure 12.1 into equal-area surface elements

sions is then

$$\frac{A_{\mathrm{cone}}}{32} = \frac{\pi(D/2)\sqrt{(D/2)^2 + H^2}}{32} = \frac{\pi(5)\sqrt{5^2 + 10^2}}{32} \cong 5.49 \quad (12.2)$$

If we choose to divide the cone into five axial divisions, then each surface element will have an area of about 1.1 mm^2, and if we divide it into six axial divisions each surface element will have an area of about 0.9 mm^2. Bearing in mind that the surface elements on the cylinder have an area of about 0.98 mm^2, we choose to divide the cone into six axial segments, as shown in Figure 12.3.

In Figure 12.3 we note that the axial locations of the six cutting planes that define the axial divisions on the conical surface are not equally spaced. This is due to the fact that the average width of each surface element on the cone decreases the closer the surface element is to the apex of the cone. The area of a conic surface element bounded in the axial direction by z_a and z_b and in the circumferential direction by $\Delta\phi$ is

$$A_c = \frac{\Delta\phi}{2}\left(R_b\sqrt{R_b^2 + z_b^2} - R_a\sqrt{R_a^2 + z_a^2}\right) \quad (12.3)$$

where R_a and R_b are the radii of the cone at z_a and z_b, respectively. Then, if H is the height of the conical surface and D is the diameter of its base, we have

$$A_c = \left[\frac{D\,\Delta\phi}{4H}\sqrt{\left(\frac{D}{2H}\right)^2 + 1}\,\right](z_b^2 - z_a^2) \cong 0.05488\,(z_b^2 - z_a^2) \quad (\mathrm{mm}^2) \quad (12.4)$$

where the constant 0.05488 is the approximate value of the lead factor for the current example. Note that in an actual MCRT analysis constants such as this would be computed and used with the highest precision available for the computer used.

Then, assuming that the conical surface, whose total surface area is given by the numerator in Equation 12.2, is to be subdivided into six equal-area axial bands, the area of each surface element on the cone must be

$$A_c = \frac{A_{cone}}{(6)(32)} = \frac{\pi(5)\sqrt{5^2 + 10^2}}{(6)(32)} \cong 0.9147 \text{ mm}^2 \qquad (12.5)$$

The values of z_a, z_b, z_c, z_d, and z_e can then be computed as

$$z_a = \sqrt{\frac{0.9127}{0.05488}} \cong 4.08 \text{ mm} \qquad (12.6)$$

$$z_b = \sqrt{\frac{0.9127}{0.05488} + z_a^2} \cong \sqrt{2}z_a \cong 5.77 \text{ mm} \qquad (12.7)$$

$$z_c = \sqrt{\frac{0.9127}{0.05488} + z_b^2} \cong \sqrt{3}z_a \cong 7.06 \text{ mm} \qquad (12.8)$$

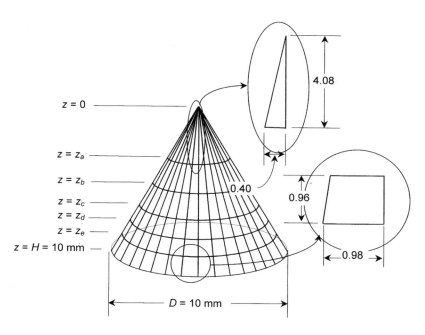

Figure 12.3 Subdivision of the conical mirror into equal-area elements whose area is approximately the same as on the cylindrical wall

$$z_d = \sqrt{\frac{0.9127}{0.05488} + z_c^2} \cong \sqrt{4}z_a \cong 8.16 \text{ mm} \qquad (12.9)$$

and

$$z_e = \sqrt{\frac{0.9127}{0.05488} + z_d^2} \cong \sqrt{5}z_a \cong 9.12 \text{ mm} \qquad (12.10)$$

We immediately see that the spacing of the axial divisions on the conical surface is very nonlinear, and that only the bottom two rings of surface elements are approximately square. In fact, the surface elements in the top ring are quasi-triangular with a height-to-base ratio of nearly ten. Thus, we see that in the case of some surfaces (segments of cones, spheres, circles, and so forth) it is not possible to have both equal-area surface elements and essentially square surface elements. Often, to gain the statistical balance resulting from having equal-area surface elements, we must make sacrifices in spatial resolution. For example, in the current analysis we find that the axial resolution near the tip of the cone is not as fine as near the base (although the circumferential resolution is better there). In the competition between the need for equal surface areas and the desire for uniform spatial resolution, the former should win out.

12.4   EXECUTING THE RAY TRACE: LOCATING THE POINT OF "EMISSION"

Having subdivided the cavity into a reasonably large number (512) of quasi-equal-area surface elements, we are now ready to trace some energy bundles. Let us begin by allowing energy bundles to enter the cavity through the opening parallel to the cavity z axis. For each entering energy bundle we want its point of origin in the plane of the opening, or aperture, to be determined randomly. We accomplish this by drawing two random numbers, R_r and R_ϕ, and then computing

$$r = \frac{D}{2}\sqrt{R_r} \qquad (12.11)$$

and

$$\phi = 2\pi R_\phi \qquad (12.12)$$

Equation 12.12 is rather obvious, but derivation of Equation 12.11 is left as an exercise for the student (see Discussion Point DP 12.6). Once r and ϕ have been determined, then

$$x = r \cos\phi \qquad (12.13)$$

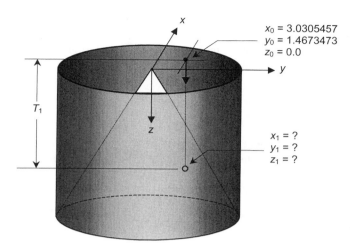

Figure 12.4 Path of an energy bundle entering the cavity through point x_0 = 3.0305457 mm, y_0 = 1.4673473 mm parallel to the cavity z axis

and

$$y = r \sin\phi \tag{12.14}$$

are the x, y coordinates of the point of "emission." For example, suppose in the problem at hand the random number pair (R_r, R_ϕ) = (0.45349261, 0.07176558) is selected. Then,

$$r = 3.3670930 \text{ mm}$$
$$\phi = 0.45091644 \, r \tag{12.15}$$

and

$$x = 3.0305457 \text{ mm}$$
$$y = 1.4673473 \text{ mm} \tag{12.16}$$

The path followed by this energy bundle as it enters the cavity is illustrated in Figure 12.4.

12.5 DETERMINE WHERE THE ENERGY BUNDLE STRIKES THE CAVITY WALLS

Where does the entering energy bundle strike the cavity walls? We know that it does not strike the cylindrical wall because it enters parallel to that wall. However, the computer does not "know" this until we tell it. Therefore, we must for the moment consider all mathematical possibilities. Recall from Chapter 11 that "back" candidates may exist, which also must be considered. The equations for the line describing the path of the energy bundle are

$$\frac{x_1 - x_0}{\ell_1} = \frac{y_1 - y_0}{m_1} = \frac{z_1 - z_0}{n_1} = T_1 \tag{12.17}$$

But we know, since the energy bundle enters parallel to the z axis, that $\ell_1 = m_1 = 0$ and $n_1 = 1.0$. Thus, Equations 12.17 become

$$x_1 = x_0 \tag{12.18}$$

$$y_1 = y_0 \tag{12.19}$$

and

$$z_1 = T_1 \tag{12.20}$$

where both z_1 and T_1 are unknown.

It is clear that we need another equation in z_1 and T_1. Available are three equations of surfaces: the equation of the x, y plane containing the cavity opening,

$$z = 0 \tag{12.21}$$

the equation of the cylindrical surface containing the cylindrical wall,

$$x^2 + y^2 - \frac{D^2}{4} = 0 \tag{12.22}$$

and the equation of the conical surface containing the conical wall of the cavity,

$$x^2 + y^2 - z^2 \tan^2\gamma = 0 \tag{12.23}$$

where γ is the half-angle of the cone,

$$\gamma = \tan^{-1}\left(\frac{D}{2H}\right) \tag{12.24}$$

We (and the computer) immediately reject the solution $z = 0$, Equation 12.21, because that is the point of origin. The computer does this by checking to see if the magnitudes of $x_1 - x_0$, $y_1 - y_0$, and $z_1 - z_0$ are all less than a predetermined tolerance δ, in which case it decides that the point x_1, y_1, z_1 is identical to the point x_0, y_0, z_0 and so rejects it. Note that it is inevitable in a ray trace involving several million rays that sooner or later a point x_1, y_1, z_1 will, in fact, lie within δ of the point x_0, y_0, z_0 without actually being identical to that point. For example, in the current problem such a situation could arise near the circle of intersection of the cylindrical wall with the conical mirror. When this occurs, an *orphan ray* is created; that is, a ray that no surface will receive. In creating our computer code we must always anticipate this possibility and compensate for it. Typically, we will have a series of logic statements designed to test the candidate points x_1, y_1, z_1 to find out which one meets the criteria of a valid "strike." This sequence must include the possibility of no valid strikes. These will be few in number and so the best way to deal with them is to return to the emitting surface and restart the trace with the next available random number. The orphan ray then appears in neither the numerator nor the denominator of the distribution factor estimates; therefore, its existence has negligible effect on the outcome.

The solution $x^2 + y^2 - D^2/4 = 0$, Equation 12.22, clearly has no intersections with the path of the energy bundle, but how does the computer "know" this? The problem is that this equation adds nothing to the search for the values of z_1 and T_1 since it contains neither of these variables. In this case, the computer code should include instructions that cause the logic flow to skip over surface subroutines for surfaces not involving the variable whose value is sought.

Finally, the equation for the third surface, Equation 12.23, is solved simultaneously with Equations 12.18 and 12.19 to obtain

$$z_1 = T_1 = \pm \sqrt{\frac{x_0^2 + x_0^2}{(D/2H)^2}} = \pm 6.7341860 \qquad (12.25)$$

Because Equation 12.25 is quadratic in z, it yields two roots, corresponding to a *forward candidate* and a *back candidate,* but which is which? Recall that the equation of a conical surface describes two branches, as illustrated in Figures 11.11 and 11.12. In the case at hand the line that includes the path of the energy bundle intersects both branches. Of the two, we know by inspection of Figure 12.4 that the intersection with the "lower" branch, corresponding to the positive root in Equation 12.25, is the forward candidate; but how does the computer know this? Recognizing that $\mathbf{n}_0 = \mathbf{k}$, the dot product

$$[(x_1 - x_0)\mathbf{i} + (y_1 - y_0)\mathbf{j} + (z_1 - z_0)\mathbf{k}] \cdot \mathbf{n}_0 = \pm 6.7341860 \qquad (12.26)$$

has a positive value using the positive root for z_1 in Equation 12.25 and a negative value when the negative root is used. Therefore, we identify the former as the forward candidate and the latter as the back candidate. Because only one forward candidate has been found using all of the available surface equations, it is the point of intersection sought. Of course, had more than one forward candidate been identified we would have selected the one yielding the smallest value of T_1.

Even though it is obvious in this case that $\mathbf{n}_0 = \mathbf{k}$, it is perhaps useful to review how we would have obtained $\mathbf{n}_0$ if it were not so obvious; after all, it might not be all that obvious to a computer incapable of inspecting a drawing. Recall Equation 11.13,

$$\mathbf{n} = \pm \frac{\nabla S}{|\nabla S|} \tag{11.13}$$

where once again $S(x, y, z) = 0$ is the equation for a surface. The "$\pm$" sign in Equation 11.13 reminds us to select the value that assures that the unit normal vector is oriented toward the interior of the cavity. In the current situation it is clear that the "+" sign is appropriate. Then for the case of the x, y plane,

$$S(x, y, z) = z = 0 \tag{12.27}$$

so that

$$\mathbf{n}_0 = \frac{\left(\dfrac{\partial}{\partial x}\mathbf{i} + \dfrac{\partial}{\partial y}\mathbf{j} + \dfrac{\partial}{\partial z}\mathbf{k}\right)z}{\left|\left(\dfrac{\partial}{\partial x}\mathbf{i} + \dfrac{\partial}{\partial y}\mathbf{j} + \dfrac{\partial}{\partial z}\mathbf{k}\right)z\right|} = \mathbf{k} \tag{12.28}$$

In summary, we now know that the first energy bundle entering the cavity strikes the walls at

$$\begin{aligned} x_1 &= 3.0305457 \text{ mm} \\ y_1 &= 1.4673473 \text{ mm} \\ z_1 &= 6.7341860 \text{ mm} \end{aligned} \tag{12.29}$$

12.6   DETERMINE THE INDEX OF THE SURFACE ELEMENT RECEIVING THE ENERGY BUNDLE

Before we can investigate whether the energy bundle is absorbed or reflected, we must know which surface element has received the energy bundle so that the correct absorptivity can be attributed to it. This naturally invokes the larger issue of numbering the surface elements and determining the index of a given surface element where an absorption event has occurred.

The details of each search to determine the surface element in which a given point lies will vary with the peculiarities of the geometry, a fact that makes this particular part of the Monte Carlo ray-trace method extremely challenging to generalize. Perhaps the most significant feature of the computer environment FELIX provided with this book is that any enclosure that can be conceived and created in a CAD environment can be imported and analyzed. If in the current example we only wanted to know whether the energy bundle intersects the conical mirror or the cylindrical absorber, we could simply determine if the radial coordinate

$$r_1 = \sqrt{x_1^2 + y_1^2} \qquad (12.30)$$

is less than or equal to the radius of the cavity, $D/2$. In the case of this first energy bundle we would deduce that the point in question lies on the conical mirror, as indicated in Figure 12.4. However, in Monte Carlo ray-trace analyses in which millions of energy bundles are traced, the point in question will occasionally be found to lie, to within the accuracy of the computer, on *both* of two (or even three) surfaces. For example, suppose the solution to Equation 12.30 was, to within the accuracy of the computer being used, $r_1 = D/2$. Note that this is the opposite of the problem, cited earlier, of the orphan ray; in the current situation we have two (or more) "homes" vying for the same "child." As before, we must anticipate this eventuality in our logic to prevent the program from "locking up" or otherwise yielding an ambiguous result. The best and most natural solution is to systematically assign the energy bundle to the first surface checked in the logic sequence that meets the criterion.

As already stated, in the case of this first energy bundle we determine that r_1 is less than $D/2$; thus, we know that the point in question lies on the conical mirror. However, we need to know in which specific surface element making up the mirror the point lies. Then if we later determine that the energy bundle is absorbed there, we can attribute the absorption event to the correct surface element index number. More to the point, it is possible that the surface optical properties of the mirror are not uniform, but rather vary from one surface element to the next. In this case the index of the surface element is needed in order to look up the local values of the surface optical properties. We might decide to number the surface elements beginning in the uppermost ring of surface elements on the conical mirror. The choice of a starting point is arbitrary, but this choice is as good as any other. We let surface element number 1 be the first element in the clockwise direction from the projection of the x axis on the cone. Then surface element number 33 is the first element on the second ring from the tip of the cone, and so forth, as illustrated in Figure 12.5.

The order of the numbering system shown in Figure 12.5 allows a single equation to be written from which the surface element index number j can be calculated; that is,

$$j = 32 \left\{ \mathrm{INT} \left[\left(\frac{z_1}{z_a} \right)^2 - \delta_1 \right] \right\} + \mathrm{INT} \left[32 \left(\frac{\phi_1}{2\pi} \right) + 1 - \delta_2 \right] \qquad (12.31)$$

The function INT[x] in Equation 12.31 returns the integer value of x; that is, if $x = 3.141592653$, INT[x] = 3. The quantities δ_1 and δ_2 are small numbers, whose values are near the smallest calculable number of the computer being used, provided to prevent Equation 12.31 from returning indices that are outside the range $1 \le j \le 192$. For example, if $z_1 = H$ (exactly), then 32 INT[$(z_1/z_2)^2$] = 192, which means that for any value of ϕ_1 the range of possible values of j will be exceeded. Similarly, when

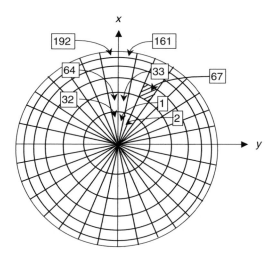

Figure 12.5 Numbering system for the surface elements on the conical surface (view looking into cavity from above)

the first term in Equation 12.31 is equal to 160, if $\phi_1 = 2\pi$ (exactly), then INT$[32(\phi_1/2\pi) + 1] = 33$ so that $j = 193$ is outside the permissible range. While the probability is low that z_1 will be exactly equal to H or that ϕ_1 will be exactly equal to 2π, when millions of energy bundles are being traced it is bound to happen sooner or later. When it does finally happen, if the constants δ_1 and δ_2 are zero an error in placement of the energy bundle will occur. Finally, if $j = 0$ we must set $j = 1$. In the case at hand we get $j = 64 + 3 = 67$. This surface element is identified in Figure 12.5.

Note that the method described above for determining the index j of the receiving surface element is a two-step process: first the receiving major surface (disk, cone or cylinder in this case) is identified and then the index number of the surface element on the receiving major surface (cone in this case) is determined. One-step processes are imaginable, especially in certain classes of enclosures. However, experience has shown that the two-step process described above is usually more practical.

12.7 DETERMINE IF THE ENERGY BUNDLE IS ABSORBED OR REFLECTED

Once the index number of the receiving surface element has been determined, the absorptivity of the surface element can be obtained from a look-up table. Better still, the surface element absorptivities can be entered as a one-dimensional subscripted variable whose subscript is the surface index number, e.g., in FORTRAN $\alpha_j =$ AL-PHA(J). Suppose that when we draw the next available random number we obtain $R_\alpha = 0.20096783$. We then compare R_α with α_j (= 0.04 in this case) and find that $R_\alpha > \alpha_j$. Therefore, the energy bundle is not absorbed and so must be reflected.

12.8   DETERMINE IF THE REFLECTION IS DIFFUSE OR SPECULAR

To determine if the reflection is diffuse or specular, we draw the next available random number, R_r, and compare its value with the specularity ratio r_s defined by Equation 11.47. Suppose $R_r = 0.73618303$. Then, since $r_s = 0.958$, we see that $R_r < r_s$ and so the reflection is specular.

12.9   DETERMINE THE DIRECTION OF THE SPECULAR REFLECTION

Recalling Equations 11.48 through 11.53, the direction cosines of the path followed by the specularly reflected energy bundle are

$$\ell_2 = \frac{V_{2,x}}{T_2} \tag{12.32}$$

$$m_2 = \frac{V_{2,y}}{T_2} \tag{12.33}$$

and

$$n_2 = \frac{V_{2,z}}{T_2} \tag{12.34}$$

where

$$\mathbf{V}_2 = \mathbf{V}_1 - 2\,(\mathbf{V}_1 \cdot \mathbf{n}_1)\,\mathbf{n}_1 = V_{2,x}\mathbf{i} + V_{2,y}\mathbf{j} + V_{2,z}\mathbf{k} \tag{12.35}$$

and

$$T_2 = \sqrt{(V_{2,x})^2 + (V_{2,y})^2 + (V_{2,z})^2} \tag{12.36}$$

In Equation 12.35

$$\mathbf{V}_1 = (x_1 - x_0)\mathbf{i} + (y_1 - y_0)\mathbf{j} + (z_1 - z_0)\mathbf{k} \tag{12.37}$$

and

$$\mathbf{n}_1 = \frac{\nabla S_1}{|\nabla S_1|} \tag{12.38}$$

the unit normal vector at the point on the conical mirror from which the energy bundle is reflected. In writing Equation 12.38 the plus sign in Equation 12.13 has been selected to ensure that the unit vector is directed into the enclosure.

The equation for the conical surface is given by Equations 12.23 and 12.24,

$$S_1(x, y, z) = x^2 + y^2 - \left(\frac{D}{2H}\right)^2 z^2 = 0 \tag{12.39}$$

Therefore,

$$\mathbf{n}_1 = \frac{2x\mathbf{i} + 2y\mathbf{j} - 2(D/2H)^2 z\mathbf{k}}{\left|2x\mathbf{i} + 2y\mathbf{j} - 2(D/2H)^2 z\mathbf{k}\right|}\Bigg|_{x_1, y_1, z_1}$$

$$= \frac{x_1/z_1}{\left(\dfrac{D}{2H}\right)\sqrt{1 + \left(\dfrac{D}{2H}\right)^2}}\mathbf{i} + \frac{y_1/z_1}{\left(\dfrac{D}{2H}\right)\sqrt{1 + \left(\dfrac{D}{2H}\right)^2}}\mathbf{j} \tag{12.40}$$

$$- \frac{D/2H}{\sqrt{1 + \left(\dfrac{D}{2H}\right)^2}}\mathbf{k}$$

With $D/2H = 0.5$, $x_1 = 3.0305457$, $y_1 = 1.4673473$, and $z_1 = 6.7341860$, Equation 12.40 becomes

$$\mathbf{n}_1 = 0.80502750\mathbf{i} + 0.38978291\mathbf{j} - 0.44721360\mathbf{k} \tag{12.41}$$

and from Equation 12.37,

$$\mathbf{V}_1 = 6.7341860\mathbf{k} \tag{12.42}$$

Then, from Equation 12.35,

$$\mathbf{V}_2 = 4.8488731\mathbf{i} + 2.3477556\mathbf{j} + 4.0405115\mathbf{k} \tag{12.43}$$

and from Equation 12.36,

$$T_2 = 6.7341860 \tag{12.44}$$

Finally, from Equations 12.32 through 12.34, we obtain

$$\ell_2 = \frac{V_{2,x}}{T_2} = \frac{4.8488731}{6.7341860} = 0.72003850 \tag{12.45}$$

$$m_2 = \frac{V_{2,y}}{T_2} = \frac{2.3477556}{6.7341860} = 0.34863243 \tag{12.46}$$

and

$$n_2 = \frac{V_{2,z}}{T_2} = \frac{4.0405115}{6.7341860} = 0.59999999 \tag{12.47}$$

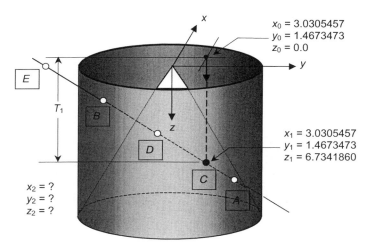

Figure 12.6   Line including the path of the energy bundle after specular reflection from the conical mirror

The line including the path described by Equations 12.45 through 12.47 is illustrated in Figure 12.6. Note here that T_2 is not (yet) the distance from point x_1, y_1, z_1 to point x_2, y_2, z_2, although later it will assume this role. This is because the coordinates of point x_2, y_2, z_2 are not yet known. Here T_2 is equal to T_1 and is simply the scaling factor needed to convert the reflected vector into a unit vector.

12.10   DETERMINE THE POINT WHERE THE ENERGY BUNDLE STRIKES THE CAVITY WALL

We now need to find the point where the reflected energy bundle next strikes the cavity wall. As human beings looking at Figure 12.6 we know that it must hit the cylindrical wall at point A, but once again the computer does not have the same intuitive grasp of the situation as does its human master. The computer must first find all of the candidates and then systematically eliminate the source point and any back candidates before settling on the nearest forward candidate. The equations of the line containing the path of the reflected energy bundle are

$$\frac{x_2 - x_1}{\ell_2} = \frac{y_2 - y_1}{m_2} = \frac{z_2 - z_1}{n_2} = T_2' \tag{12.48}$$

three equations in four unknowns: x_2, y_2, z_2, and T_2', where now T_2' is the distance from point x_1, y_1, z_1 to point x_2, y_2, z_2. The fourth equation is, of course, the equation of the surface for the various candidate surfaces.

Three candidate surfaces are available: the cylindrical wall, the conical mirror, and the disk opening. Thus, we have

$$x_2 = x_1 + \ell_2 T_2' \tag{12.49}$$

$$y_2 = y_1 + m_2 T_2' \tag{12.50}$$

$$z_2 = z_1 + n_2 T_2' \tag{12.51}$$

$$x_2^2 + y_2^2 - \frac{D^2}{4} = 0 \tag{12.52}$$

$$x_2^2 + y_2^2 - z_2^2 \left(\frac{D}{2H} \right)^2 = 0 \tag{12.53}$$

and

$$z_2 = 0 \tag{12.54}$$

We obtain candidate values of T_2' by substituting Equations 12.49, 12.50, and 12.51 into Equations 12.52, 12.53, and 12.54. When we do this we get, respectively,*

$$T_2' = 2.0411337, -10.458866 \text{ mm} \tag{12.55}$$
$$T_2' = 0.0, -6.1219874 \text{ mm} \tag{12.56}$$

and

$$T_2' = -11.223643 \text{ mm} \tag{12.57}$$

Substituting these values into Equations 12.49, 12.50, and 12.51 yields the coordinates of the candidate points,

$$\left. \begin{array}{l} x_2 = 4.5002405 \text{ mm} \\ y_2 = 2.1789527 \text{ mm} \\ z_2 = 7.9588662 \text{ mm} \end{array} \right\} \quad \text{(Point } A) \tag{12.58}$$

---

* Eight significant figures are shown here to emphasize the fact that the maximum number of significant figures available on the computer being used should be retained to minimize round-off error, which can accumulate into a significant error after hundreds of thousands or even millions of rays have been traced. Of course, final results would be expressed in far fewer significant figures in an engineering calculation.

$$\left.\begin{array}{l} x_2 = -4.5002405 \text{ mm} \\ y_2 = -2.1789527 \text{ mm} \\ z_2 = 0.45886650 \text{ mm} \end{array}\right\} \quad \text{(Point } B) \quad\quad\quad (12.59)$$

$$\left.\begin{array}{l} x_2 = x_1 \\ y_2 = y_1 \\ z_2 = z_1 \end{array}\right\} \quad \text{(Point C)} \quad\quad\quad (12.60)$$

$$\left.\begin{array}{l} x_2 = -1.3775209 \text{ mm} \\ y_2 = -0.66697604 \text{ mm} \\ z_2 = 3.0609936 \text{ mm} \end{array}\right\} \quad \text{(Point D)} \quad\quad\quad (12.61)$$

and

$$\left.\begin{array}{l} x_2 = -5.0509094 \text{ mm} \\ y_2 = -2.4455786 \text{ mm} \\ z_2 = 0.0 \text{ mm} \end{array}\right\} \quad \text{(Point E)} \quad\quad\quad (12.62)$$

respectively. We must now test each of these candidates systematically, without invoking our human prejudices, to discover which is the true point.

Let us first consider point E in Figure 12.6, whose coordinates are given in Equation 12.62. This is an interesting point because we have two bases for rejecting it: (1) it lies outside the cavity ($x_2^2 + y_2^2 > D^2/4$) and (2) it is a "back" candidate; that is

$$[(x_2 - x_1)\mathbf{i} + (y_2 - y_1)\mathbf{j} + (z_2 - z_1)\mathbf{k}] \cdot \mathbf{n}_1 < 0 \quad\quad\quad (12.63)$$

Now we do not have a priori knowledge of either of these two conditions, but the fact that both disqualify the point in question raises an interesting question. How do we know which test to apply to a given point? The answer is that we apply all tests to all points until only one point remains. Thus, point E will be eliminated by whichever of the two tests to which it is first subjected. The suggested hierarchy of test cycles, based on their simplicity of application, is

1. Eliminate the source point by testing all points to find the one for which x_2, y_2, z_2 is within radius δ of x_1, y_1, z_1, where δ is slightly larger than the smallest calculable real number of the computer being used.
2. Eliminate the back candidates by applying the test represented by Equation 12.63 to all remaining points.

3. Sort through the surviving forward candidates to find the one with the smallest magnitude of T.

Note that by following this order we never need to test a point to see if it lies outside the cavity.

Point C is eliminated as a candidate by the first test, but the other points survive this test. Points B, D, and E are all three eliminated by the second test since

$$
\begin{aligned}
&[(-4.5002405 - 3.0305457)\mathbf{i} + (-2.1789527 - 1.4673473)\mathbf{j} \\
&+(0.45886650 - 6.7341860)\mathbf{k}]\cdot(0.80502750\mathbf{i} + 0.38978291\mathbf{j} \\
&-0.44721360\mathbf{k}) = (-7.5307862\mathbf{i} - 3.6463000\mathbf{j} \\
&-6.2753195\mathbf{k})\cdot(0.80502750\mathbf{i} + 0.38978291\mathbf{j} \\
&-0.44721360\mathbf{k}) = (-6.0624900 - 1.4212654 + 2.8064082) \\
&= -4.6773472 < 0
\end{aligned}
\tag{12.64}
$$

for point B,

$$
\begin{aligned}
&[(-1.3775209 - 3.0305457)\mathbf{i} + (-0.66697604 - 1.4673473)\mathbf{j} \\
&+(3.0609936 - 6.7341860)\mathbf{k}]\cdot(0.80502750\mathbf{i} + 0.38978291\mathbf{j} \\
&-0.44721360\mathbf{k}) = (-4.4080666\mathbf{i} - 2.13432334\mathbf{j} \\
&-3.6731924\mathbf{k})\cdot(0.80502750\mathbf{i} + 0.38978291\mathbf{j} \\
&-0.44721360\mathbf{k} = (-3.5486148 - 0.83192276 + 1.6427016) \\
&= -2.7378360 < 0
\end{aligned}
\tag{12.65}
$$

for point D, and

$$
\begin{aligned}
&[(-5.0509094 - 3.0305457)\mathbf{i} + (-2.4455786 - 1.4673473)\mathbf{j} \\
&+(0.0 - 6.7341860)\mathbf{k}]\cdot(0.80502750\mathbf{i} + 0.38978291\mathbf{j} \\
&-0.44721360\mathbf{k}) = (-8.0814551\mathbf{i} - 3.9129259\mathbf{j} \\
&-6.7341860\mathbf{k})\cdot(0.80502750\mathbf{i} + 0.38978291\mathbf{j} \\
&-0.44721360\mathbf{k}) = (-6.5057936 - 1.5251916 + 3.0116195) \\
&= -5.0193656 < 0
\end{aligned}
\tag{12.66}
$$

for point E. That is, all three are demonstrated to be back candidates. Thus, by default point A is the point of intersection. However, it still needs to be checked, as the computer program would do automatically. Thus,

$$[(4.5002405 - 3.0305457)\mathbf{i} + (2.1789527 - 1.4673473)\mathbf{j}$$

$$+(7.9588662 - 6.7341860)\mathbf{k}]\cdot(0.80502750\mathbf{i} + 0.38978291\mathbf{j}$$

$$-0.44721360\mathbf{k}) = (1.4696948\mathbf{i} + 0.7116054\mathbf{j}$$

$$+1.2246802\mathbf{k})\cdot(0.80502750\mathbf{i} + 0.38978291\mathbf{j} \tag{12.67}$$

$$-0.44721360\mathbf{k}) = (1.1831447 + 0.27737162 - 0.54769364)$$

$$= 0.91282268 > 0$$

thereby confirming that point *A*, as the only forward candidate in this case, is the correct point of intersection.

12.11   DETERMINE THE INDEX NUMBER OF THE SURFACE ELEMENT RECEIVING THE ENERGY BUNDLE

The next step is to determine the index number of the surface element where the energy bundle impacts. Recall that this is a two-step process in which we first determine the receiving major surface: the disk, the cone or the cylinder in the current example. Thus we begin by computing the radius of the impact point,

$$r_2 = \sqrt{x_2^2 + y_2^2} = \sqrt{(4.5002405)^2 + (2.1789527)^2}$$

$$= \sqrt{24.999994} = 4.9999999 \cong 5.0 \tag{12.68}$$

This result* is necessary but insufficient for the point to lie on the cylinder. To verify that this point does indeed lie on the cylindrical surface we must also check to verify that $0 < z_2 = 7.9588662 < 10.0$, which is of course the case.

Now that we know that point *A* lies on the cylindrical surface, in order to compute the index number of the surface element that it strikes, we can access the function statement

$$j = 192 + 32 \text{ INT}(z_2) + \text{INT}\left[32\left(\frac{\phi_2}{2\pi}\right) + 1 - \delta_2\right] \tag{12.69}$$

where

$$\phi_2 = \tan^{-1}\left(\frac{y_2}{x_2}\right) \tag{12.70}$$

---

* The calculations in this chapter were done using an 8-place hand calculator. This result, 4.9999999 rather than 5.0000000, illustrates how round-off error can accumulate. For this reason it is essential that double precision be used in MCRT applications.

In the case at hand,

$$\phi_2 = \tan^{-1}\left(\frac{2.1789527}{4.5002405}\right) = 0.45091643r \tag{12.71}$$

and

$$j = 192 + 32 \text{ INT}(7.9588662)$$
$$+ \text{INT}\left[32\left(\frac{0.450916427}{2\pi}\right) + 1 - \delta_2\right] = 419 \tag{12.72}$$

12.12   DETERMINE IF THE ENERGY BUNDLE IS ABSORBED OR REFLECTED

At this point we once again make use of the vector $\alpha_j = $ ALPHA(J) to obtain the absorptivity of surface element 419. The next available random number is then drawn and its value compared with α_j. If $R_\alpha < \alpha_j$, the energy bundle is absorbed and the counter N_{ij} is incremented; otherwise the energy bundle is reflected. For example, if $R_\alpha = 0.92673011 > 0.9$, then the energy bundle is reflected, in which case the next step is to determine whether the reflection is diffuse or specular.

12.13   DETERMINE IF THE REFLECTION IS DIFFUSE OR SPECULAR

Suppose the next available random number has a value of 0.97200923 (beating all odds!). Then, $R_r > r_s$ and the reflection is diffuse.

12.14   DETERMINE THE DIRECTION OF THE DIFFUSE REFLECTION

What are the direction cosines of the reflected energy bundle? Suppose the next two available random numbers are 0.79364701 and 0.38923567. Then

$$\phi_2 = 2\pi(0.79364701) = 4.9866312 \ r \quad (285.7 \text{ deg}) \tag{12.73}$$

and

$$\theta_2 = \sin^{-1}\left(\sqrt{0.38923567}\right) = 0.67370726 \ r \quad (38.6 \text{ deg}) \tag{12.74}$$

Now recall that these two angles are with respect to local $\mathbf{n}, \mathbf{t}_1, \mathbf{t}_2$ coordinates. Therefore we must express $\mathbf{n}, \mathbf{t}_1$, and $\mathbf{t}_2$ in the global $\mathbf{i}, \mathbf{j}, \mathbf{k}$ coordinates and then use Equations 11.27, 11.28 and 11.29 (with the "*e*" subscript replaced by "*r*") to express the vector in the reflected direction, $V_{r,2}$, in the global coordinate system.

The equation for the cylindrical surface is

$$S(x, y, z) = x^2 + y^2 - \left(\frac{D}{2}\right)^2 = 0 \tag{12.75}$$

so that, from Equation 11.16,

$$\mathbf{n}_2 = -\frac{2x_2}{D}\mathbf{i} - \frac{2y_2}{D}\mathbf{j} = -0.2(x_2\mathbf{i} + y_2\mathbf{j})$$

$$= -0.90004810\mathbf{i} - 0.43579054\mathbf{j} \tag{12.76}$$

from Equation 11.24,

$$\mathbf{t}_{1,2} = \frac{\mathbf{n}_2 \times \mathbf{i}}{|\mathbf{n}_2 \times \mathbf{i}|} = \frac{-0.2(x_2\mathbf{i} + y_2\mathbf{j}) \times \mathbf{i}}{|-0.2(x_2\mathbf{i} + y_2\mathbf{j}) \times \mathbf{i}|} = \frac{0.2y_2\mathbf{k}}{|0.2y_2\mathbf{k}|} = \mathbf{k} \tag{12.77}$$

and from Equation 11.26,

$$\mathbf{t}_{2,2} = \mathbf{n}_2 \times \mathbf{t}_{1,2} = -0.2(y_2\mathbf{i} - x_2\mathbf{j}) = -0.43579054\mathbf{i} + 0.90004810\mathbf{j} \tag{12.78}$$

As expected, $\mathbf{n}_2$, $\mathbf{t}_{1,2}$, and $\mathbf{t}_{2,2}$ all have magnitudes of unity. These unit vectors are illustrated in Figure 12.7.

Now that the unit normal and tangent vectors have been determined, we use Equations 11.30, 11.31, and 11.32 (once again, with the "*e*" subscript replaced by "*r*") to find the *x, y* and *z* components of the direction of reflection of the energy bundle in the global coordinate system. When we do this we obtain

$$V_{x,3} = -0.44167618 \tag{12.79}$$

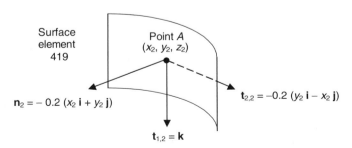

Figure 12.7 The unit normal and tangent vectors at point *A*

from Equation 11.30,

$$V_{y,3} = -0.88112130 \qquad (12.80)$$

from Equation 11.31, and

$$V_{z,3} = 0.16895970 \qquad (12.81)$$

from Equation 11.32. The reflected ray is shown in Figure 12.8.

Note that because $\mathbf{V}_{r,3}$ is itself a unit vector, so that $T_{r,3} = 1.0$. Then, from Equations 11.51, 11.52, and 11.53 we have

$$\ell_3 = \frac{V_{3,x}}{T_3} = V_{3,x} = -0.44167618 \qquad (12.82)$$

$$m_3 = \frac{V_{3,y}}{T_3} = V_{3,y} = -0.88112130 \qquad (12.83)$$

and

$$n_3 = \frac{V_{3,z}}{T_3} = V_{3,z} = 0.16895970 \qquad (12.84)$$

The line containing the path of this diffusely reflected energy bundle is shown in Figure 12.9.

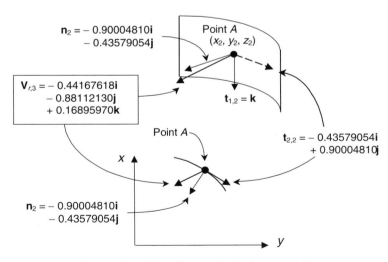

Figure 12.8 The diffuse reflection from point A

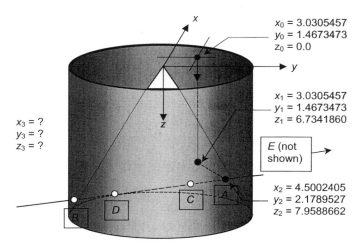

$x_0 = 3.0305457$
$y_0 = 1.4673473$
$z_0 = 0.0$

$x_1 = 3.0305457$
$y_1 = 1.4673473$
$z_1 = 6.7341860$

$x_3 = ?$
$y_3 = ?$
$z_3 = ?$

E (not shown)

$x_2 = 4.5002405$
$y_2 = 2.1789527$
$z_2 = 7.9588662$

Figure 12.9 Path of the energy bundle after diffuse reflection from the cylindrical wall

12.15 FIND THE POINT WHERE THE DIFFUSELY REFLECTED ENERGY BUNDLE STRIKES THE CAVITY WALL

Our task is now to find the point where the diffusely reflected ray strikes the cavity wall. Once again, for the human observer looking at Figure 12.9 it is obvious that this is point C (on the far outside surface of the cone), but the computer must sift through all possible candidates to find the correct one.

We begin by solving the three equations for the line containing the path

$$x_3 = x_2 + \ell_3 T_3' \qquad (12.85)$$

$$y_3 = y_2 + m_3 T_3' \qquad (12.86)$$

and

$$z_3 = z_2 + n_3 T_3' \qquad (12.87)$$

simultaneously with each of the equations for the three surfaces making up the enclosure, Equations 12.52, 12.53, and 12.54 (only now with subscripts "3"). When we do this we obtain the following candidate points:

$$\left. \begin{aligned} x_3 &= x_2 \\ y_3 &= y_2 \\ z_3 &= z_2 \end{aligned} \right\} \quad (\text{Point } A) \qquad (12.88)$$

$$\left.\begin{array}{l} x_3 = 0.94704419 \\ y_3 = -4.9094915 \\ z_3 = 9.3181131 \end{array}\right\} \quad \text{(Point } B) \tag{12.89}$$

$$\left.\begin{array}{l} x_3 = 3.9405783 \\ y_3 = 1.0624553 \\ z_3 = 8.1729605 \end{array}\right\} \quad \text{(Point } C) \tag{12.90}$$

$$\left.\begin{array}{l} x_3 = 1.1693609 \\ y_3 = -4.4659811 \\ z_3 = 9.2330676 \end{array}\right\} \quad \text{(Point } D) \tag{12.91}$$

and

$$\left.\begin{array}{l} x_3 = 25.307448 \\ y_3 = 43.684273 \\ z_3 = 0.0 \end{array}\right\} \quad \text{(Point } E) \tag{12.92}$$

Point A is first eliminated by a routine that establishes that it is within a small diameter δ of the source point. Point E is eliminated by noting that the sign of the dot product

$$[(x_3 - x_2)\mathbf{i} + (y_3 - y_2)\mathbf{j} + (z_3 - z_2)\mathbf{k}] \cdot \mathbf{n}_2 \tag{12.93}$$

is negative. This leaves three forward candidates: points B, C, and D. The values of T'_3 for these three points have already been computed in the process of obtaining the coordinates of the three forward candidates. They are

$$T'_{3,B} = 8.0447995 \tag{12.94}$$

$$T'_{3,C} = 1.2671324 \tag{12.95}$$

and

$$T'_{3,D} = 7.5414518 \tag{12.96}$$

After sorting these for length we (and the computer) conclude that point C in Figure 12.9, on the cone, is the point where the reflected energy bundle hits the cavity wall.

The index number of point C in Figure 12.9 may be computed from Equation 12.31 once ϕ_3 has been computed. Therefore,

$$\phi_3 = \tan^{-1}\left(\frac{y_3}{x_3}\right) = \tan^{-1}\left(\frac{1.0624553}{3.9405783}\right) = 0.26335681r \qquad (12.97)$$

and so

$$j = 32\left\{\mathrm{INT}\left[\left(\frac{z_3}{z_a}\right)^2 - \delta_1\right]\right\} + \mathrm{INT}\left[32\left(\frac{\phi_3}{2\pi}\right) + 1 - \delta_2\right]$$

$$= 32\left\{\mathrm{INT}\left[\left(\frac{8.1729605}{4.0780916}\right)^2 - \delta_1\right]\right\} \qquad (12.98)$$

$$+ \mathrm{INT}\left[32\left(\frac{0.26335681}{2\pi}\right) + 1 - \delta_2\right]$$

$$= 128 + 2 = 130$$

The surface element containing this point is shown in Figure 12.10.

12.16   DETERMINE IF THE ENERGY BUNDLE IS ABSORBED OR REFLECTED

We now ask if the energy bundle is absorbed or reflected. Suppose the next available random number has a value of 0.00162758. Then, recalling that the absorptivity α of

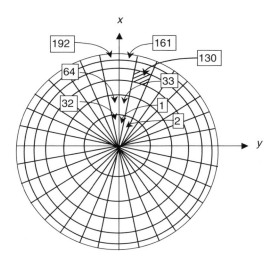

Figure 12.10   Numbering system for the surface elements on the conical surface (view looking into the cavity from above) showing surface element 130

the conical mirror is 0.04, we see that this energy bundle is one of the rare ones striking the mirror that is in fact absorbed. (This is just as well, because nothing of pedagogical importance is to be learned by considering another specular or diffuse reflection.) Therefore, we increment the counter $N_{0,130}$ by unity and return to the beginning of the analysis where another energy bundle is launched into the cavity parallel to the z axis through the opening from a randomly selected location in the plane of the opening.

12.17   COMPUTE THE ESTIMATE OF THE DISTRIBUTION FACTOR MATRIX

When a sufficiently large number of energy bundles has been traced in this manner, we estimate the elements of the distribution factor matrix using

$$D_{oj} \cong \frac{N_{oj}}{N_o} \qquad (12.99)$$

where N_o is the total number of energy bundles entering the cavity through the opening ("o") and N_{oj} is the number of those absorbed by surface element j.

Team Projects

TP 12.1   Use FELIX to work the extended problem whose solution is outlined in this chapter. In particular, compute the effective absorptivity of the cavity for collimated incident radiation. Also obtain a plot of the area-weighted distribution factors, D_{oj}/A_j, on the cavity surfaces, where the index "o" indicates the opening. (This result is shown in Figure 12.11. The corresponding effective absorptivity is 0.988. This is a significant improvement over the absorptivity—0.90—of the absorber layer itself.)

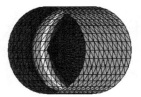

Figure 12.11   Distribution of absorbed radiation on the notional thermal radiation detector of Figure 12.1

TP 12.2 Use FELIX to compute the effective emissivity of the laboratory black-body of Team Project TP 2.2. Assume that the reflectivity $\rho_e = \rho_e^d + \rho_e^s$ of the interior wall of the elbow is $0.01 + 0.09$, the reflectivity $\rho_f = \rho_f^d + \rho_f^s$ of the floor is $0.09 + 0.01$, and the reflectivity $\rho_a = \rho_a^d + \rho_a^s$ of the aluminum foil is $0.05 + 0.90$.

TP 12.3 Use FELIX to determine the boil-off rate of the liquid nitrogen used to maintain the cylindrical wall of the vacuum chamber in Figure 12.12 at 77 K if the hemispherical dome temperature is 300 K and the floor is perfectly insulated.

Discussion Points

DP 12.1 Look up the original paper on effective absorptivity ("apparent emissivity," actually), Reference 1. For the geometries studied in Reference 1, what is the general influence of a high specularity ratio on the apparent emissivity? Can you explain this?

DP 12.2 The last sentence in the first paragraph of Section 12.2 states, "It is a curious and useful property of cavities that we can estimate their effective absorptivity with greater accuracy than we know the surface properties themselves!" Why is this true?

DP 12.3 In the second paragraph of Section 12.2 it is stated that "we must subdivide the cavity into surface elements whose temperatures and heat fluxes can at least in principle be different from one another" in order to study self-contamination of the detector. Please explain why we must subdivide the cavity to accomplish this study.

DP 12.4 Just below Equation 12.1 it is stated that "if we use Equation 12.1 to estimate some of the distribution factors for an enclosure, then we forfeit the possibility of using Equation 11.6 to estimate the global precision of the distribution factor estimates." Why is this true?

DP 12.5 In the paragraph just below Equation 12.10 it is stated that "In the competition between the need for equal surface areas and the desire for uniform spatial resolution, it should always be the former that wins out." Why is this true? Use the example of the conical mirror to justify this claim.

DP 12.6 Derive Equation 12.11.

DP 12.7 Referring to the discussion in the paragraph just below Equation 12.24 concerning the radius δ, how would you choose the value of δ?

DP 12.8 Compose a segment of computer code that accomplishes the task described in the second paragraph below Equation 12.24.

DP 12.9 The paragraph immediately after Equation 12.24 describes a procedure for dealing with orphan rays. Consider the alternative of simply defaulting to the emission point as the receiving point when no point tested is chosen. Also, are there other ways of creating an orphan ray than the one described in this paragraph?

DP 12.10 Compose a segment of computer code that accomplishes the task described in the first sentence beginning with the word "However" below Equation 12.30.

DP 12.11 Derive Equation 12.31.

DP 12.12 Verify that $j = 67$ in Figure 12.5.

DP 12.13 Derive Equation 12.69.

DP 12.14 Discuss the significance of the absorbed heat flux distribution on the absorber wall shown in Figure 12.11 (see Team Project TP 12.1). What does the distribution of absorbed radiation say about the design of the detector? Can you recommend any dimensional changes that might improve the design?

Problems

A certain vacuum chamber, shown in Figure 12.12, consists of three major surfaces: a hemispherical dome, a cylindrical wall, and a circular floor. All surfaces of the cavity have the same surface properties, $\rho = \rho^d + \rho^s = 0.5 + 0.2$. The chamber is to be subdivided into surface elements that are as nearly as possible one centimeter on a side in order to carry out a Monte Carlo ray-trace analysis.

Part A: Subdivide the enclosure into surface elements. Begin by dividing the cylindrical wall in the axial (z) direction into 100 1.0-cm-high rings.

P 12.1   (a) How many circumferential (ϕ) divisions do we need on the cylindrical wall?

(b) What are their arc lengths?

(c) What is the surface area (cm^2) of each of these surface elements?

P 12.2   (a) How many circumferential (ϕ) divisions do we need on the circular floor?

(b) Into how many radial (r) divisions should we divide the circular floor?

(c) What are the radii of the circles defining each of these divisions?

(d) What is the area of each surface element on the circular floor?

P 12.3   (a) Into how many axial (z) divisions should we divide the hemispherical dome?

(b) What are the z coordinates of each cutting plane?

(c) What is the area of each surface element on the hemispherical dome?

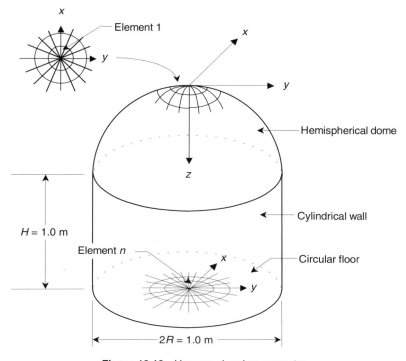

Figure 12.12   Vacuum chamber geometry

P 12.4   What is the total number of surface elements n into which the chamber has been divided?

Part B: Number the surface elements. We number the surface elements starting with the first surface element, in the clockwise direction from the x axis, in the first ring of surface elements at the top of the hemispherical dome, as shown in Figure 12.12. The last (nth) surface element is the last element in the *inner* ring at the center of the circular floor, as shown in the figure.

P 12.5   Derive an equation for the index j of a surface element on the hemispherical dome in terms of the ϕ, z coordinates of a point on the surface element.

P 12.6   Derive an equation for the index j of a surface element on the cylindrical surface in terms of the ϕ, z coordinates of a point on the surface element.

P 12.7   Derive an equation for the index j of a surface element on the circular floor in terms of the r, ϕ coordinates of a point on the surface element.

Part C: Perform the ray trace. The sequence of random numbers listed below is available. Please draw these IN ORDER as you respond to the questions below. (Draw a line through each random number as you use it so you will not use the same number twice.) Trace a diffusely emitted ray from element 1.

1. 0.63021418   4. 0.30441737   7. 0.89473127
2. 0.01041773   5. 0.97237214   8. 0.77271371
3. 0.21123241   6. 0.00197432   9. 0.46372978

P 12.8   What are the x,y,z coordinates of the first emission point on surface element 1?

P 12.9   What are the direction cosines of the first energy bundle emitted from surface element 1?

P 12.10   What are the x,y,z coordinates of the point where the first energy bundle emitted from surface element 1 strikes the chamber?

P 12.11   What is the index number j of the surface element where the first energy bundle emitted from surface element 1 intercepts the chamber?

P 12.12   Is the first energy bundle emitted from surface element 1 absorbed or reflected?

P 12.13   If the first energy bundle emitted from surface element 1 is reflected, is the reflection diffuse or specular?

P 12.14   If the first energy bundle emitted from surface element 1 is reflected, what are the direction cosines of the reflected energy bundle?

REFERENCE

1. Sparrow, E. M., and S. H. Lin, "Absorption of thermal radiation in V-groove cavities," *International Journal of Heat and Mass Transfer,* Vol. 5, 1962, pp. 1111–1115.

13

THE DISTRIBUTION FACTOR FOR NONDIFFUSE, NONGRAY, SURFACE-TO-SURFACE RADIATION

In Chapter 11 we defined the radiation distribution factor and indicated how it might be used to formulate the problem of radiation heat transfer among the surfaces of an enclosure. The Monte Carlo ray-trace (MCRT) method was then introduced as a means for estimating its value. An extended example of estimating the distribution factor matrix for the case of a diffuse–specular, gray enclosure was presented in Chapter 12. In this chapter we develop the concept of the band-averaged spectral radiation distribution factor, which is the most general tool available for treating radiation exchange among surfaces.

13.1 THE BAND-AVERAGED SPECTRAL RADIATION DISTRIBUTION FACTOR

In Chapter 11 we defined the band-averaged spectral radiation distribution factor, D_{ijk}, as the fraction of power emitted in wavelength interval $\Delta\lambda_k$ by surface element i that is absorbed by surface element j due to direct radiation and to all possible reflections. Once again, attention is directed to the differences between this definition and those for the configuration and exchange factors. In particular, the reader is invited to contrast the words "emitted," "element," and "absorbed," which appear in the definition of the distribution factor, with the words "leaving," "surface," and "arrives at," which appear in the standard definitions of the configuration and exchange factors. In fact, we will discover in Chapter 14 that the distribution factor is defined and used for both surface and volume elements. However, before taking up the case of radiant exchange among volume elements we will first fully develop and demonstrate the method as applied to radiant exchange among surface elements.

13.2 USE OF THE BAND-AVERAGED SPECTRAL RADIATION DISTRIBUTION FACTOR FOR THE CASE OF SPECIFIED SURFACE TEMPERATURES

We begin by considering the situation in which the temperatures are specified for all n surface elements of an enclosure. The surfaces are directional emitters and absorbers and bidirectional reflectors, and the related emissivities, absorptivities, and reflectivities are band-averaged spectral in the sense illustrated in Figure 13.1. This is the most general, and therefore the most difficult, problem in surface-to-surface radiation analysis. The need for methods to solve this class of problem has been severely limited in the past by the lack of (bi)directional, spectral surface property data. It may be anticipated that the availability of the tool developed in this chapter will encourage experimentalists to compile and publish more of this kind of data in the open literature.

Consider the general enclosure, shown in Figure 13.2, which features blockage of the view among at least some of the surface elements. The surface elements are directionally emitting and absorbing and bidirectionally reflecting, and the surface properties vary with wavelength. For the moment we assume that all surface temperatures are known, and we seek the unknown surface net heat fluxes. We assume further that the enclosure has been subdivided into sufficiently small surface elements that the temperature and net heat flux of each may be considered uniform to an acceptable degree of accuracy. No further assumptions are required.

The power (W) absorbed by surface element i may be computed directly from the definition of the band-averaged spectral radiation distribution factor; that is,

$$Q_{i,a} = \sum_{k=1}^{K} \sum_{j=1}^{n} \varepsilon_{jk} e_b(\Delta\lambda_k, T_j) A_j D_{jik}, \qquad 1 \le i \le n \qquad (13.1)$$

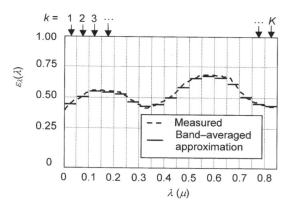

Figure 13.1 Wavelength dependence of a hypothetical spectral emissivity, as measured and the corresponding band-averaged approximation

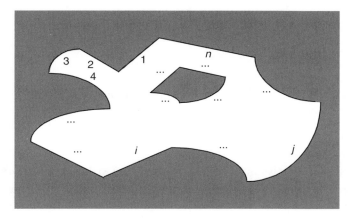

Figure 13.2 A general enclosure some of whose directional, spectral surface elements are at least partially obscured from each other by blockage

where k is the index of the K wavelength bands and i and j are the indices of the n surface elements comprising the enclosure. The quantity $e_b(\Delta\lambda_k, T_j)$ in Equation 13.1 is the blackbody emissive power (W/m^2) contained in the wavelength interval $\Delta\lambda_k$; that is,

$$e_b(\Delta\lambda_k, T_j) \equiv \int_{\lambda_k}^{\lambda_k + \Delta\lambda_k} e_{b\lambda}(\lambda, T_j)\, d\lambda \tag{13.2}$$

The quantity ε_{jk} in Equation 13.1 is the hemispherical, band-averaged spectral emissivity of surface element j in wavelength band k (we note that D_{jik} contains the required directional information).

Reciprocity holds for the band-averaged spectral radiation distribution factor; that is,

$$\varepsilon_{ik} A_i D_{ijk} = \varepsilon_{jk} A_j D_{jik} \tag{13.3}$$

Then the flux (W/m^2) absorbed by element i may be found by introducing Equation 13.3 into Equation 13.1 and dividing through by A_i, yielding

$$\frac{Q_{i,a}}{A_i} = q_{i,a} = \sum_{k=1}^{K} \sum_{j=1}^{n} \varepsilon_{ik}\, e_b(\Delta\lambda_k, T_j) D_{ijk}, \qquad 1 \le i \le n \tag{13.4}$$

Finally, the *net* heat flux from element i is given by

$$q_i \equiv q_{i,e} - q_{i,a} = \sum_{k=1}^{K} \sum_{j=1}^{n} \varepsilon_{ik} e_b(\Delta\lambda_k, T_j)(\delta_{ijk} - D_{ijk}), \qquad 1 \le i \le n \tag{13.5}$$

where now δ_{ijk}, the Kronecker delta function, is defined

$$\delta_{ijk} \equiv \begin{cases} 1, & i = j \\ 0, & i \neq j \end{cases}, \quad \text{all } k \tag{13.6}$$

This formulation provides at least four distinct advantages over the band-averaged spectral net exchange formulation for diffuse–specular enclosures, which is the most general problem that can be solved using the net exchange formulation or other formulations based on geometric optics. They are

1. The formulation is fully directional (as opposed to diffuse–specular).
2. There are no matrices to invert, at least in the case of specified surface temperatures, and in the case where N of the n surfaces have specified net heat fluxes, the order of the matrix to be inverted is only N rather than n.
3. The formulation permits routine analysis of enclosures involving partial or full blockage of the line-of-sight between surface elements. In fact, distribution factors between surface element pairs having a partially or fully blocked view of each other are as easy to compute as distribution factors for surface element pairs that view each other without blockage.
4. For most enclosures of practical interest, the distribution factors are significantly easier to compute using the MCRT method than are configuration or exchange factors, required in the net exchange formulation, using standard methods.

13.3 CALCULATION OF (BI)DIRECTIONAL, BAND-AVERAGED SPECTRAL RADIATION DISTRIBUTION FACTORS FOR THE CASE OF SURFACE-TO-SURFACE EXCHANGE

We now undertake the task of computing the (bi)directional, band-averaged spectral radiation distribution factor matrix D_{ijk} for the generic enclosure of Figure 13.2. In the present development we will follow the general logic of Figure 11.5, adding only the new information that permits directionality and wavelength dependence to be taken into consideration when computing the distribution factor.

13.4 DETERMINE THE DIRECTION OF EMISSION

We first consider how determination of the direction of emission, described in step 2 in Chapter 11 and illustrated in Section 12.14 for the case of a diffuse reflection, must be modified to take into account directional emission. It is assumed at this point that a directional, spectral emissivity model exists for each of the n surfaces. This model could be based on experimental measurements, on theory, or on a mixture of the two (i.e., a semi-empirical model). This is equivalent to assuming that a reasonable approximation of $\varepsilon'_\lambda(\lambda, \theta, \phi)$ exists for each surface. It is emphasized that such data currently exist in the open literature for only a relatively small number of surfaces of practical interest. One important exception is in the domain of atmospheric radiation. Atmospheric scientists seek to measure what they refer to as bidirectional reflectivity functions ("BDRFs"), or angular distribution models ("ADMs"), for earth-emitted radiation and solar radiation reflected from the earth–atmosphere system as viewed from space. BDRFs are needed to convert intensity (which atmospheric scientists refer to as "radiance"), observed from orbiting radiometric instruments, into outward-directed flux at the top of the atmosphere. This need has resulted in the creation and publication of increasingly accurate tables of BDRFs corresponding to an ever-widening range of scene types (e.g., partly cloudy conditions over the ocean, etc.). Much of the bidirectional, spectral reflectivity data for surfaces of engineering interest were compiled in support of infrared low-observable ("IR stealth") programs, and so are not generally available to the public. This is also true of similar data measured at considerable cost and considered to be company proprietary in certain sectors of private industry. This lamentable state of affairs has led to a rather healthy growth of the goniometry industry, whose products are used to measure bidirectional, spectral reflectivity.

To use the MCRT method to compute radiation distribution factors for directional enclosures, an "engine" is needed whose properties are suggested in Figure 13.3. In this version of the engine the user provides as inputs three random numbers: R_λ, R_θ, and R_ϕ. Unless otherwise specifically stated, whenever random numbers are mentioned in this book it should be assumed that they are uniformly distributed between zero and unity, as defined in Appendix D and illustrated in Figure 13.4. In this latter figure, the random number corresponding to i has the *value* $i/(m + 1)$ and the corresponding *probability* of $1/m$ for all i, where m is the total number of random numbers in the set and $1 \le i \le m$. The engine illustrated in Figure 13.3 then returns the corresponding values of λ, θ, and ϕ for the emitted energy bundle.

How does one go about designing, in a numerical environment, a spectral emission phase-function generator? The spectral emission phase-function generator must possess one essential property: if a sufficiently large number of three-member sets of random numbers $(R_\lambda, R_\theta, R_\phi)$ is introduced to the input of the generator, the corresponding three-member sets (λ, θ, ϕ), when appropriately plotted, will yield $\varepsilon'_\lambda(\lambda, \theta, \phi)$ to an acceptable degree of accuracy.

Let us imagine that there exists an integer K such that the wavelength subintervals $\Delta\lambda_1, \Delta\lambda_2, \ldots, \Delta\lambda_k, \ldots, \Delta\lambda_K$ divide up the wavelength interval between λ_0 and λ_K in

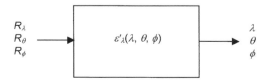

Figure 13.3   The spectral emission phase-function generator

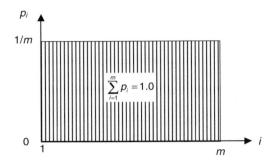

Figure 13.4 Probability distribution of m random numbers uniformly distributed between zero and unity

a way that, for any specified wavelength interval $\Delta\lambda_k$, the directional, spectral emissivity is, to an acceptable approximation, a function of direction θ, ϕ only. The wavelengths λ_0 and λ_K represent practical limits, with λ_0 usually greater than zero and λ_K always less than infinity. Specification of the size and distribution of the wavelength intervals to be used is a judgment call and depends on (1) the inherent quality of the directional, spectral emissivity model or data available, (2) the required accuracy of the result sought, and (3) the computer resources available.

To this point we have been very careful to use the term "energy bundle" rather than the term "ray" or "photon" when referring to the entity being traced. The *path* of an energy bundle can be properly referred to as a "ray" and it is in this sense that we speak of a Monte Carlo "ray trace." However, in the case of a band-averaged spectral analysis, the entity being traced must be a quantity of energy, or power, whose value is properly weighted with respect to the Planck blackbody radiation distribution function.

We begin by computing and storing the two-dimensional matrix

$$f_{ik} \equiv \frac{\int_{\lambda_{k-1}}^{\lambda_k} e_{b\lambda}(\lambda, T_i) \, d\lambda}{\sigma T_i}, \qquad i = 1, 2, \ldots, n, \quad k = 1, 2, \ldots, K \qquad (13.7)$$

The elements of this matrix have the property that, to an acceptable accuracy,

$$\sum_{k=1}^{K} f_{ik} = 1.0, \qquad i = 1, 2, \ldots, n \qquad (13.8)$$

Then, for each emission event from surface element i we draw a random number R_k and search f_{ik} for the value of k such that

$$\sum_{1}^{k-1} f_{ik} < R_k \leq \sum_{1}^{k} f_{ik} \qquad (13.9)$$

The effect of this process is to cause surface element i to emit a number of energy bundles in band k proportional to the amount of energy in that band. Then, when a sufficiently large number of energy bundles has been emitted from surface i, the band-averaged spectral distribution factor for any band k is well approximated by

$$D_{ijk} \cong \frac{N_{ijk}}{N_{ik}} \qquad (13.10)$$

In Equation 13.10, N_{ik} is the number of energy bundles emitted from surface element i in wavelength band k, and N_{ijk} is the number of these absorbed by surface element j. The process just described for estimating D_{ijk} makes it clear why we refer to "energy bundles" rather than "rays" or "photons." It is essential that each emission event from a given surface involve the same amount of energy, otherwise some energy bundles will be weighted more heavily than others and the results obtained will be biased.

Once the wavelength interval, or band index k, for a given emitted energy bundle is known, we then must access a corresponding directional, band-averaged spectral emissivity function $\varepsilon'_k(\Delta\lambda_k, \theta, \phi)$. For each wavelength band k of interest it is usually possible to obtain functions of the form

$$\theta = f_k(R_\theta) \qquad (13.11)$$

and

$$\phi = g_k(R_\phi) \qquad (13.12)$$

based on regressions of available data. For each emission event a set of random numbers (R_θ, R_ϕ) is drawn and Equations 13.11 and 13.12 are used to determine the direction of emission. Then, for a statistically significant number of emissions an acceptable approximation of the directional distribution of radiation emitted in a given wavelength band will be obtained. The following example illustrates this approach.

Example Problem 13.1

Representations of the directional, spectral emissivity are generally available in one of two forms: tabular or symbolic. Tabular data can usually be converted to a symbolic representation by application of suitable regression techniques. Therefore, we consider the case where the directional emissivity can be expressed as a function of the zenith and azimuth angles, θ and ϕ. Consider the directional, spectral emissivity of platinum at 2.0 μm as predicted using the theory of Section 4.5 and plotted in Figure 4.7. In this case emission is assumed to be axisymmetric and so directional emissivity does not vary with azimuth angle ϕ. Therefore, Equation 11.10 is used for Equation 13.12. The directional emissivity can be interpreted in terms of the relative probability that an energy bundle will be emitted in a particular direction. It can be

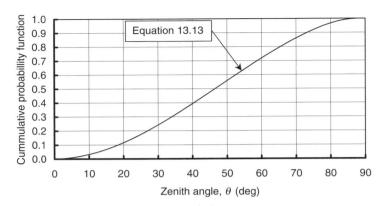

Figure 13.5 Relationship between a random number and the zenith angle of emission for the theoretical directional, spectral emissivity of Figure 4.7

shown (see Problem P 13.1) that the relative probability $p(\theta) = \varepsilon'_\lambda(\theta)\cos\theta \sin\theta$ and a random number R_θ uniformly distributed between zero and unity are related according to

$$R_\theta = \frac{\int_0^\theta p(\theta')\, d\theta'}{\int_0^{\pi/2} p(\theta')\, d\theta'} \tag{13.13}$$

In the case of a diffuse emitter, application of Equation 13.13 leads to Equation 11.11 (see Problem P 13.2). In the present case the relationship obtained by application of Equation 13.13 is plotted in Figure 13.5.

How can the result depicted in Figure 13.5 be used to recover the zenith angle of emission θ from the random number R_θ? Two possibilities come to mind:

1. A regression of appropriate order could be fitted to the data forming the curve of Figure 13.5 to produce an estimate of the function f_k of Equation 13.11.
2. A search algorithm could be set up to locate the value of θ corresponding to R_θ in the table represented by the curve in Figure 13.5.

The former approach is numerically more efficient, while the latter approach is potentially more accurate. As is often the case, the ultimate choice is an engineering decision based on the tradeoff between available computer resources and time, on the one hand, and desired accuracy, on the other.

Once the direction of emission θ, ϕ has been determined in local $\mathbf{n}$, $\mathbf{t}_1$, $\mathbf{t}_2$ coordinates, it is converted to global coordinates and the energy bundle is traced to a surface within the enclosure, as described in Chapter 11 and illustrated in Chapter 12.

13.5 DETERMINE WHETHER THE ENERGY BUNDLE IS ABSORBED OR REFLECTED

As always, when the energy bundle arrives at a surface its disposition must be determined: is it absorbed or reflected? At this point it is noted that information about the wavelength of origin (i.e., the value of the index k) of the energy bundle must be carried along with it so that the correct wavelength-dependent surface property values can be attributed to the surfaces it encounters. As a practical matter a computer program such as FELIX could be executed separately for each wavelength interval of interest. In each execution of the code the directional surface properties would be independent of wavelength, and the results obtained from each run would be suitably combined in a postprocessing step. In this scenario the problem of carrying along the value of the index k is sidestepped. However, the task remains of properly weighting the influence of the different wavelength intervals in the subsequent heat transfer analysis.

We now consider the usual case where the directional absorptivity depends only on incident zenith angle, and so begin by calculating the angle of incidence to the surface element in the local $\mathbf{n}, \mathbf{t}_1, \mathbf{t}_2$ coordinate system. This additional step, required for bidirectional analyses and so not considered in Chapters 11 and 12, is not difficult (see Problem P 13.3). Consistent with Kirchhoff's law we interpret the directional, band-averaged spectral emissivity as an absorptivity. We draw the next available random number, R_a. Then, if R_a is less than the value computed using Equation 13.13, with $p(\theta) = \alpha'(\theta) \cos\theta \sin\theta$, the energy bundle is absorbed; otherwise, it is reflected.

13.6 IF REFLECTED, DETERMINE THE DIRECTION OF REFLECTION

If it is determined that the energy bundle is reflected, an engine called a *band-averaged spectral scattering phase-function generator*, depicted schematically in Figure 13.6, is invoked. The *band-averaged spectral scattering phase function* $p_k(\theta_i, \phi_i, \theta_r, \phi_r)$ is the probability that an energy bundle in wavelength interval $\Delta\lambda_k$ incident from a specified direction θ_i, ϕ_i will be reflected in a given direction θ_r, ϕ_r. It is related to a uniformly distributed random number R_k such that

$$R_k(\theta_i, \phi_i, \theta_r, \phi_r) \equiv \int_{\phi=0}^{\phi_r} \int_{\theta=0}^{\theta_r} dp_k(\theta_i, \phi_i, \theta_r^*, \phi_r^*) \tag{13.14}$$

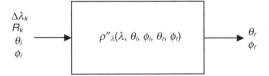

Figure 13.6 The spectral scattering phase-function generator

where

$$dp_k(\theta_i, \phi_i, \theta_r, \phi_r) = \frac{di_{k,r} \cos\theta_r d\omega_r}{\displaystyle\int_{2\pi} di_{k,r} \cos\theta_r d\omega_r} \tag{13.15}$$

or, with the introduction of Equation 3.42,

$$dp_k(\theta_i, \phi_i, \theta_r, \phi_r) = \frac{\rho_k''(\theta_i, \phi_i, \theta_r, \phi_r) \cos\theta_r d\omega_r}{\displaystyle\int_{2\pi} \rho_k''(\theta_i, \phi_i, \theta_r, \phi_r) \cos\theta_r d\omega_r} \tag{13.16}$$

The definition of the scattering phase function permits an analogy to be drawn between bidirectional reflection and directional emission. An equation similar to Equation 13.13 can be written and a graph similar to the one in Figure 13.5 can be constructed, except that now two independent variables, θ_r and ϕ_r, are involved. For a given wavelength band k and a given direction of incidence θ_i, ϕ_i, Equation 13.14 implies a two-dimensional surface of local height $R_k(\theta_i, \phi_i, \theta_r, \phi_r)$ above the θ_r, ϕ_r plane, where $0 \leq \phi_r \leq 2\pi$ and $0 \leq \theta_r \leq \pi/2$.

With the introduction of Equation 13.16, Equation 13.14 may be written

$$R_k(\theta_i, \phi_i, \theta_r, \phi_r) = \frac{\displaystyle\int_{\phi_r=0}^{\phi_r} \int_{\theta_r=0}^{\theta_r} \rho_k''(\theta_i, \phi_i, \theta_r^*, \phi_r^*) \cos\theta_r^* d\omega_r}{\displaystyle\int_{\phi_r=0}^{2\pi} \int_{\theta_r=0}^{\pi/2} \rho_k''(\theta_i, \phi_i, \theta_r, \phi_r) \cos\theta_r d\omega_r} \tag{13.17}$$

Equation 13.17 can be well approximated by the numerical form

$$R_k(\theta_i, \phi_i, \theta_r, \phi_r) = \frac{\displaystyle\sum_{i=1}^{i_\phi} \sum_{i=1}^{i_\theta} \rho_k''(\theta_i, \phi_i, \theta_{r,j}, \phi_{r,i}) \cos\theta_{r,j}}{\displaystyle\sum_{i=1}^{N_\phi} \sum_{j=1}^{N_\theta} \rho_k''(\theta_i, \phi_i, \theta_{r,j}, \phi_{r,i}) \cos\theta_{r,j}} \tag{13.18}$$

where N_ϕ and N_θ are, respectively, the number of ϕ and θ divisions into which the reflection hemisphere has been divided and i_ϕ and j_θ are the indices corresponding to the angular coordinates of a particular direction of reflection. Note that the form of Equation 13.18 assumes that the reflection hemisphere has been divided into equal-solid-angle patches. When used with the appropriate curve fit or interpolating polynomial, Equation 13.18 can yield a continuum of reflection directions just as Equation 13.13 can yield a continuum of emission directions.

If, for a given θ_r, ϕ_r direction, the inner sum (over θ_r^*) is carried out completely for each increment of the outer sum (over $\phi_{r,z}^*$) until θ_r, ϕ_r is reached, a pseudo-one-dimensional series will be created in which the index n, corresponding to a given combination of θ_r and ϕ_r for a given wavelength band k and a given incident direc-

Table 13.1   Illustration (hypothetical) of using scattering phase-function data to obtain the directional coordinates of a bidirectional, spectral reflection using a sequence of uniformly distributed random numbers

| | $k = 3$, $\theta_i = 28$ deg, $\phi_i = 0$ deg | | |
| | ϕ_r | θ_r | |
n	(deg)	(deg)	$R_k(\theta_i, \theta_r, \phi_r)$
1	0	5	0.00163783
2	0	15	0.00198945
3	0	25	0.00219375
4	0	35	0.00230023
5	0	45	0.00249355
6	0	55	0.00268351
7	0	65	0.00281045
8	0	75	0.00299795
9	0	85	0.00314562
10	15	5	0.00339456
11	15	15	0.00351946
12	15	25	0.00369365
13	15	35	0.00391123
14	15	45	0.00414156
15	15	55	0.00440345
16	15	65	0.00459688

tion θ_i, ϕ_i, is a monotone increasing function of $R_k(\theta_i, \phi_i, \theta_r, \phi_r)$. Of course, the order of summation is arbitrary. Table 13.1 shows how the resulting series would be organized. Once this pseudo-one-dimensional series is established, the random number $R_k(\theta_i, \phi_i, \theta_r, \phi_r)$ can then be used exactly as described in Example Problem 13.1 to recover the direction of a bidirectional reflection from the value of a single random number. It is noted that in principle the function $R_k(\theta_i, \phi_i, \theta_r, \phi_r)$ must be known for each incident direction θ_i, ϕ_i and for each wavelength band k; however, in most cases a different function will be needed only for each incident zenith angle θ_i due to axisymmetry.

Typical results using this approach are illustrated in Figure 13.7. Figure 13.7a shows measured bidirectional reflectivity data for two coats of Krylon true blue interior/exterior paint (#1910) on an aluminum substrate at 3.28 μm when $\theta_i = 20$ deg and $\phi_i = 0$ deg [1], and Figure 13.7b shows simulated data for the same situation obtained using the MCRT method [2]. Figure 13.8, generated using Equation 13.18, relates the value of the index n, corresponding to the reflected direction, to a random number R_k. In the MCRT experiment ten million energy bundles, each incident to the sample at the prescribed angle and reflected at an angle governed by a look-up table represented graphically by Figure 13.8, were collected in the appropriate solid-angle bins. The global difference between the original data and the simulated data is much less than 0.1 percent, and the experiment required about 35 min to run on a Pentium III-based PC.

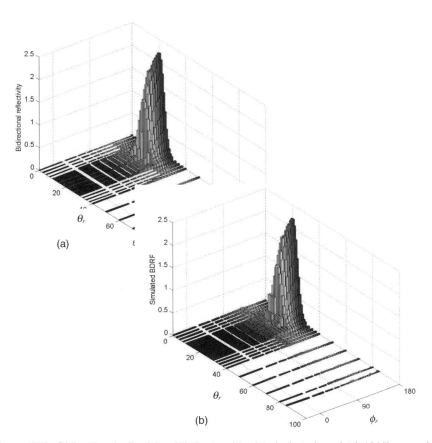

Figure 13.7 Bidirectional reflectivity of Krylon true blue interior/exterior paint (#1910) on an aluminum substrate at 3.28 μm when $\theta_i = 20$ deg and $\phi_i = 0$ deg [1] (a) as measured [1] and (b) from an MCRT simulation [2].

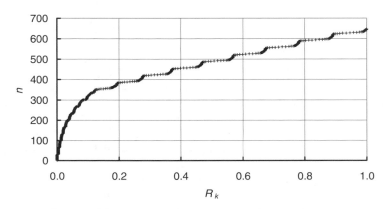

Figure 13.8 Index of scattering angle θ_r, ϕ_r as a function of the value of a random number R_k for blue Krylon paint at 3.28 μm with $\theta_i = 20$ deg and $\phi_i = 0$ deg

13.7   USE OF THE BAND-AVERAGED SPECTRAL RADIATION DISTRIBUTION FACTOR FOR THE CASE OF SOME SPECIFIED SURFACE NET HEAT FLUXES

How do we handle problems in which the surface net heat flux is specified for one or more surfaces? We face two complications:

1. The formulation results in an explicit expression for the surface net heat flux in terms of the surface temperatures. The form of the resulting equations (see Equation 13.5) is such that we cannot explicitly solve them for the surface temperatures in terms of the surface heat fluxes. Therefore, as in Section 11.10, we will be forced to invert a square matrix of order N, where N is the number of surface elements having a specified net heat flux condition.
2. The surface net heat flux is an inherently "total" quantity; that is, it cannot be expressed in terms of band-averaged spectral components. This means that we will be forced to undertake an iterative procedure.

13.7.1   Formulation

When N of the n surface elements have specified surface net heat fluxes we must invert an N-by-N matrix to recover the N unknown surface temperatures. We begin by splitting the right-hand side of Equations 13.5 into two summations,

$$
q_i = \sum_{k=1}^{K} \left[\sum_{j=1}^{N} \varepsilon_{ik} e_b(\Delta\lambda_k, T_j)(\delta_{ijk} - D_{ijk}) \right.
$$

$$
\left. - \sum_{j=N+1}^{n} \varepsilon_{ik} e_b(\Delta\lambda_k, T_j) D_{ijk} \right], \qquad 1 \le i \le N
$$

(13.19)

The first N terms in brackets on the right-hand side involve the N unknown temperatures, while the remaining $n - N$ terms involve only known quantities. It is convenient to rewrite Equation 13.19

$$
\sum_{k=1}^{K} \sum_{j=1}^{N} \varepsilon_{ik} e_b(\Delta\lambda_k, T_j)(\delta_{ijk} - D_{ijk}) = q_i + Q_i, \qquad 1 \le i \le N \qquad (13.20)
$$

where

$$
Q_i \equiv \sum_{k=1}^{K} \sum_{j=N+1}^{n} \varepsilon_{ik} e_b(\Delta\lambda_k, T_j) D_{ijk}, \qquad 1 \le i \le N \qquad (13.21)
$$

Thus, the right-hand side of Equations 13.20 is a known vector of length N. Equation 13.20 represents N equations in N unknown surface temperatures. The system is determinate but the solution of the system for the unknown temperatures is not easy.

Still it is pointed out that in contrast to the net exchange formulation only N (rather than $n \geq N$) equations need be solved.

Once the N equations have been solved for the N unknown surface temperatures, Equations 13.5 may be used $n - N$ times to compute the $n - N$ unknown surface net heat fluxes,

$$q_i = \sum_{k=1}^{K} \sum_{j=1}^{n} \varepsilon_{ik}\, e_b(\Delta\lambda_k, T_j)(\delta_{ijk} - D_{ink}), \qquad N + 1 \leq i \leq n \qquad (13.5a)$$

13.7.2   Solution Strategy

Let us begin by considering the special (but not all that unusual) case, where $N = 1$; that is, where only one surface has a specified net heat flux. In this case, Equations 13.20 may be written

$$f(T_1) = \sum_{k=1}^{K} \varepsilon_{1k}\, e_b(\Delta\lambda_k, T_1)(1 - D_{11k}) - q_1$$

$$- \sum_{k=1}^{K} \sum_{j=2}^{n} \varepsilon_{1k} e_b(\Delta\lambda_k, T_j) D_{1jk} = 0 \qquad (13.22)$$

We seek an algorithm capable of solving this transcendental equation for the root T_1. One very robust algorithm for finding the roots of a transcendental equation, the half-interval search method, is illustrated in Figure 13.9 and described below.

A half-interval search for a root involves systematically incrementing T_1 through a sequence of equally spaced values such that $T_1^{(m+1)} = T_1^{(m)} + \Delta T_1$, where ΔT_1 is a constant whose value determines the accuracy of the solution. For each value of T_1

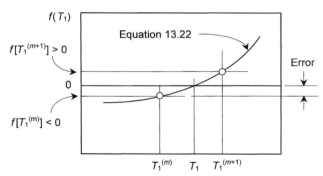

Figure 13.9   Illustration of the half-interval search method for finding the roots of a transcendental equation

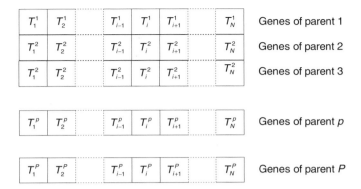

Figure 13.10   Illustration of the "genes" making up one generation of the population of P possible solution vectors

Equation 13.22 is used to compute the corresponding value of $f(T_1)$. At each step the product

$$P_{m+1} = f[T_1^{(m+1)}]f[T_1^{(m)}] \tag{13.23}$$

is computed and the sign of its value tested. If (as illustrated in Figure 13.9) P_{m+1} is negative this means that

$$T_1^{(m)} < T_1 < T_1^{(m+1)} \tag{13.24}$$

The search can now be stopped and the value of $T_1^{(m)}$ taken as an acceptable estimate for T_1, or it may be continued between $T_1^{(m)}$ and $T_1^{(m+1)}$ with a smaller value of ΔT_1, say $\Delta T_1 = \Delta T_1/10$.

Now consider the more difficult case where $N > 1$; that is, where more than one surface element has a specified net heat flux. We now have N equations (Equations 13.20) to solve simultaneously using an iterative approach. Standard techniques exist for accomplishing this, but perhaps the most robust among these are the so-called *genetic algorithms* [3–5].

As the name implies, genetic algorithms involve an analogy with the evolution of the species. Briefly, an initial population is set up whose individual members have "genes" represented by a complete set of unknown temperatures. The genetic content of each member of the original population, illustrated in Figure 13.10, is established randomly, using a pseudo-random number generator. That is, upper and lower bounds on the temperature range within the cavity are established and then each temperature making up the genes of each individual in the original population is computed as

$$T_i = T_{min} + R_T(T_{max} - T_{min}) \tag{13.25}$$

where R_T is the next available random number.

The next step is to test the genetic content of each member of the original population to see which potential "parents" have the "best" genes. This may be accomplished by introducing each set of candidate surface temperatures into Equations 13.20 and then computing the corresponding value of a penalty function. A convenient penalty function is the root-mean-square (rms) difference between the left-hand side (LHS) and the right-hand side (RHS) of the equation; that is,

$$\text{rms} = \sqrt{\frac{1}{N} \sum_{i=1}^{N} (\text{LHS}_i - \text{RHS}_i)^2} \tag{13.26}$$

All members of the original population are now ranked in ascending order of their value of rms, i.e., with the parents having the smallest values of rms at the top of the list.

Next, "marriages" are arranged between randomly paired couples from the upper half of the list (members of the lower half are, alas, not permitted to mate). A marriage consists of using the genetic information of the parents to create "offspring" whose genes are improved in the sense that they will, when tested, produce an even smaller value of RMS. The simplest approach is to average the two values of each surface temperature provided by the two parents. However, faster convergence is obtained if the parent with the "better" genes is weighed more heavily in the averaging process. Also, experience has shown that judicious use of random "mutations" can lead to more rapid improvement of the gene pool while avoiding "local" minima in the penalty function. In any case it is best if each marriage leads to two nonidentical offspring. This assures that the size of the gene pool will not diminish from one generation to the next while at the same time maximizing its genetic diversity.

Each successive generation of offspring is treated in the same manner as the original population, with an attendant increase in the quality of the gene pool. This procedure is continued until the relative root-mean-square difference between the LHS vector and the RHS vector

$$\text{rms} = \frac{\sqrt{\dfrac{1}{N} \sum_{i=1}^{N} (\text{LHS}_i - \text{RHS}_i)^2}}{\dfrac{1}{N} \sum_{i=1}^{N} \text{RHS}_i} \tag{13.27}$$

is less than some predetermined value, say 0.01.

The advantage of genetic algorithms in this application is that, in contrast to other methods, they do not require numerical estimation of partial derivatives of the left-hand side of Equations 13.20 with respect to each of the unknown surface temperatures. Also, they are extremely robust in that they are unconditionally stable.

13.8 SUMMARY

With the new material presented in this chapter and the material previously presented in Chapters 11 and 12, we are now able to perform a fully bidirectional, band-averaged spectral analysis of an enclosure for which we have appropriate bidirectional, spectral reflectivity data. We are in no way inconvenienced by partial or total blockage of the view of one surface element by another; this simply is not an issue. While the method elaborated in Part III of the book works perfectly well for diffuse, gray analyses, it is easier to use than the more traditional net exchange formulation of Part II. More to the point, it is equally easy to use to solve far more general problems such as those involving (bi)directional, wavelength-dependent surface properties. There remains the issue of how to handle participating media. That is, how do we account for emission, absorption, and scattering of radiation by media, such as gases, that may fill the enclosure? In Chapter 14 we will see that the MCRT method is equally adept at treating this class of problem.

Team Projects

TP 13.1   Consider the thermal radiation detector studied in Chapter 12. Suppose that the mirror was polished platinum (see Figure 4.7 and Example Problem 13.1) and the cylindrical detector was a thick dielectric having an index of refraction of 1.5 in the wavelength range of interest, the near infrared (use $\lambda = 2.0$ μm in the analysis). What is the effective absorptivity of the cylindrical wall of the cavity for collimated radiation at 2.0 μm normally incident to the detector aperture? Modify the MCRT code developed in TP 12.1 by including directional, spectral absorptivity for both surfaces. Assume that the reflectivity of both surfaces is ninety-five-percent specular ($r_s = 0.95$).

Discussion Points

DP 13.1   Compare and contrast the exchange factor of Part II of this book with the band-averaged spectral distribution factor introduced in this chapter. How are they similar and how are they different? The discussion should utilize the vocabulary of the discipline.

DP 13.2   The discussion of Figure 13.2 emphasizes the role of blockage. What difficulties would blockage provoke in the net-exchange formulation? What is meant by a "partial" configuration factor or exchange factor? How might one go about computing a partial configuration factor?

DP 13.3   Describe how the hemispherical, band-averaged spectral emissivity ε_{ik} in Equation 13.1 might be obtained from the traditional, spectral emissivity.

DP 13.4   A list of four advantages of the MCRT method over the net-exchange method appears in Section 13.2. Please discuss the second advantage. Is it clear to you why the matrix to be inverted in the case of N specified net heat fluxes is only of order N? Please explain (*Hint:* Review Section 13.7).

DP 13.5   Compare the spectral emission phase-function generator of Figure 13.3 with the spectral scattering phase-function generator of Figure 13.6. After studying these two phase-function generators, can you think of a modification of the former that would require only two random numbers rather than three as an input?

DP 13.6   How would the search algorithm suggested in possibility number 2 of Example Problem 13.1 be structured?

DP 13.7   Produce a new logic block diagram to replace Figure 11.5 that takes into account the new material in this chapter.

DP 13.8   How might the information about the wavelength of the energy bundle at its origin (i.e., the value of the index k) be carried with a given energy bundle as it navigates through the enclosure? (The inherent structure of object-oriented programming languages such as C++, the language used internal to FELIX, makes this relatively easy.)

DP 13.9   Consider Table 13.1. How many rows would be needed to represent the bidirectional, band-averaged spectral scattering phase function if $K = 10$ (ten bands) and the same increment size, 5 deg, were used for θ_i as for θ_r? How does this number constrain the minimum number of significant figures of the random number generator used?

DP 13.10 Can you think of some practical problems where only one surface element of an enclosure has a specified net heat flux and all other surface elements have specified temperatures?

DP 13.11 Go to the library and look up information on genetic algorithms. Then prepare a brief discussion of "crossover rules."

DP 13.12 It is pointed out near the end of Section 13.7.2 that an advantage of genetic algorithms is that they avoid the need to numerically estimate partial derivatives. (Why is it desirable to avoid numerical estimation of partial derivatives?) What alternative approaches to parameter estimation do you know about that do require the numerical estimation of partial derivatives? Prepare a brief discussion of one of these alternative approaches.

Problems

P 13.1 Show formally that in the context of directional emission, $p(\theta) = \varepsilon'_\lambda \cos\theta \sin\theta$ and a random number R_θ uniformly distributed between zero and unity are related according to Equation 13.13.

P 13.2 Show that Equation 13.13 yields Equation 11.11 in the case of a diffuse surface.

P 13.3 Describe, with appropriate equations and figures, a procedure for determining the zenith angle of incidence, in local $\mathbf{n}$, $\mathbf{t}_1$, $\mathbf{t}_2$ coordinates, of an energy bundle arriving at a surface element.

P 13.4 For the extended example in Chapter 12, find the zenith angle of incidence for points C and A in Figure 12.6, and for point C in Figure 12.9.

P 13.5 Derive a suitable regression formula based on the data plotted in Figure 13.5 to produce an estimate of the function f_k of Equation 13.11. You will first have to create a spreadsheet to compute the data for Figure 13.5.

P 13.6 Derive Equation 13.18 from Equation 13.17. How would Equation 13.18 be modified if the reflection hemisphere was not divided into equal-solid-angle patches?

P 13.7 Design a spectral emission phase-function generator like the one shown in Figure 13.3 for which azimuthal symmetry is not assumed. Your design should include the appropriate computer code.

P 13.8 Design a spectral scattering phase-function generator like the one shown in Figure 13.6. Your design should include the appropriate computer code.

REFERENCES

1. Beecroft, M. T., and P. R. Mattison, "Design review of an in-situ bidirectional reflectometer," *Proceedings of the SPIE on Scattering and Surface Roughness,* Vol. 3141, Zu-Han Gu and Alexei A. Maradudin (eds.), 1997, pp. 196–208 (data provided by the authors).
2. Smith, D., J. R. Mahan, M. T. Beecroft, and P. R. Mattison, "Bidirectional reflectivity in the Monte Carlo environment," Paper No. 4540-68, 8th International Symposium on Remote Sensing, Conference 4540A on Sensors, Systems, and Next-Generation Satellites V, Toulouse, France, 17–21, September 2001.
3. Chambers, L., *Practical Handbook of Genetic Algorithms,* CRC Press, Boca Raton, FL, 1995.
4. Goldberg, D., *Genetic Algorithms in Search, Optimization, and Machine Learning,* Addison-Wesley, New York, 1989.
5. Davis, L., *Handbook of Genetic Algorithms,* Van Nostrand Reinhold, New York, 1991.

14

THE MCRT METHOD APPLIED TO RADIATION IN A PARTICIPATING MEDIUM

In Chapter 6 we learned that radiation propagating through certain semitransparent media is subject to attenuation by absorption and scattering, and that its intensity can be augmented by emission and scattering. In this chapter we will find that these phenomena are readily treated using the Monte Carlo ray-trace method. In fact, the MCRT method is particularly well suited for treating radiation within an enclosure filled with a participating medium, especially when coupling exists between surface and volume radiation.

14.1 THE ENCLOSURE FILLED WITH A PARTICIPATING MEDIUM

We now consider an enclosure filled with a participating medium. The enclosure volume is suitably subdivided into volume elements and the enclosure walls are subdivided into compatible surface elements, as illustrated in Figure 14.1. As in the case of surface-to-surface radiation, we seek a matrix of distribution factors D_{ijk} but where now i and j are the indices of either surface *or* volume elements, and k is the index of a wavelength interval $\Delta\lambda_k$ within which the various optical properties can be considered uniform. We have already developed the concept of radiation distribution factors between surface elements when the intervening medium is nonparticipating. We will now develop the radiation distribution factor between volume elements of a participating medium, and between a surface element and a volume element.

In Figure 14.1 the lines separating white from gray represent real surfaces and the lines drawn through the white region represent imaginary surfaces that divide the enclosure into volume elements. The shape of the volume elements is completely arbitrary as far as the MCRT formulation is concerned. However, as explained in Section 14.2.3, certain restrictions are placed on the size and aspect ratio of the volume ele-

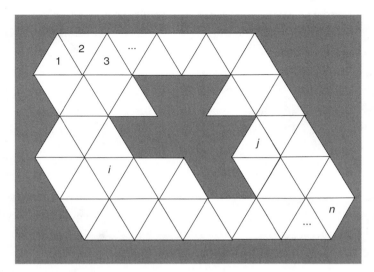

Figure 14.1   An enclosure containing a participating medium subdivided into compatible surface and volume elements

ments. A radiative analysis is frequently performed on the basis of the results of an earlier computational fluid dynamics analysis used to establish the flow, temperature, and species concentration field. In this case it may be convenient and acceptable to use the same mesh already defined for the computational fluid dynamic (CFD) analysis. Figure 14.2 compares the measured and predicted infrared images of a Boeing 747 in flight. The predicted image is based on a CFD analysis whose output has been post processed using an MCRT analysis based on the method developed in this chapter [1]. The CFD mesh used has over 11 million mesh points! Figure 14.3 compares measured and predicted infrared images of the exhaust plume from an auxiliary power unit. In this case data obtained by pressure and temperature probing of the plume were used as the input to the MCRT model [2, 3]. In the following development it is assumed that the enclosure has already been appropriately divided into surface and volume elements and that a suitable surface and volume element numbering system has been established.

14.2 THE MCRT FORMULATION FOR ESTIMATING THE DISTRIBUTION FACTORS

The logic flowchart for estimating the band-averaged spectral distribution factors in the case of an enclosure filled with a participating medium is shown in Figure 14.4. The logic flow begins after the enclosure has been divided into surface and volume elements and the first volume element has been selected from which energy bundles are to be released. Normally, all of the surface and volume elements are treated se-

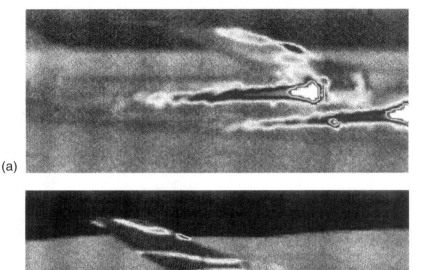

(a)

(b)

Figure 14.2 Comparison of (a) measured and (b) predicted infrared images of a Boeing 747 in flight [1]

quentially as source elements. The function and design of each block is presented in the following subsections.

Consider the volume element ΔV of a participating medium, shown in Figure 14.5. For purposes of this development ΔV is considered to be a rectangular "solid," but this is by no means a limitation of the method; just as in the case of the surface-to-surface development, we are not limited to rectangular volume elements. The steps for implementing the Monte Carlo ray-trace method in the case of emission by a volume element are as follows (refer to Figure 14.4).

14.2.1 Step 1: Determine the Emission Site Within the Volume Element

As a concession to clarity the development followed here assumes rectangular-solid volume elements. The actual shape of the volume elements used will always be dictated by the enclosure geometry. Tetrahedral volume elements are often used in general formulations, such as in the commercial version of FELIX, because they conform well to a wide range of geometries.

Let x_0, y_0, and z_0 be the coordinates of the corner of the volume element that identify the volume element. A relationship analogous to Equations 12.31 and 12.69 could be used to compute the volume element index number from knowledge of x_0, y_0, and z_0. Then let Δx_0, Δy_0, and Δz_0 be the x, y and z dimensions of ΔV. We now draw three random numbers: R_x, R_y, and R_z. Then the coordinates of the point of emission within ΔV are

$$x_0' = x_0 + R_x \Delta x_0 \qquad (14.1)$$

$$y_0' = y_0 + R_y \Delta y_0 \qquad (14.2)$$

and

$$z_0' = z_0 + R_z \Delta z_0 \qquad (14.3)$$

14.2.2 Step 2: Determine the Direction of Emission in the Global Coordinate System

We begin by drawing three random numbers: R_ℓ, R_m, and R_n. We then shift their range of operation from $(0, 1)$ to $(-1, 1)$; that is,

$$R_\ell' = 2R_\ell - 1 \qquad (14.4)$$

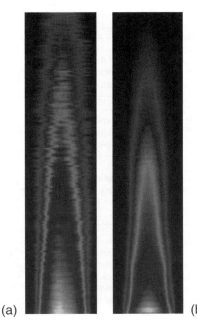

(a)　　　　　　　　　　　　　　(b)

Figure 14.3 Comparison of (a) measured and (b) predicted infrared images of an auxiliary power unit exhaust plume [2, 3]

$$R'_m = 2R_m - 1 \tag{14.5}$$

and

$$R'_n = 2R_n - 1 \tag{14.6}$$

Note that it is usually computationally more efficient to program Equations 14.4 through 14.6 as

$$R' = R + R - 1 \tag{14.7}$$

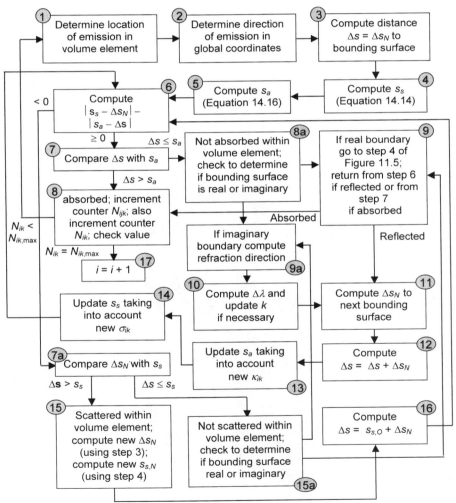

Figure 14.4 Logic flowchart for estimating the distribution factors for an enclosure containing a participating medium

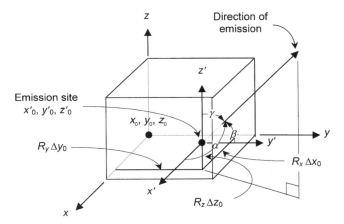

Figure 14.5   Emission of an energy bundle from a volume element

The direction cosines ℓ_1, m_1, and n_1 may then be computed as

$$\ell_1 \equiv \cos \alpha = \frac{R'_\ell}{\sqrt{(R'_\ell)^2 + (R'_m)^2 + (R'_n)^2}} \tag{14.8}$$

$$m_1 \equiv \cos \beta = \frac{R'_m}{\sqrt{(R'_\ell)^2 + (R'_m)^2 + (R'_n)^2}} \tag{14.9}$$

and

$$n_1 \equiv \cos \gamma = \frac{R'_n}{\sqrt{(R'_\ell)^2 + (R'_m)^2 + (R'_n)^2}} \tag{14.10}$$

Equations 14.8 through 14.10 represent a set of three numbers having the following properties:

1. All three members of the set are random and are randomly related to each other
2. All three members of the set are in the open interval $(-1, 1)$
3. The square root of the sum of the squares of the members of the set is equal to unity

That is, ℓ_1, m_1, and n_1 represent the direction cosines, in the global coordinate system, of a randomly emitted energy bundle in 4π-space.

14.2.3   Absorption of Thermal Radiation by a Volume Element

If an energy bundle does not escape the volume element where it is emitted, it is said to be *self-absorbed.* Self-absorption will generally occur if the volume element is optically thick. If volume elements are sized such that they are optically thick in a wavelength interval of interest, then only radiation from the volume elements at the walls will penetrate to the walls of the enclosure. This situation leads to gross inefficiencies in the MCRT implementation and unnecessarily compromises the accuracy of the solution obtained. On the other hand, if the volume element is optically thin it is probably too small for efficient treatment. Experience has verified the intuitively attractive idea that volume elements should be chosen so that they are neither optically thick nor optically thin for a given wavelength interval; that is, they should be chosen such that $\kappa_{ik}L \approx 1$, where L is the characteristic length of the volume element. For example, for a rectangular solid volume element we could use $L = (L_x + L_y + L_z)/3$ for elements that are relatively cubic or $L = [(L_x)(L_y)(L_z)]^{1/3}$ for rectangular elements that are elongated or flattened in one direction.

When subdividing a particular medium into volume elements, every effort should be made to assure that L_x, L_y, and L_z are not greatly different from each other. If one dimension is either significantly longer or significantly shorter than the other two, then the volume element will be optically thicker or thinner in that direction than in the other two. This leads to inefficiencies and inaccuracies in the resulting MCRT implementation.

Clearly, it is not possible to use optimally sized ($\kappa_{ik}L \approx 1$) volume elements when the value of the (band-averaged) spectral absorption coefficient varies widely over the wavelength range of interest. In this case the volume elements should be sized so that as much of the energy as possible in the Planck blackbody radiation distribution function is exchanged among volume elements that are of moderate optical thickness. A scheme for accomplishing this goal would be one in which the volume elements corresponding to the more weakly absorbing bands are built up from smaller "building block" volume elements that are themselves sized for the strongest absorbing bands. Then the number of these building blocks used to form a given volume element would depend on the wavelength interval. Wavelength-dependent meshing of a medium is yet to be fully explored in the literature.

14.2.4 Step 3: Compute the Distance Δs to the Bounding Surface in the Direction ℓ_1, m_1, and n_1

Now, the volume element shown in Figure 14.5 has six faces: one at $x = x_0$, one at $x = x_0 + \Delta x_0$, one at $y = y_0$, and so forth. Three of these faces can be immediately eliminated as potential candidates for intersection with the path of the emitted energy bundle based on the signs of the direction cosines. For example, when ℓ_1, m_1, and n_1 are positive, the path of the energy bundle cannot intersect the faces at $x = x_0$, $y = y_0$, and $z = z_0$, respectively (this is the situation illustrated in Figure 14.5). Then each of the three following sets of two equations in two unknowns describes a possible intersection:

$$\frac{x_0 + \Delta x_0 - x_0'}{\ell_1} = \frac{y_1 - y_0'}{m_1} = \frac{z_1 - z_0'}{n_1} \tag{14.11a}$$

if the intersection is with the face at $x = x_0 + \Delta x_0$,

$$\frac{x_1 - x_0'}{\ell_1} = \frac{y_0 + \Delta y_0 - y_0'}{m_1} = \frac{z_1 - z_0'}{n_1} \qquad (14.11b)$$

if the intersection is with the face at $y = y_0 + \Delta y_0$, and

$$\frac{x_1 - x_0'}{\ell_1} = \frac{y_1 - y_0'}{m_1} = \frac{z_0 + \Delta z_0 - z_0'}{n_1} \qquad (14.11c)$$

if the intersection is with the face at $z = z_0 + \Delta z_0$. Note that all three versions of Equations 14.11 provide a candidate coordinate set (x_1, y_1, z_1); the planes containing the faces at $x = x_0 + \Delta x_0$, $y = y_0 + \Delta y_0$, and $z = z_0 + \Delta z_0$ are infinite in extent and the path of the emitted energy bundle in general will intersect all three. However, only one of the three intersections will be in the range of a face of the volume element. It is emphasized that Equations 14.11 represent the special case where ℓ_1, m_1, and n_1 are positive. It is left as an exercise for the student (see Problem P 14.1) to write versions of Equations 14.11 corresponding to cases where one (or more) of the direction cosines is negative.

All three versions of Equations 14.11 are solved to obtain the three candidate coordinate sets x_1, y_1, z_1. Then for each candidate point of intersection the distance Δs_1 from the emission site to the point of intersection,

$$\Delta s_1 = \sqrt{(x_1 - x_0')^2 + (y_1 - y_0')^2 + (z_0 - z_0')^2} \qquad (14.12)$$

is computed. The correct intersection x_1, y_1, z_1 is the one corresponding to the minimum value of Δs_1.

A word of caution is appropriate here. Just as in the case of finding intersections with surface elements (see Chapters 11 and 12), it is possible (though highly unlikely) that two or even three candidate points of intersection will yield exactly the same value of Δs_1 to within the accuracy of the processor being used. This event corresponds to the path of an energy bundle intersecting the volume element along an edge or in a corner. Remember that millions or even billions of energy bundles may be traced through hundreds or even thousands of volume elements in a given application, so that very small individual probabilities can result in rather significant joint probabilities. The logic of the algorithm that applies this test must take this possibility into consideration.

Another equally unlikely but nevertheless possible scenario is that an energy bundle will be emitted, to within the calculation accuracy of the processor being used, exactly parallel to one or even two of the faces. In this case one or even two of the direction cosines will be zero, and the program logic must be able to handle this eventuality. For example, Equations 14.11 could be set equal to a new variable T as suggested in Chapter 11 (see Equation 11.39 and the related discussion).

14.2.5   Step 4: Compute the Free Path of the Energy Bundle Before It Will Suffer a Scattering Event

When the value of Δs_1 has been obtained, we then draw another uniformly distributed random number R_s to determine the distance this particular energy bundle can be expected to travel before being scattered. Equation 6.30 can be used (see Problem P14.2) to establish that the probability that an energy bundle whose energy is confined to wavelength interval $\Delta\lambda$ will travel distance s_s before being scattered is

$$p_{\Delta\lambda}(s_s) = 1 - e^{-\sigma_{\Delta\lambda} s_s} \tag{14.13}$$

Setting $p_{\Delta\lambda}(s_s)$ equal to R_s and solving for the scatter-free path length we obtain

$$s_s = \frac{-1}{\sigma_{ik}} \ell n(1 - R_s) \tag{14.14}$$

where now σ_{ik} is the band-averaged spectral scattering coefficient. Note that the quantity s_s is the *total* distance that this particular energy bundle is destined to travel before being scattered, even if it suffers reflections from solid surfaces and travels through several volume elements before covering this distance.

14.2.6   Step 5: Compute the Free Path of the Energy Bundle Before It Is Absorbed

The *spectral absorptance* associated with a path length s_a is defined (see Problem P 14.3)

$$a_{\Delta\lambda}(s_a) = 1 - e^{-\kappa_{\Delta\lambda} s_a} \tag{14.15}$$

The spectral absorptance can be interpreted as the probability that an energy bundle whose energy lies in wavelength interval $\Delta\lambda$ will be absorbed after it has traveled a distance s_a. We can then draw the next available random number R_a and set it equal to the band-averaged spectral absorptance and, after minor algebraic manipulation, obtain the absorption-free path length

$$s_a = \frac{-1}{\kappa_{ik}} \ell n(1 - R_a) \tag{14.16}$$

14.2.7   Step 6: Compare the Scatter-free Path Length, s_s, with the Absorption-free Path Length, s_a

Is the energy bundle scattered before it is absorbed? To answer this question we compare the scatter-free path length s_s with the absorption-free path length s_a. If absorption occurs before (or simultaneous with) a scattering event, we continue to step 7; otherwise, we continue to step 7a.

14.2.8   Step 7: Compare the Distance from the Current Location to the Wall, Δs, with the Absorption-free Path Length, s_a

Does the energy bundle escape from the current volume element without being absorbed? To answer this question we compare Δs with s_a. If Δs is greater than s_a, the energy bundle is absorbed, in which case control is transferred to step 8; otherwise, we continue to step 8a.

14.2.9   Step 7a: Compare the Distance from the Current Location to the Wall, Δs, with the Scatter-free Path Length, s_s

Does the energy bundle escape from the current volume element without being scattered? To answer this question we compare Δs with s_s. If Δs is greater than s_s, the energy bundle is scattered (step 15), and therefore changes direction within the current volume element; otherwise, it can reach the bounding surface without being scattered (step 16).

14.2.10   Step 8: The Energy Bundle Is Absorbed; Increment the Counters N_{ijk} and N_{ik} and Check the Value of the Latter

In this step we increment two counters: the counter N_{ijk} that keeps track of the number of energy bundles emitted from volume element i that are absorbed by volume element j in wavelength interval $\Delta\lambda_k$,

$$N_{ijk} = N_{ijk} + 1 \tag{14.17}$$

and the counter N_{ik} that keeps track of the total number of energy bundles emitted from volume element i in wavelength interval $\Delta\lambda_k$. If in the case of Equation 14.17, $j = i$, that is if the energy bundle is still in the emitting volume element when it is absorbed, it is said to be *self-absorbed*. We compare the value of N_{ik} to a preset limit. If N_{ik} has reached the limit value, we move on to the next $(i + 1\text{st})$ source element and restart the process (step 17). Otherwise, we return to step 1 and emit another energy bundle from source element i.

14.2.11   Step 8a: The Energy Bundle Is Not Absorbed and So Reaches a Bounding Surface; Determine if the Bounding Surface Is Real (i.e., Solid) or Imaginary

We now know that the energy bundle has reached a bounding surface of the current volume element without being absorbed, although it has perhaps undergone one or more scattering events that have changed its direction and added to its total nonabsorbed path length, Δs. Note that other logic strings lead to step 8a that include the possibility of scattering. The quantity Δs computed in steps 3, 12, and 17 is the total distance traveled by an energy bundle since suffering its last scattering event.

At this juncture it is interesting to point out an important but often overlooked effect of scattering: it lengthens the path of an energy bundle through a volume element when s_s is less than s_a. This means that a scattered energy bundle is more likely to be absorbed within a volume element than a nonscattered energy bundle. The net result of scattering is to increase the absorptance of a volume element by increasing the mean path length through it.

Is the bounding surface that the energy bundle has reached real (solid) or imaginary? Equations like Equations 12.31 and 12.69 are used to determine the index number of the surface reached and, depending on the value of the index number, a look-up table provides the surface properties.

14.2.12 Step 9: The Boundary Is Real; Determine if the Energy Bundle Is Absorbed or Reflected

We have arrived at a solid wall of the enclosure. Is the energy bundle absorbed or re-flected? To answer this question control is transferred to step 4 of the logic flowchart for the MCRT method applied to enclosures without a participating medium, Figure 11.5. If the energy bundle is reflected, either diffusely or specularly, it reenters the current logic flowchart at step 11 from step 6 (d or s) of Figure 11.5. If the energy bundle is absorbed by the wall element, we return control to step 8 of the logic flowchart in Figure 14.4.

14.2.13 Step 9a: The Boundary Is Imaginary; Determine the Change of Direction of the Energy Bundle as It Crosses the Interface Between Adjacent Volume Elements

Here we invoke Snell's law, Equation 4.13, as interpreted using the notation of Figure 14.6; that is,

$$\frac{\sin\theta_1}{\sin\theta_2} = \frac{n_2}{n_1} \tag{14.18}$$

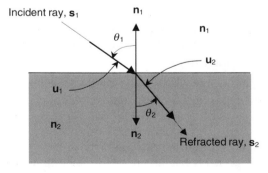

Figure 14.6 Illustration of Snell's law as applied to the path of an energy bundle crossing the boundary of a volume element

where the angles θ_1 and θ_2 are with respect to the surface normal on the two sides of the boundary at the point of intersection. The new direction corresponding to θ_2 must be expressed in terms of direction cosines in the global coordinate system. Figure 14.6 is an adaptation of Figure 4.2 to the problem under consideration. In the figure and the related analysis care should be taken not to confuse three quantities:

$\mathbf{n}_{1,2}$ = the unit normal vector (**bold**)

$n_{1,2}$ = the index of refraction

$n_{1,2}$ = the direction cosine with respect to the z direction (*italics*)

In most cases the context will be sufficient to distinguish among the three uses of the letter "n."

In the following discussion we assume that the volume element under consideration is aligned such that the unit normal vector $\mathbf{n}_1$ on the face of interest is aligned with one of the unit vectors along the coordinate axes, i.e., $\mathbf{n}_1 = \pm\,\mathbf{i},\,\pm\,\mathbf{j},$ or $\pm\,\mathbf{k}$. This will often be the case with thoughtful definition of the coordinate axes in a given problem. Also, $\mathbf{n}_2 = -\mathbf{n}_1$ as indicated in Figure 14.6.

We begin by computing θ_1 from the relation

$$\mathbf{u}_1\cdot\mathbf{n}_1 = -\cos\theta_1 \tag{14.19}$$

where

$$\mathbf{u}_1 \equiv \frac{\mathbf{s}_1}{|\mathbf{s}_1|} = \ell_1\mathbf{i} + m_1\,\mathbf{j} + n_1\mathbf{k} \tag{14.20}$$

For example, if $\mathbf{n}_1 = -\mathbf{j}$, then $\mathbf{u}_1\cdot\mathbf{n}_1 = -m_1 = -\cos\theta_1$, or $\theta_1 = \cos^{-1}(-m_1)$, or finally $\theta_1 = \beta_1$, the angle measured from the positive-y axis to $\mathbf{u}_1$.

We then compute θ_2 from Snell's law and knowledge of the indices of refraction n_1 and n_2,

$$\theta_2 = \sin^{-1}\left[\left(\frac{n_1}{n_2}\right)\sin\theta_1\right] \tag{14.21}$$

But from Figure 14.6 we see that $\mathbf{u}_2\cdot\mathbf{n}_2 = \cos\theta_2$, where

$$\mathbf{u}_2 \equiv \frac{\mathbf{s}_2}{|\mathbf{s}_2|} = \ell_2\mathbf{i} + m_2\,\mathbf{j} + n_2\,\mathbf{k} \tag{14.22}$$

Now in the current example $\mathbf{n}_2 = -\mathbf{n}_1 = \mathbf{j}$. Therefore, $m_2 = \cos\theta_2$, where θ_2 is given by Equation 14.21.

The remaining two direction cosines, ℓ_2 and n_2, may be obtained by geometrical construction, as illustrated in Figure 14.7. When the unit vectors $\mathbf{u}_1$ and $\mathbf{u}_2$ are pro-

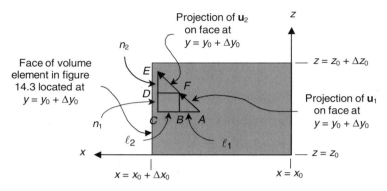

Figure 14.7   Geometry for determining ℓ_2 and n_2

jected on the plane at $y + \Delta y_0$, the magnitudes of their projections are $\sin\theta_1$ and $\sin\theta_2$, respectively. Noting that the triangles ABF and FDE are similar, we can write

$$\frac{\ell_1}{\ell_2} = \frac{n_1}{n_2} = \frac{\sin\theta_1}{\sin\theta_2} = \frac{n_2}{n_1} \tag{14.23}$$

where particular care must be exercised not to confuse the direction cosine n with the index of refraction n in reading the equation. The final equality expressed by Equation 14.23 is Snell's law, Equation 14.18. After appropriate manipulation of Equations 14.23 we have, for the present example,

$$\ell_2 = \left(\frac{n_1}{n_2}\right)\ell_1 \tag{14.24}$$

and

$$n_2 = \left(\frac{n_1}{n_2}\right)n_1 \tag{14.25}$$

It must be emphasized that these results for ℓ_2, m_2, and n_2 are not general but rather are specific to the rectangular solid volume element of the present example. However, the principles embodied in this example are general and can easily be adapted to other geometries.

14.2.14   Step 10: Compute $\Delta\lambda$ and Update the Index k if Necessary

It is important to bear in mind that when an energy bundle crosses into a volume element whose index of refraction n is different, e.g., when $n_2 \neq n_1$, its wavelength changes according to

$$\nu = \frac{c_1}{\lambda_1} = \frac{c_2}{\lambda_2} \tag{14.26}$$

that is, the frequency ν is constant across the interface. Invoking the definition of the index of refraction, $n \equiv c_0/c$, where c_0 is the speed of light in a vacuum, we have for the wavelength shift as the energy bundle crosses from the emitting volume element where $n = n_1$ to the neighboring volume element where $n = n_2$,

$$\frac{\lambda_2 - \lambda_1}{\lambda_1} = \frac{\Delta\lambda}{\lambda_1} = \frac{n_1}{n_2} - 1 \tag{14.27}$$

Recall that we are formulating a band-averaged analysis in which the spectral absorption coefficient is considered constant in a given wavelength interval $\Delta\lambda_k$. However, the wavelength shift provoked by the change in index of refraction from one volume element to the next may take us outside the range of $\Delta\lambda_k$. In this case we continue tracing the energy bundle taking into account the local spectral absorption coefficient based on the new wavelength interval when evaluating and using the absorptance $a_{\Delta\lambda}$ of the path length. Now, for a gas the index of refraction n is proportional to the mass density ρ (kg/m^3), so that

$$\frac{\Delta\lambda}{\lambda_0} = \frac{-\Delta\rho/\rho_0}{\Delta\rho/\rho_0 + 1} \qquad \text{(gas)} \tag{14.28}$$

and for an ideal gas ($p = \rho RT$), $\Delta\rho/\rho_0 = -\Delta T/T_0$, so that

$$\frac{\Delta\lambda}{\lambda_0} = \frac{\Delta T/T_0}{1 - \Delta T/T_0} \qquad \text{(ideal gas)} \tag{14.29}$$

This means that a 10-percent increase in temperature of an ideal gas over the life path of an energy bundle would translate into an 11-percent increase in wavelength.

14.2.15   Step 11: Compute the Nonscattered Path Length to the Next Bounding Surface

We are now in a new volume element. Our next step is to determine which bounding surface the energy bundle will strike next, assuming it is not scattered or absorbed in the current volume element. The procedure is exactly as already described in step 3, and so is not repeated here.

14.2.16   Step 12: Compute the Total Path Length to the Next Bounding Surface from the Last Scattering Event

The energy bundle may suffer an absorption or scattering event before reaching the bounding surface toward which it is headed. To check this possibility we sum the path length segments since the last scattering event and then transfer control to step 14. This total path length is the cumulative path length from the point of emission, including the projected distance across the current volume element, Δs_N; that is,

$$\Delta s = \Delta s_O + \Delta s_N \tag{14.30}$$

In Equation 14.30, Δs_O refers to the "old" cumulative path length traveled before reaching the current volume element.

14.2.17   Step 13: Update the Value of the Absorption-free Path Length Taking into Account the Possibility of a New Value of the Band-Averaged Spectral Absorption Coefficient

If step 11 was entered from step 10 and the band interval $\Delta\lambda_k$ changed from k_1 to k_2 as a result of refraction, then we must update the value of the absorption-free path length. We begin by using Equation 14.16 to compute s_{a2} corresponding to the new value of κ_{ik}. We then must recognize that part of the current absorption-free path was traveled in band interval k_1 so that the updated value of the absorption-free path length is a pathlength-weighted average of the two values of s_a; that is,

$$s_a = \frac{\Delta s_1 s_{a1} + \Delta s_N s_{a2}}{\Delta s_1 + \Delta s_N} \tag{14.31}$$

where Δs_1 is the absorption-free path length actually followed by the energy bundle as it traveled in band interval k_1, and Δs_N is the straight-line distance to the boundary of the current volume element.

14.2.18   Step 14: Update the Value of the Scatter-free Path Length Taking into Account the Possibility of a New Value of the Band-Averaged Spectral Scattering Coefficient

Once again, if step 11 was entered from step 10 and the band interval $\Delta\lambda_k$ changed as a result of refraction, then we must update the value of the scatter-free path length. The procedure is similar to that for the absorption-free path length in the previous section. For example, if the first segment of the scatter-free path was traveled in band interval k_1 and the second segment in band interval k_2, then the updated value of the scatter-free path length is

$$s_s = \frac{\Delta s_1 s_{s1} + \Delta s_N s_{s2}}{\Delta s_1 + \Delta s_N} \tag{14.32}$$

where the symbols have an interpretation similar to those in Equation 14.31, except that now scattering is involved rather than absorption.

14.2.19   Step 15: Scattering Has Occurred in the Volume Element; Compute the New Direction and Distance from the Scattering Location to the Wall After Scattering Within the Volume Element

If the energy bundle was scattered within the current volume element we need to determine its new direction, an updated location where it would strike a bounding surface, and a new total distance traveled from its point of emission to this surface. To determine the direction of emission we need to access a band-averaged spectral scattering phase function generator. Such an engine could be based on either the Mie scattering or the Rayleigh scattering model or on some other appropriate model, including one based on available data.

The design of a band-averaged spectral scattering phase function generator would proceed along the lines already described in Section 13.6 for the case of bidirectional reflection from a surface. The details are left as an exercise for the student. In general, we will draw the next one or two available random numbers and—along with the incident direction cosines, the volume element index i, and the wavelength band index k—enter the band-averaged spectral phase function generator, from which we will exit with the direction cosines of the new direction. We then return to step 3 where the distance from the scattering location to the bounding surface is computed. Next control is transferred to step 4 where a new value of the scatter-free path s_s is computed. Finally, control is passed to step 16, where the total distance traveled since emission is computed.

14.2.20 Step 16: Update the Total Distance Traveled Since Emission, Assuming the Energy Bundle Reaches the Next Bounding Surface Without Being Absorbed

We have had a scattering event but the energy bundle still has not been absorbed. Beyond this point we must keep track of two accumulating distances: the distance traveled by the energy bundle since emission, assuming that it reaches the boundary of the current volume element without being absorbed, and the distance traveled since the last scattering event, once again assuming that the energy bundle reaches the bounding surface without being scattered. The former distance is given by

$$\Delta s = s_s + \Delta s_N \qquad (14.33)$$

and the latter is simply Δs_N. After execution of step 16, control is transferred to step 6.

14.2.21 Step 15a: Scattering Has Not Occurred in the Volume Element; Check to See if the Boundary in the Scattering Direction Is Real or Imaginary

When we reach this point we know that the energy bundle can reach the volume element boundary from its current location without being absorbed or scattered. We now need to check to see if that boundary is real or imaginary (see Section 14.2.11). If the boundary is real we transfer control to step 9, and if it is imaginary we transfer control to step 9a.

14.2.22 Estimation of the Band-Averaged Spectral Distribution Factor

All of the emission, scattering, and absorption issues have now been covered and there remains only the task of computing the estimate of the distribution factor matrix. As in the case of surface-to-surface radiation we have,

$$D_{ijk} \cong \frac{N_{ijk}}{N_{ik}} \tag{14.34}$$

where N_{ijk} is the number of energy bundles emitted from volume element i in wavelength interval $\Delta\lambda_k$, which are absorbed by volume element j, and N_{ik} is the number of energy bundles emitted from volume element i in the same wavelength interval.

The distribution factor defined by Equation 14.34 is used in almost exactly the same manner as the surface-to-surface distribution factor. It may be interpreted as the sensitivity of the radiative power absorbed by volume element j in wavelength interval k to radiation emitted from volume element i in the same wavelength interval.

14.3 USE OF BAND-AVERAGED SPECTRAL RADIATION DISTRIBUTION FACTORS IN A PARTICIPATING MEDIUM

We are generally interested in analyzing radiation through participating media bounded by connected surfaces that define an enclosure. In such situations radiation exchange necessarily occurs among a mixture of surface and volume elements. Suppose that for $1 \leq i \leq N$ the elements are surface elements bounding an enclosure, and for $N + 1 \leq i \leq n$ the elements are volume elements within the enclosure. Then, from the definition of the distribution factor, the power (W) absorbed by (surface or volume) element i is

$$Q_{i,a} = \sum_{k=1}^{K} \sum_{j=1}^{N} \varepsilon_{jk} \, \Delta A_j \, e_b(\Delta\lambda_k, T_j) \, D_{jik}$$
$$+ 4\pi \sum_{k=1}^{K} \sum_{j=N+1}^{n} \kappa_{jk} \, \Delta V_j \, i_b(\Delta\lambda_k, T_j) \, D_{jik} \tag{14.35}$$

The first term on the right-hand side of Equation 14.35 represents radiation absorbed from the surface elements of the enclosure, and the second term represents radiation absorbed from the volume elements of the enclosure.

The net power from element i is

$$Q_i = Q_{i,e} - Q_{i,a} \tag{14.36}$$

where $Q_{i,e}$ is the power emitted from the surface or volume element,

$$Q_{i,e} = \begin{cases} \displaystyle\sum_{k=1}^{K} \varepsilon_{ik} \, \Delta A_i \, e_b(\Delta\lambda_k, T_i), & 1 \leq i \leq N \\[2ex] \displaystyle 4\pi \sum_{k=1}^{K} \kappa_{ik} \, \Delta V_i \, i_b(\Delta\lambda_k, T_i), & N + 1 \leq i \leq n \end{cases} \tag{14.37}$$

With the introduction of the Kronecker delta function, Equation 13.6, Equations 14.35, 14.36, and 14.37 can be combined to yield

$$
Q_i =
\begin{cases}
\displaystyle\sum_{k=1}^{K}\left[\sum_{j=1}^{N}\varepsilon_{jk}\Delta A_j e_b(\Delta\lambda_k, T_j)(\delta_{jik} - D_{jik})\right.\\
\qquad\qquad\left. - 4\pi\sum_{j=N+1}^{n}\kappa_{jk}\Delta V_j\, i_b(\Delta\lambda_k, T_j)D_{jik}\right], \qquad 1 \le i \le N \\[2ex]
-\displaystyle\sum_{k=1}^{K}\left[\sum_{j=1}^{N}\varepsilon_{jk}\Delta A_j\, e_b(\Delta\lambda_k, T_j)D_{jik}\right.\\
\qquad\qquad\left. - 4\pi\sum_{j=N+1}^{n}\kappa_{jk}\Delta V_j\, i_b(\Delta\lambda_k, T_j)(\delta_{jik} - D_{jik})\right], \qquad N+1 \le i \le n
\end{cases}
\tag{14.38}
$$

It can be shown (see Problems P 14.4 and P 14.5) that the reciprocity relation between two volume elements i and j is

$$
\kappa_{ik}\Delta V_i D_{ijk} = \kappa_{jk}\Delta V_j D_{jik}
\tag{14.39}
$$

and the reciprocity relation between a surface element i and a volume element j is

$$
\varepsilon_{ik}\Delta A_i D_{ijk} = 4\,\kappa_{jk}\Delta V_j D_{jik}
\tag{14.40}
$$

Recognizing that $e_b = \pi i_b$ and introducing Equations 14.39 and 14.40, Equations 14.38 can be written

$$
q_i = \sum_{k=1}^{K}\varepsilon_{ik}\left[\sum_{j=1}^{N}e_b(\Delta\lambda_k, T_j)(\delta_{ijk} - D_{ijk}) - \sum_{j=N+1}^{n}e_b(\Delta\lambda_k, T_j)D_{ijk}\right],
$$
$$
1 \le i \le N \quad (14.41)
$$

and

$$
\dot q_i = -4\sum_{k=1}^{K}\kappa_{ik}\left[\sum_{j=1}^{N}e_b(\Delta\lambda_k, T_j)D_{ijk} - \sum_{j=N+1}^{n}e_b(\Delta\lambda_k, T_j)(\delta_{ijk} - D_{ijk})\right],
$$
$$
N+1 \le i \le n
$$
$$
\tag{14.42}
$$

Note that Equation 14.41 defines a net *heat flux* (W/m²) from a surface element, while Equation 14.42 defines a net *volumetric cooling rate* (W/m³) of a volume element. In Equation 14.35 and Equations 14.37 through 14.42 the band-averaged spectral emissivity ε_{jk} and the band-averaged spectral absorption coefficient κ_{jk} are the values averaged over the wavelength interval $\Delta\lambda_k$, as suggested by Equations 6.8 or 6.8a.

Equations 14.41 and 14.42 are explicit expressions for the net heat transfer *from* a given surface or volume element given the temperature distribution on the enclosure walls and within the participating medium filling the enclosure. It is emphasized that, if all of the temperatures are specified, there is no matrix to invert; the net heat transfers can be computed in a spreadsheet environment once the band-averaged spectral distribution factors have been estimated. Equations 14.41 and 14.42 are extremely useful because it is often the case that the temperature field is completely known and the wall heat transfer is desired, for example, in the prediction of the infrared signature of the hot parts and exhaust plume of a jet engine.

14.4 EVALUATION OF UNKNOWN TEMPERATURES WHEN THE NET HEAT TRANSFER IS SPECIFIED FOR SOME SURFACE AND/OR VOLUME ELEMENTS

Situations are inevitable in which the net heat transfer is prescribed for some or all of the surface or volume elements. A typical example might be the thermal design of a gas turbine combustor, in which case radiation combines with heat convection to produce one of the most difficult heat transfer analysis problems imaginable. In this situation we seek to predict the combustor liner temperature distribution from estimates of the volumetric heat release distribution subject to candidate liner cooling schemes. Our approach to this type of problem is almost identical to that described in Section 13.7.

The first step is to recast Equations 14.41 and 14.42 so that the terms representing surface and volume elements having specified net heat transfer are separated from those representing surface and volume elements having a specified temperature; that is,

$$
q_i = \sum_{k=1}^{K} \varepsilon_{ik} \left[\sum_{j=1}^{N_1} e_b(\Delta\lambda_k, T_j)(\delta_{ijk} - D_{ijk}) + \sum_{j=N_1+1}^{N} e_b(\Delta\lambda_k, T_j)(\delta_{ijk} - D_{ijk}) \right.
$$
$$
\left. - \sum_{j=N+1}^{n_1} e_b(\Delta\lambda_k, T_j)D_{ijk} - \sum_{j=n_1+1}^{n} e_b(\Delta\lambda_k, T_j)D_{ijk} \right], \qquad 1 \le i \le N
$$

(14.43)

and

$$
\dot{q}_i = -4 \sum_{k=1}^{K} \kappa_{ik} \left[\sum_{j=1}^{N_1} e_b(\Delta\lambda_k, T_j)D_{ijk} + \sum_{j=N_1+1}^{N} e_b(\Delta\lambda_k, T_j)D_{ijk} \right.
$$
$$
- \sum_{j=N+1}^{n_1} e_b(\Delta\lambda_k, T_j)(\delta_{ijk} - D_{ijk})
$$
$$
\left. - \sum_{j=n_1+1}^{n} e_b(\Delta\lambda_k, T_j)(\delta_{ijk} - D_{ijk}) \right], \qquad N+1 \le i \le n
$$

(14.44)

In Equations 14.43 and 14.44 the first N_1 surface elements have specified net heat fluxes and the first $n_1 - N$ volume elements have specified net cooling rates. Next we rewrite Equations 14.43 and 14.44 with the unknown quantities moved to the left-hand side (LHS) and the known quantities moved to the right-hand side (RHS),

$$\sum_{k=1}^{K} \varepsilon_{ik} \left[\sum_{j=1}^{N_1} e_b(\Delta\lambda_k, T_j)(\delta_{ijk} - D_{ijk}) - \sum_{j=N+1}^{n_1} e_b(\Delta\lambda_k, T_j)D_{ijk} \right]$$

$$= q_i - \sum_{k=1}^{K} \varepsilon_{ik} \left[\sum_{j=N_1+1}^{N} e_b(\Delta\lambda_k, T_j)(\delta_{ijk} - D_{ijk}) \right. \tag{14.45}$$

$$\left. - \sum_{j=n_1+1}^{n} e_b(\Delta\lambda_k, T_j)D_{ijk} \right], \qquad 1 \le i \le N$$

and

$$-4\sum_{k=1}^{K} \kappa_{ik} \left[\sum_{j=1}^{N_1} e_b(\Delta\lambda_k, T_j)D_{ijk} - \sum_{j=N+1}^{n_1} e_b(\Delta\lambda_k, T_j)(\delta_{ijk} - D_{ijk}) \right]$$

$$= \dot{q}_i + 4\sum_{k=1}^{K} \kappa_{ik} \left[\sum_{j=N_1+1}^{N} e_b(\Delta\lambda_k, T_j)D_{ijk} \right. \tag{14.46}$$

$$\left. + \sum_{j=n_1+1}^{n} e_b(\Delta\lambda_k, T_j)(\delta_{ijk} - D_{ijk}) \right], \qquad N+1 \le i \le n$$

We are now in a position to use a genetic algorithm to compute the unknown temperatures exactly as described in Section 13.7.2.

Finally, it is noted that even though Equations 14.45 and 14.46 explicitly involve only the band-averaged spectral emissivity and absorption coefficient, scattering effects are fully accounted for, coming into the analysis through the distribution factors.

Team Projects

TP 14.1 Design a band-averaged spectral phase function generator for solar radiation scattered by rain droplets based on Problems P 6.6 through P 6.8. Allow the diameters of the spherical scattering particles to be normally distributed with a specified mean and standard deviation (the mean diameter should be large compared to the wavelength of the scattered radiation for the scattering model to be valid). Let the wavelength dependence of the index of refraction correspond to that for water. Apply the phase function generator to the problem of generating a rainbow.

TP 14.2 Consider a cylindrical plume of hot gas escaping from a jet engine nozzle. Suppose that for a certain wavelength in the infrared the plume optical diameter is twenty. Our task is to predict the infrared image of a length of the

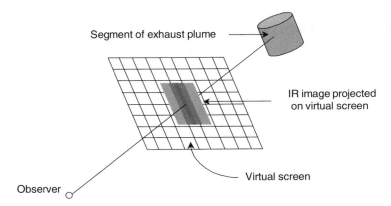

Segment of exhaust plume

IR image projected
on virtual screen

Virtual screen

Observer

Figure 14.8 Geometry for Team Project TP 14.2

plume one diameter long when viewed from a point at right angles to the plume, as shown in Figure 14.8. Assume that the radial temperature distribution in kelvins is reasonably well approximated by $300 + 1000[1 - (2r/D)^2]$, where r is the radial optical coordinate and D is the optical diameter of the plume. Assume that the axial variation in temperature can be neglected over one diameter of length in this part of the plume. We will also neglect scattering. The infrared image is to be projected on the "virtual screen" depicted in the figure. [*Hint:* Use an "inverse ray trace" in which energy bundles are traced through the virtual screen into the plume from the observer point and reciprocity is then used to interpret the distribution factors obtained.]

Discussion Points

DP 14.1 Derive an equation analogous to Equations 12.31 and 12.69 for computing the index number of the volume element of Figure 14.5 from knowledge of the corner coordinates x_0, y_0, z_0.

DP 14.2 How would Equations 14.1 through 14.3 change if the emitting volume element were a cylindrical segment?

DP 14.3 How would Equations 14.1 through 14.3 change if the emitting volume element were a spherical segment?

DP 14.4 In Section 14.2.3 two relationships are suggested for determining the characteristic dimension of rectangular-solid volume elements. One of these is claimed to be preferable for volume elements that are elongated or flattened in one direction. Why? Study the behavior of these two relations and prepare a discussion of their suitability for use in defining the characteristic dimension of a rectangular-solid volume element.

DP 14.5   Critique the sentence in the second paragraph of Section 14.2.3, "This leads to inefficiencies and inaccuracies in the resulting Monte Carlo implementation."

DP 14.6   Consider the statement at the end of the second paragraph in Section 14.2.3, "Wavelength-dependent meshing of a medium is yet to be fully explored in the literature." Briefly discuss how you would attack this problem as, say, a doctoral dissertation.

DP 14.7   How would you implement the logic implied in the second and third sentences in Section 14.2.4?

DP 14.8   How would you translate the "word of caution" beginning the last paragraph in Section 14.2.4 into a computer algorithm?

DP 14.9   How would you modify Equations 14.11 as suggested in the last paragraph in Section 14.2.4?

DP 14.10   What is the ramification of the last sentence in Section 14.2.5? That is, how would you use the quantity s_s between the point where the energy bundle is emitted and the point where it is finally scattered, assuming that it is scattered before it is absorbed?

DP 14.11   The mechanism described in the second paragraph of Section 14.2.11 may be the source of a phenomenon called *anomalous absorption,* which is currently a subject of some controversy in the atmospheric radiation community. The lack of closure between visible radiative flux data obtained from spaced-based and ground-based observations seems to indicate that clouds absorb more visible (solar) radiation than predicted by their "water path length." Some authorities claim that radiative transfer analysis that includes the increased path length due to scattering does not completely account for observed anomalous absorption. Others question whether or not the radiative transfer analysis referred to in this claim was done correctly (as described in the current chapter). Briefly outline a doctoral dissertation aimed at settling this controversy. What would be some of the essential features of such a study?

DP 14.12   In Equation 14.21, if $n_1 > n_2$, (i.e., if medium 1 is more dense than medium 2), we have the mathematical possibility that $\sin\theta_2$ will be greater than unity, which is, of course, meaningless. How should this eventuality be interpreted physically? (Please see Reference 4.)

DP 14.13   How would you apply the principles of Section 14.2.13 to the geometry of Discussion Point DP 14.2 if, for example, the energy bundle crosses the curved face at $r_0 + \Delta r$?

DP 14.14   How would you design the band-averaged spectral scattering phase function generator described in the second and third paragraphs of Section 14.2.19? Your response to this question should include an appropriate equation (see Team Project TP 14.1).

DP 14.15 Team Project TP 14.2 invokes the very important concept of the inverse (or reverse) ray trace. Prepare a brief discussion of this concept. When should it be used?

Problems

P 14.1 Rewrite Equations 14.11 for the case where all three direction cosines of the emitted energy bundle path are negative.

P 14.2 Begin with Equation 6.30 and derive Equation 14.13. In what sense is $p_{\Delta\lambda}(s_s)$ a probability?

P 14.3 Derive Equation 14.15.

P 14.4 Derive the reciprocity relation between two volume elements, Equation 14.39.

P 14.5 Derive the reciprocity relation between a surface element i and a volume element j, Equation 14.40.

REFERENCES

1. Turk, J. A., *Acceleration Techniques for the Radiative Analysis of General Computational Fluid Dynamics Solutions Using Reverse Monte-Carlo Ray Tracing,* Ph.D. Dissertation, Department of Aerospace & Ocean Engineering, Virginia Polytechnic Institute and State University, Blacksburg, VA, December 1994.
2. Nelson, E. L., *Temperature, Pressure and Infrared Image Survey of an Axisymmetric Heated Exhaust Flow,* Ph.D. Dissertation, Department of Mechanical Engineering, Virginia Polytechnic Institute and State University, Blacksburg, VA, December 1994.
3. Nelson, E. L., J. R. Mahan, L. D. Birckelbaw, J. A. Turk, D. A. Wardwell, and C. E. Hange, *Temperature, Pressure, and Infrared Image Survey of an Axisymmetric Heated Exhaust Plume,* National Aeronautics and Space Administration Technical Memorandum 110382, February 1996.
4. Lindley, C. A., *Practical Ray Tracing,* Wiley, New York, p. 96.

15

STATISTICAL ESTIMATION OF UNCERTAINTY IN THE MCRT METHOD

The Monte Carlo ray-trace (MCRT) method is based on a probabilistic interpretation of the radiative properties of surface and volume elements, and the radiation distribution factor is itself a probability. Therefore, the uncertainty of results obtained using the method is expected to be predictable using standard statistical methods. Specifically, we should be able to use statistical inference to state, to a specified level of confidence, the uncertainty of a result obtained. The chapter begins with a brief review of probability and statistics, after which the principles of statistical inference are applied to the MCRT method. Finally, a formal structure is presented for the experimental design of MCRT algorithms.

15.1 STATEMENT OF THE PROBLEM

The *uncertainty* in the value of a physical quantity is generally taken to be the difference between the true and observed values of the quantity, where the true value is unknown. A procedure is sought that provides an estimate, with a stated level of confidence, of the uncertainty of a result obtained using the Monte Carlo ray-trace method in radiation heat transfer analysis.

The problem of estimating the uncertainty of a result obtained using the MCRT method consists of three steps:

1. Quantify the uncertainty of the model that uses the distribution factors assuming that the true values of the distribution factors are known.
2. Quantify the uncertainty of the estimates of the distribution factors.
3. Combine the two components of the uncertainty in a statistically valid manner.

413

Although we will not attack the problem in the order listed above, this is the logical sequence for understanding error propagation through a MCRT-based analysis.

15.2   STATISTICAL INFERENCE

The MCRT method allows us to obtain an *estimate* of the distribution factor based on a *sample* of a large *population*. In principle, the "true" value of the distribution factor is based on an infinite population of uniformly distributed random numbers, while the estimate is based on a large but finite sample of this population. The validity of the method is based on the hypothesis that the statistical behavior of the sample reflects that of the population from which it is drawn. This also turns out to be one of the underlying hypotheses of *statistical inference*. Therefore, we are encouraged to seek means of applying statistical inference to results obtained using the MCRT method.

The distribution of a population has a *mean* μ and *standard deviation* σ, neither of which is known in most applications of practical interest. A third property of a population distribution is the *proportion* π having a certain characteristic. The proportion property divides a distribution into two parts (male and female, old and young, heads and tails, and so forth). It is sometimes convenient to think of the total number of energy bundles emitted from surface or volume element i as being divided into two groups: those absorbed by surface or volume element j, and those *not* absorbed by surface or volume element j. The true value of the distribution factor D_{ij}^t (which is never known) is the *probability* that an energy bundle emitted by a given surface or volume element will be absorbed by some other specified surface or volume element. Because they are probabilities, all of the distribution factors from a given surface or volume element sum to unity over the enclosure. Thus, if D_{ij}^t is the probability that an energy bundle emitted from surface or volume element i will be absorbed by surface or volume element j, then $1 - D_{ij}^t$ is the probability that the energy bundle will be absorbed elsewhere in the enclosure. Thus, we have the pair of possible results "absorbed by j" and "not absorbed by j." It is in this sense that the radiation distribution factor can be thought of as a proportion.

The *sample mean m,* the *sample standard deviation s,* and the *sample proportion p* are *estimators* of, respectively, the population mean μ, the population standard deviation σ, and the population proportion π. Statistical inference in the present context consists of using a sample of the population to obtain values of m, s, and p whose deviations from μ, σ, and π are bounded with a known level of confidence. The validity of statistical inference increases with the size of the sample as long as the sample remains small compared to the population from which it is drawn. This is because the statistical distribution of the estimators becomes increasingly normal under these sample size restrictions, which clearly apply in the MCRT environment.

In the realm of statistics the term *confidence* bears a formal relationship to the probability that the value of a parameter will fall within a stated interval. The *degree of confidence* is normally stated in terms of a *confidence interval*. For example, a 95-percent confidence interval means that the probability of the true value of a parameter lying within a specified interval is at least 0.95.* The relevant statistical tool for computing confidence intervals for the mean μ is the *Student's t statistic,*

$$t = \frac{m - \mu}{s/\sqrt{N}} \qquad (15.1)$$

where the sample standard deviation s is

$$s = \sqrt{\frac{1}{N - 1} \sum_{i=1}^{N} (x_i - m)^2} \qquad (15.2)$$

and N is the number of observations in the sample. If the sample size is sufficiently large (but still small compared to the population size), the Student's t statistic is related to the *normal probability distribution function,*

$$p(z) = \frac{1}{\sigma\sqrt{2\pi}} e^{-(z-\mu)^2/2\sigma^2} \qquad (15.3)$$

as indicated in Figure 15.1. The total area under the curve is unity, and the shaded area under the curve, bounded by $\pm t$, is the probability $P(t)$. In the context of Equation 15.1, $P(t)$ is interpreted as the level of confidence that the true mean μ lies on the interval

$$m - t\frac{s}{\sqrt{N}} \le \mu \le m + t\frac{s}{\sqrt{N}} \qquad (15.4)$$

---

* Formally, statisticians prefer to say that the proposition that the true value lies within the specified range holds with 95-percent confidence.

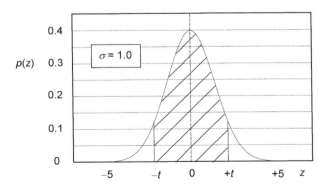

Figure 15.1 The normal probability distribution function showing the values of *t* corresponding to a given confidence interval

Example Problem 15.1

A certain pseudo-random number generator (PRNG) is claimed to provide random numbers uniformly distributed between zero and unity. We would like to know with 95-percent confidence the upper and lower limits on the mean value of the random numbers provided by this PRNG. We will base our estimate on the limited sample of 35 random numbers whose values are plotted in Figure 15.2. The mean value of the sample is found to be $m = 0.4676$ with a sample standard deviation $s = 0.2995$. The value of the t statistic as a function of confidence interval is commonly tabulated in statistics texts. For a 95-percent confidence interval with a sufficiently large population ($N > 30$) we find $t = 1.960$. Then with a 95-percent level of confidence we can state that the true mean of the random numbers provided by the candidate PRNG lies in the interval

$$m - 1.960 \frac{s}{\sqrt{N}} \leq \mu \leq m + 1.960 \frac{s}{\sqrt{N}} \tag{15.5}$$

or

$$0.4676 - 1.960 \frac{0.2995}{\sqrt{35}} \leq \mu \leq 0.4676 + 1.960 \frac{0.2995}{\sqrt{35}} \tag{15.6}$$

or finally

$$0.3684 \leq \mu \leq 0.5668 \tag{15.7}$$

This is not really a very satisfying result because (1) the sample mean (0.4676) deviates significantly from its expected value (0.5000), and (2) the range of values where we can be 95-percent confident of finding the true mean of the PRNG is fairly broad. Based on this analysis we might be tempted to conclude that our PRNG is biased toward the low end of its range, i.e., toward zero. However, that might be a hasty conclusion in the sometimes-surprising world of statistics, as we shall see in Example Problem 15.2.

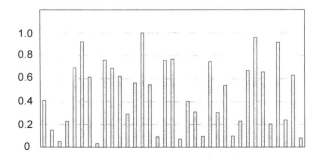

Figure 15.2   A sample of 35 pseudo-random numbers provided by the PRNG

15.3 HYPOTHESIS TESTING FOR POPULATION MEANS

In Example Problem 15.1 we learned only that we can be 95-percent confident that the PRNG provides random numbers whose true mean value is in the range 0.3684 $\leq \mu \leq$ 0.5668. Presumably we would want to know more than this about the performance of the PRNG. For example, how *significant* is the deviation of the sample mean (0.4676) from the expected value (0.5000)? The statistical technique of *hypothesis testing* is used to answer this question.

We may perform one-sided or two-sided hypothesis tests. In the current scenario a *lower one-sided test* is appropriate. The *hypothesis under test,* designated H_0, is stated in such a way that no action is taken if it is accepted. For example, in the current situation we suspect that the PRNG of Example Problem 15.1 is biased toward the low end of the distribution. If this is so the action contemplated is to reject it for use in our MCRT studies. Therefore, the hypothesis under test is

$$H_0: \mu \geq 0.5000 \tag{15.8}$$

If the test fails, the hypothesis is rejected and we accept instead the *alternative hypothesis,*

$$H_1: \mu < 0.5000 \tag{15.9}$$

and conclude that, to a stated level of significance, the PRNG is biased toward the lower limit. However, if the test is passed we accept H_0 and conclude that, to the stated level of significance, the PRNG is not biased toward the lower end of the range.

Two levels of error can occur in hypothesis testing [1]:

1. Type I error in which we reject H_0 when it is in fact true.
2. Type II error in which we accept H_0 when H_1 is true.

The type I error—rejecting a true hypothesis—is generally considered more serious than the type II error—accepting a false hypothesis. For this reason, the test is designed to minimize the incidence of type I errors, that is, to minimize taking action when no action is called for.

The *significance level* α is the upper bound on the probability of rejecting H_0 when it is, in fact, true. Then we set out to test H_0 to a stated level of significance, say 5 percent ($\alpha = 0.05$). The basis of our test is once again the Student t statistic. We use Equation 15.1 to compute the value of t. We then reject H_0 if t is less than or equal to (in the algebraic sense) a *critical value* based on the normal probability distribution function shown in Figure 15.3. Specifically, the critical value of t (<0 in this example) delineates an area (shown shaded in Figure 15.3) under the normal probability distribution function curve, between $-\infty$ and t, equal to α.

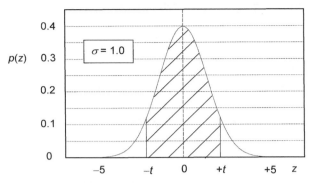

Figure 15.3   The normal probability distribution function showing the critical value of t for a one-sided hypothesis test

Example Problem 15.2

Is the PRNG of Example Problem 15.1 biased toward the low end?
To answer this question we test the hypothesis

$$H_0: \mu \geq 0.5000 \tag{15.10}$$

The values for the sample mean m ($= 0.4676$), the expected mean μ ($= 0.5000$), the sample standard deviation s ($= 0.2995$), and the sample size N ($= 35$) are introduced into Equation 15.1. When this is done there results $t = -0.6400 > -1.645$, where the critical value (-1.645), read from a table, is the value of t shown in Figure 15.3. Therefore, we accept the hypothesis under test and conclude that the PRNG is unbiased to a 5-percent level of significance. This significance level is usually associated with "ordinary" concern about making a type I error, while a 1-percent significance level is associated with "extreme" concern and a 10-percent significance level with "mild" concern.

It should be noted here that a larger sample size than $N = 35$ in Example Problems 15.1 and 15.2 would have presumably yielded a value of m closer to 0.5000, a smaller sample standard deviation, and a narrower 95-percent confidence interval. Ten experiments were conducted in which 350 random numbers were drawn from the same PRNG in each experiment. The results are summarized in Table 15.1. The mean of the mean,

$$\langle m \rangle = \frac{1}{10} \sum_{i=1}^{10} m_i \tag{15.11}$$

for the data given in Table 15.1 is 0.5099. The upper and lower limits listed in Table 15.1 are computed using Equation 15.5. These results, based on 3500 random numbers, support the conclusion of the hypothesis test, based on a sample of only 35 ran-

Table 15.1 Summary of results obtained using the PRNG to obtain ten sets of 350 random numbers each

i	m_i	s_i	Lower Limit	Upper Limit
1	0.5322	0.3011	0.5006	0.5637
2	0.5056	0.2813	0.4762	0.5351
3	0.5419	0.2753	0.5131	0.5707
4	0.4967	0.2792	0.4675	0.5260
5	0.5139	0.2884	0.4837	0.5441
6	0.4954	0.2861	0.4654	0.5254
7	0.5042	0.2784	0.4750	0.5333
8	0.5036	0.2966	0.4725	0.5347
9	0.5077	0.2904	0.4773	0.5381
10	0.4975	0.2908	0.4671	0.5280

dom numbers, that the PRNG is not biased toward the low end. It should be remarked that 3500 is still a very small number of random numbers in the context of a MCRT analysis.

15.4 CONFIDENCE INTERVALS FOR POPULATION PROPORTIONS

It has already been established that the true value of the distribution factor D_{ij}^t is a *population proportion*. We would like to know, to a stated level of confidence, the maximum width of the interval containing the true value of the distribution factor (corresponding to an infinite number of energy bundles traced) and the estimate obtained using a large but finite number of energy bundles.

We recall that population proportions come in pairs (heads and tails, male and female, etc.). Such paired variables are often referred to as *Bernoulli variables*. Population proportions follow a *binomial probability distribution*,

$$P(k) = \binom{N}{k} \pi^k (1 - \pi)^{N-k} \tag{15.12}$$

where

$$\binom{N}{k} = \frac{N!}{k!(N - k)!} \tag{15.13}$$

In Equation 15.12, π is the population proportion (a number between zero and unity), and in Equations 15.12 and 15.13 N is the number of observations, or "experiments," and k is the number of distinguishable ways a particular outcome of the experiment can occur. As an example, what is the probability that two "fair" coins, when tossed on the floor, will end up one heads and one tails? In this example $\pi = 0.5$ (equal chance of landing heads up), $N = 2$ (two independent coins), and $k = 1$ (H/T is indistinguishable from T/H). The probability of getting one heads and one tails (in no particular order) is then

$$P(\text{HT or TH}) = \left(\frac{2!}{1!0!}\right)(0.5)^1(1 - 0.5)^1 = \frac{1}{2} \tag{15.14}$$

Now if a population is *randomly* sampled for proportion and the sample size is sufficiently large, the statistical distribution of the sample proportion p will be normal even though the underlying distribution is binomial. This means that hypothesis testing for proportion can be done using essentially the same statistical tools as used in the case of the population mean. The only difference is that now our estimate of the sample standard deviation is based on the standard deviation of a binomial distribution, that is,

$$s = \frac{\sigma}{N^2} \tag{15.15}$$

The standard deviation for a binomial distribution in Equation 15.15 may be approximated

$$\sigma = \sqrt{N\pi(1 - \pi)} \cong \sqrt{Np(1 - p)} \tag{15.16}$$

in which case

$$s \cong \sqrt{\frac{p(1 - p)}{N}} \tag{15.17}$$

The equivalent of the t statistic in establishing population proportion confidence intervals is the W statistic,

$$W = \frac{p - \pi}{\sqrt{p(1 - p)/N}} \tag{15.18}$$

where the probability distribution of W is approximately normal. Then the confidence interval for the population proportion π is

$$p - W_c\sqrt{\frac{p(1 - p)}{N}} \leq \pi \leq p + W_c\sqrt{\frac{p(1 - p)}{N}} \tag{15.19}$$

where the critical value of W, W_c, is determined exactly as in the case of population mean. So, for example, for a 95-percent confidence interval with $N > 30$, $W_c = 1.960$.

Example Problem 15.3

An MCRT study based on the emission of 100,000 energy bundles leads to an estimate for the total distribution factor from surface i to surface j of $D_{ij}^e = 0.00263$ (that is 263 of the 100,000 energy bundles emitted from surface element i were absorbed by surface element j). Compute the 95-percent confidence interval for this result.

SOLUTION

We use Equation 15.18 with $p = 0.00263$, $W_c = 1.960$, and $N = 100,000$ and obtain

$$0.00231 \le D_{ij}^t \le 0.00295$$

We conclude with a 95-percent level of confidence that the true value of the distribution factor (obtained by tracing an infinite number of energy bundles) lies between 0.00231 and 0.00295. That is, to a 95-percent level of confidence,

$$D_{ij}^t = 0.00263 \pm 0.00032, \quad \text{or } \mathrm{D}_{ij}^t = 0.00263 \pm 12 \text{ percent}$$

The somewhat disappointing result of Example Problem 15.3 immediately leads us to make several important observations. First, the result was obtained without any consideration of the details of the enclosure, its geometry, or its surface optical properties other than the value of the distribution factor itself. Second, for the investment of 100,000 energy bundles one might reasonably expect to get a narrower 95-percent confidence interval than ± 12 percent. Finally, it is clear from the form of Equation 15.19 that interplay occurs between the value of the distribution factor estimate, $p = D_{ij}^e$, and the number of energy bundles emitted. To see this clearly, let us rewrite Equation 15.19 in terms of the fractional width of the confidence interval,

$$1 - W_c \sqrt{\frac{1 - D_{ij}^e}{ND_{ij}^e}} \le \frac{D_{ij}^t}{D_{ij}^e} \le 1 + W_c \sqrt{\frac{1 - D_{ij}^e}{ND_{ij}^e}} \tag{15.20}$$

where now D_{ij}^t is the (unknown) true value of the distribution factor and D_{ij}^e is the MCRT estimate based on tracing n energy bundles. It is clear from this form of the confidence interval expression that we want to minimize the quantity

$$W_c \sqrt{\frac{1 - D_{ij}^e}{ND_{ij}^e}} \tag{15.21}$$

to minimize the departure of D_{ij}^t/D_{ij}^e from unity. For sufficiently large sample sizes N the critical value of the W statistic is fixed once we decide our level of confidence. Therefore, all that remains to work with is the quantity

$$f(N, D_{ij}^e) = \sqrt{\frac{1 - D_{ij}^e}{ND_{ij}^e}} \tag{15.22}$$

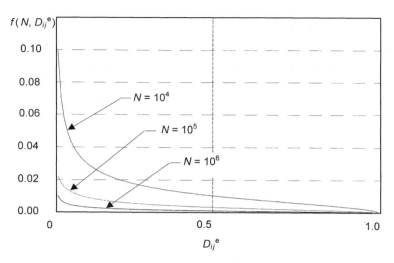

Figure 15.4   Behavior of $f(N, D_{ij}^e)$ defined by Equation 15.22

whose behavior is shown in Figure 15.4. Figure 15.4 may be interpreted as showing the fractional uncertainty in the distribution factor per critical value of the W statistic for a given number of energy bundles traced. It is clear from the figure and from the form of Equation 15.22 that for very small values of the distribution factor the uncertainty can be very large. This is because the absorbing surface, surface j, is increasingly *undersampled* by the Monte Carlo ray-trace method as the value of the distribution factor decreases. Figure 15.4 also verifies the expectation that increasing the number of energy bundles traced reduces the uncertainty in the value of the distribution factor. While this result is not surprising, it is significant to note that the improvement achieved by increasing N is much greater for small values of the distribution factor than for large values.

15.5   EFFECTS OF UNCERTAINTIES IN THE ENCLOSURE GEOMETRY AND SURFACE OPTICAL PROPERTIES

In the previous section it was tacitly assumed that the estimates for the distribution factors were based on exactly known enclosure geometry and surface optical properties. We refer to confidence intervals obtained under that assumption as *inherent confidence intervals*. That is, they are inherent to the statistical nature of the MCRT method and do not take into account our degree of confidence in the model upon which the Monte Carlo ray trace is based. An imperfect model will lead to imperfect results for the distribution factors whose degree of imperfection goes beyond that implied by the width of the inherent confidence interval.

Now, in practice, enclosure dimensions and surface optical properties carry their own uncertainties. Recall that the uncertainty in a quantity is the (unknown) difference between the true and observed values. At this point it is convenient to distinguish between the *inherent uncertainty* in the distribution factor estimate related to the inherent confidence interval developed in the previous section, and the *aggravated uncertainty* related to the uncertainties in the enclosure geometry and surface optical properties. All uncertainties are themselves random variables and so are subject to statistical treatment. We seek a formalism that would allow the inherent uncertainty to be combined with the aggravated uncertainty in a statistically meaningful way.

15.6   SINGLE-SAMPLE VERSUS MULTIPLE-SAMPLE EXPERIMENTS

Section 15.4 deals with what are called *single-sample experiments* because the statistical analysis leading to the uncertainty interval is based on only a single sample of size N. Now consider a situation in which M samples of a population are obtained, where each sample consists of N observations. Such experiments are called *multiple-sample experiments*. Now for a single-sample experiment the population mean μ and standard deviation σ are related to the *mean of the sample means* μ_m and the *standard deviation of the sample means* σ_m according to

$$\mu_m = \mu \tag{15.23}$$

and

$$\sigma_m = \frac{\sigma}{\sqrt{N}} \tag{15.24}$$

These same quantities can be estimated from a multiple-sample experiment in which M independent sample means are obtained as

$$m_m = \frac{1}{M} \sum_{i=1}^{M} m_i \tag{15.25}$$

and

$$s_m = \sqrt{\frac{1}{M-1} \sum_{i=1}^{M} (m_i - m_m)^2} \tag{15.26}$$

Similarly, the *mean of the sample proportions* μ_p and the *standard deviation of the sample proportions* σ_p are given by

$$\mu_p = \pi \tag{15.27}$$

and

$$\sigma_p = \sqrt{\frac{\pi(1 - \pi)}{N}} \tag{15.28}$$

These latter two quantities can be estimated from a multiple-sample experiment, in which M independent sample proportions are obtained, as

$$m_p = \frac{1}{M} \sum_{i=1}^{M} p_i \tag{15.29}$$

and

$$s_p = \sqrt{\frac{m_p(1 - m_p)}{MN}} \tag{15.30}$$

Note that in the special case where $M = 1$ (i.e., a single-sample experiment) Equation 15.30 reduces to Equation 15.17.

Now recall that an estimate of σ_p (Equation 15.17) was used in the single-sample experiment as the basis for establishing the inherent confidence interval in Section 15.4. In the case of a multiple-sample experiment we use Equation 15.29 as the estimate of μ_p, and Equation 15.30 as the estimate of σ_p to establish the corresponding confidence interval,

$$m_p - W_c s_p \leq \pi \leq m_p + W_c s_p \tag{15.31}$$

It is noted that as a practical matter an M-sample experiment in the MCRT sense is completely equivalent to a single-sample experiment with $N' = MN$. Therefore, it should not be surprising that the confidence interval obtained using Equation 15.19 with N' observations is the same as that obtained using Equation 15.31 using M samples, each consisting of N observations. Then what is gained by performing a multiple-sample experiment (other than the obvious fact that the uncertainly interval is smaller)? In the next section we will see that multiple-sample experiments provide a convenient means for evaluating the aggravated uncertainty of the distribution factor estimate.

15.7   EVALUATION OF AGGRAVATED UNCERTAINTY

In the previous section it was established that for a multiple-sample experiment having M samples, each consisting of N observations, the uncertainty interval expression is equivalent to that for a single-sample experiment consisting of $N' = MN$ observations. However, it was tacitly assumed in the development that the aggravated uncertainty was zero, or at least nothing was said about the possible effect of uncertainties in the enclosure dimensions or optical properties. But suppose these quantities are themselves known only to within stated uncertainty intervals. In fact, Equations 15.29 through 15.31 still apply except that now the variation in the estimator for the population proportion p from experiment to experiment can include the effects of the uncertainties in geometry and optical properties.

Suppose that, during the execution of the Monte Carlo ray trace, a random perturbation is applied to the local (directional, band-averaged spectral) absorptivity before this latter is used to determine whether or not the energy bundle is absorbed. (Of course, a similar argument can be used for volume elements.) Similarly, if the energy bundle is found to be reflected, the (bidirectional, band-averaged spectral) reflectivity could also be randomly perturbed before being applied. In both cases the random perturbation would follow a normal probability distribution, Equation 15.3, whose mean is identified with the mean value of the local property.

Consider the directional, band-averaged spectral absorptivity. Two successive normally distributed perturbed values of $\alpha'_{ik}(\theta, \phi)$ may be related to its mean value $\mu'_{ik}(\theta, \phi)$ and its standard deviation $\sigma'_{ik}(\theta, \phi)$ according to [2]

$$\alpha'_{ik,1}(\theta,\phi) = \mu'_{ik}(\theta,\phi) + \sigma'_{ik}(\theta,\phi)\, \sqrt{2\, \ell n(1/R_1)}\, \cos(2\pi R_2) \qquad (15.32a)$$

and

$$\alpha'_{ik,2}(\theta,\phi) = \mu'_{ik}(\theta,\phi) + \sigma'_{ik}(\theta,\phi)\, \sqrt{2\, \ell n(1/R_1)}\, \sin(2\pi R_2) \qquad (15.32b)$$

where R_1, and R_2 are two successive random numbers whose values are uniformly distributed between zero and unity. An algorithm implementing Equations 15.32 would be built in to the directional, band-averaged spectral absorptivity engine. Note that the use of Equations 15.32 presupposes the availability of statistically meaningful estimates of the mean and standard deviation of the directional, band-averaged spectral absorptivity (or absorption coefficient in the case of a volume element) for each of the surface (and volume) elements of the enclosure.

A similar treatment can be applied to the dimensions of the surfaces making up the enclosure. The exact procedure would depend on the details of the enclosure, but one can imagine writing an equation similar to Equations 15.32 for any critical dimension. Obviously, a great deal of judgment is involved here. How sensitive is the distribution factor estimate to a given dimension? The only way to answer this question is to carry out the appropriate MCRT experiment. In a large, complicated enclosure this could be prohibitively expensive.

We now are faced with two equivalent choices:

1. We can perturb the dimensions and optical properties *each time* a surface or volume element is visited during a single-sample experiment consisting of $N' = MN$ observations (energy bundles traced).

2. We can maintain the same set of perturbed values of dimensions and optical properties throughout each of *M* MCRT experiments consisting of *N* observations (energy bundles traced) each, changing the perturbed values from one experiment to the next.

The result obtained using the two approaches would be the same for a given value of $N' = MN$. However, the second approach is recommended because it provides superior control over the sampling statistics, which becomes important at the experimental design stage. For example, the statistics of the distribution of the population proportion estimate *p* over the *M* samples provide important information about the validity and convergence of the MCRT model. In any case, it is clear that either approach provides a statistically valid means for the simultaneous evaluation of *both* the inherent and aggravated uncertainties.

This approach for estimating the aggravated uncertainty of the distribution factors is potentially very costly. First, we must assure that the number of energy bundles emitted in a given experiment is statistically significant. To accomplish this we use the procedure described in Section 15.4 for single-sample experiments. Then we must repeat the experiment a sufficiently large number of times *M* to obtain a statistically meaningful distribution of population proportion estimates *p*. There now remains the question of, How are the uncertainties in the distribution factor estimates used to determine the uncertainties in the temperatures and net heat transfer rates obtained using the distribution factors?

15.8 UNCERTAINTY IN TEMPERATURE AND HEAT TRANSFER RESULTS

The celebrated 1953 article by Kline and McClintock [3] provides a widely accepted formalism for relating the uncertainty of an experimental result, ω_R, to the uncertainties of the component measurements used to calculate the result. The procedure, which is equally applicable to physical experiments and numerical experiments of the type considered in this book, may be summarized as follows.

Suppose a result *R* is related to a series of *n* measurement variables X_i, $i = 1, 2, \ldots, n$, according to

$$R = R(X_1, X_2, \ldots, X_n) \tag{15.33}$$

Then, following the formalism of Kline and McClintock,

$$\omega_R = \pm \sqrt{\left(\frac{\partial R}{\partial X_1} \omega_{X_1}\right)^2 + \left(\frac{\partial R}{\partial X_2} \omega_{X_2}\right)^2 + \cdots + \left(\frac{\partial R}{\partial X_n} \omega_{X_n}\right)^2} \tag{15.34}$$

in which the ω_i are the uncertainties of the component measurement variables X_i. Two observations are appropriate at this point. First, to use Equation 15.34, all of the component uncertainties must be stated with the same level of confidence. Implied is that they are each random variables and so are normally distributed. Second, the existence of an explicit analytical expression relating the result to the component measurements, i.e., Equation 15.33, is implied. This requirement is needed to allow the partial derivatives ("sensitivities") in Equation 15.34 to be evaluated. For problems amenable to solution by the MCRT method, the explicit analytical expressions depend on the complexity of the problem being solved. In the following discussion, application of the method of Kline and McClintock is illustrated for the case of a total, diffuse–specular analysis as developed in Chapter 11 and illustrated in Chapter 12.

We consider the case where a gray, diffuse–specular enclosure has specified net heat fluxes on surface elements 1, 2, . . ., N and specified temperatures on surface elements $N + 1$, $N + 2$, ..., n. In this situation Equation 11.67 for the unknown surface temperatures becomes

$$T_i = \sqrt[4]{\frac{1}{\sigma} \sum_{j=1}^{N} \left\{ [\psi_{ij}]^{-1} \left[q_j - \varepsilon_j \sum_{i=N+1}^{n} \sigma T_j^4 \left(\delta_{ji} - D'_{ji} \right) \right] \right\}}, \qquad 1 \le i \le N \quad (15.35)$$

where

$$\psi_{ij} = \varepsilon_i \sum_{j=1}^{N} (\delta_{ij} - D'_{ij}), \qquad 1 \le i \le N \tag{11.65}$$

and Equation 11.59 can be rewritten for the unknown surface heat fluxes

$$q_i = \varepsilon_i \sum_{j=1}^{n} \sigma T_j^4 (\delta_{ij} - D'_{ij}), \qquad N + 1 \le i \le n \tag{15.36}$$

In Equations 15.35, 11.65, and 15.36, and in the rest of this chapter, it is understood that D'_{ij} is the *estimate* of the total, diffuse–specular radiation distribution factor.

It is assumed in the following development that each of the "known" surface temperatures T_i has associated with it an uncertainty ω_T and that each "known" surface net heat flux q_i has associated with it an uncertainty ω_q. This is in addition to the uncertainties ω_ε associated with the emissivities ε and the uncertainties ω_D associated with the distribution factors D'_{ij}. Further, it is assumed that the confidence interval is the same, usually 95 percent, on all of these uncertainties.

Because of its form, Equation 15.35 presents a particularly nasty challenge in determining the sensitivities $\partial T_i / \partial T_{i>N}$ and $\partial T_i / \partial q_j$ and especially $\partial T_i / \partial D'_{ji}$ and $\partial T_i / \partial \varepsilon_j$. Evaluation of the sensitivities $\partial q_i / \partial D'_{ij}$, $\partial q_i / \partial \varepsilon_i$, and $\partial q_i / \partial T_j$ on the other hand is straightforward. When analytical expressions for these sensitivities have been obtained, their values are computed for each surface or volume element for which unknown surface temperatures and surface net heat fluxes are to be found.

15.9 APPLICATION TO THE CASE OF SPECIFIED SURFACE TEMPERATURES

When Equation 15.34 is used, with appropriate expressions for the sensitivities listed above, in an attempt to compute the uncertainties in surface temperature and surface net heat flux, formidable mathematical and numerical complications can arise. Before continuing, it is interesting to apply some simplifying assumptions that do not unduly compromise the statistical validity of the analysis but that reveal the relative importance of uncertainties in the surface temperatures and uncertainties in the distribution factors.

Let us consider the case in which all of the surface temperatures are specified, with appropriate uncertainties. In this situation Equation 15.36 becomes

$$q_i = \varepsilon_i \sum_{j=1}^{n} \sigma T_j^4 (\delta_{ij} - D'_{ij}), \qquad 1 \le i \le n \tag{15.36a}$$

and Equation 15.34 becomes

$$\omega_{q_i} = \pm \sqrt{\left(\frac{\partial q_i}{\partial \varepsilon_i} \omega_{\varepsilon_i}\right)^2 + \left(\frac{\partial q_i}{\partial T_j} \omega_{T_j}\right)^2 + \left(\frac{\partial q_i}{\partial D'_{ij}} \omega_{D'_{ij}}\right)^2}, \qquad 1 \le i \le n \tag{15.37}$$

where the sensitivities in Equation 15.37 are

$$\frac{\partial q_i}{\partial \varepsilon_i} = \sum_{j=1}^{n} \sigma T_j^4 (\delta_{ij} - D'_{ij}) \tag{15.38}$$

$$\frac{\partial q_i}{\partial T_j} = 4\varepsilon_i \sum_{j=1}^{n} \sigma T_j^3 (\delta_{ij} - D'_{ij}) \tag{15.39}$$

and

$$\frac{\partial q_i}{\partial D'_{ij}} = -\varepsilon_i \sum_{j=1}^{n} \sigma T_j^4 \tag{15.40}$$

Upon substitution of these sensitivities into Equation 15.37 there results

$$\omega_{q_i} =$$

$$\pm \sqrt{\left(\omega_{\varepsilon_i} \sum_{j=1}^{n} \sigma T_j^4 (\delta_{ij} - D'_{ij})\right)^2 + \left(4\varepsilon_i \sum_{j=1}^{n} \sigma T_j^3 (\delta_{ij} - D'_{ij})\omega_{T_j}\right)^2 + \left(\varepsilon_i \sum_{j=1}^{n} \sigma T_j^4 \omega_{D'_{ij}}\right)^2}$$

$$\tag{15.41}$$

Finally, it is convenient to express the result represented by Equation 15.41 as a fraction of the flux emitted from surface element i; that is,

$$\frac{\omega_{qi}}{\varepsilon_i \sigma T_i^4} = \pm \left\{ \left(\frac{\omega_{\varepsilon_i}}{\varepsilon_i} \right)^2 \left[1 - \sum_{j=1}^{n} \left(\frac{T_j}{T_i} \right)^4 D_{ij}' \right]^2 \right.$$

$$- \sum_{j=1}^{n} \left(\frac{T_j}{T_i} \right)^3 \left(\frac{\omega_{T_j}}{T_i} \right) D_{ij}' \Bigg]^2 \qquad + 16 \Bigg[\qquad\qquad \frac{\omega_{T_i}}{T_i} \tag{15.42}$$

$$+ \left[\sum_{j=1}^{n} \left(\frac{T_j}{T_i} \right)^4 \left(\frac{\omega_{D_{ij}'}}{D_{ij}'} \right) D_{ij}' \right]^2 \Bigg]^{1/2} \Bigg\}$$

Note that Equations 15.38 through 15.42 all can be written for each surface element i.

The result represented by Equations 15.41 and 15.42 is exact; that is, no simplifying assumptions have been invoked to this point. Now to provide an estimate of the contribution to the uncertainty of q_i associated with the uncertainty in the distribution factors, consider the special case of an isothermal enclosure with $T_j = T$, a constant. To this we add the reasonable assumption that the uncertainty in the temperature is the same for all surface elements; that is, $\omega_{T_j} = \omega_T$, a constant. This allows the temperature and its uncertainty to be factored out of the summations in Equations 15.41 and 15.42. Then noting that

$$\sum_{j=1}^{n} (\delta_{ij} - D_{ij}') = 1 - \sum_{j=1}^{n} D_{ij}' = 0, \qquad i = 1, 2, \ldots, n \tag{15.43}$$

Equation 15.42 reduces to

$$\frac{\omega_{qi}}{\varepsilon_i \sigma T^4} = \pm \sum_{j=1}^{n} \left(\frac{\omega_{D_{ij}'}}{D_{ij}'} \right) D_{ij}' = \pm \sum_{j=1}^{n} \omega_{D_{ij}'} = \pm n \langle \omega_{D_i'} \rangle \tag{15.44}$$

where $\langle \omega_{D_i'} \rangle$ is the mean value of the uncertainty in the distribution factors from surface element i. Then Equation 15.44 establishes that the uncertainty of the surface net heat flux for surface element i, relative to the heat flux emitted from surface element i, is equal to the simple sum of the uncertainties of the distribution factors from surface element i to the n surface elements of the enclosure.

It is interesting to note that, for the special case described by Equation 15.44 ($T_i = T_j = T$, a constant), the uncertainty in the net heat flux from surface element i is independent of the uncertainty in the temperature itself. Of greater interest, however, is that the uncertainty in q_i, relative to the flux emitted from surface element i, is n times the mean uncertainty in the distribution factors from surface element i to the n surface elements of the enclosure. Thus, if the mean uncertainty of the n distribution factors is 1 percent (i.e., $\langle \omega_{D_i'} \rangle = 0.01 \langle D_i' \rangle = 0.01/n$), the uncertainty in q_i is 1 percent of $\varepsilon_i \sigma T^4$. Of course, the result represented by Equation 15.44 is fictitious almost to a fault and so must be interpreted with care. After all, if the enclosure is indeed isothermal, the radiation flux arriving at and leaving (emitted plus reflected) each of the surface elements must be the same, σT^4. This means that the net heat flux from each surface element is identically zero and thus has no associated uncertainty. The importance of Equation 15.44 is that it provides a lower bound on the uncertainty of the surface net heat flux corresponding to a given mean uncertainty in the distribution factors: that is, the uncertainty in the heat transfer result can be no less than the mean uncertainty in the distribution factors. A more realistic view may be obtained as follows.

Returning to Equation 15.42, suppose we make the reasonable assumption that the fractional (percentage) uncertainties of temperature, emissivity, and distribution factor are the same for all surface elements of the enclosure; that is, that $\omega_{T_j}/T_j = \omega_T/T$, $\omega_{\varepsilon_j}/\varepsilon_i = \omega_\varepsilon/\varepsilon$, and $\omega_{D'_{ij}}/D'_{ij} = \omega_{D'}/D'$, where each is a generally different constant. In this case Equation 15.42 becomes

$$
\frac{\omega_{q_i}}{\varepsilon_i \sigma T_i^4} = \pm \left\{ \left[\left(\frac{\omega_\varepsilon}{\varepsilon} \right)^2 + \left(4\frac{\omega_T}{T} \right)^2 \right] \left[1 - \sum_{j=1}^{n} \left(\frac{T_j}{T_i} \right)^4 D'_{ij} \right]^2 \right.
$$
$$
\left. + \left(\frac{\omega_{D'}}{D'} \right)^2 \left[\sum_{j=1}^{n} \left(\frac{T_j}{T_i} \right)^4 D'_{ij} \right]^2 \right\}^{1/2}
$$

(15.45)

Equation 15.45 clearly illustrates the role played by the *temperature spread* T_j/T_i in the uncertainty of the net heat flux result. We have already shown (Equation 15.44) that in an isothermal enclosure ($T_j/T_i = 1$ for all combinations of i and j) the uncertainty of the net heat flux from surface element i, relative to the flux emitted from surface element i, is the same as the relative uncertainty in the distribution factors. It is clear that as the temperature distribution in the enclosure becomes more uniform, the summations in Equation 15.45 approach unity and thus the approximation represented by Equation 15.44 becomes increasingly valid.

In summary, Equation 15.42 can be used to estimate, to a stated level of confidence, the uncertainty in the net heat flux from surface element i if all of the surface temperatures and emissivities, and the surface-to-surface distribution factors, are known and their uncertainties are known to the same stated level of confidence. The somewhat simpler Equation 15.45 can be used when the relative uncertainties of the surface temperatures, emissivities, and the distribution factors are the same in each category. Finally, as the temperature distribution in the enclosure becomes more and more uniform, the uncertainty in the distribution factors dominates the uncertainty in the net heat flux result. The development for the case where surface net heat flux is specified for one or more surfaces is left as an exercise (see Team Project TP 15.1).

15.10   EXPERIMENTAL DESIGN OF MCRT ALGORITHMS

We are now ready to apply the statistical principles developed in this chapter to the design of MCRT algorithms. Here the term "design" is employed in the restricted sense used by statisticians; that is, we seek the number of surface elements n and the number of energy bundles N traced per surface element required to attain a specified uncertainty in the surface net heat flux results to a stated level of confidence.

We continue to consider the special case of a gray, diffuse–specular enclosure for which the temperature is specified on all of the surface elements, and we proceed under the assumption that the relative uncertainties in surface temperature and surface emissivity, and the desired uncertainty in surface net heat flux relative to the local emitted flux, have been specified in advance, all to the same level of confidence. In the present development we assume further that the relative uncertainties are the same within a given category; that is the development will be based on Equation 15.45.

Experimental design in the present context may be divided into three steps: (1) determine the number of surface elements n into which the enclosure must be divided, (2) determine the relative uncertainty in the distribution factors consistent with obtaining the desired uncertainty in the surface net heat fluxes, and (3) determine the number of energy bundles that must be traced per surface element to achieve the required uncertainty in the distribution factors.

We begin the process of experimental design by solving Equation 15.45 for the mean value of the relative uncertainty of *all* of the distribution factors, ω'_D/D'; that is,

$$
\frac{\omega_{D'}}{D'} = \pm \left[\left\langle \sum_{j=1}^{n} \left(\frac{T_j}{T} \right)^4 D' \right\rangle \right]^{-1} \left\{ \left(\left\langle \frac{\omega_q}{\varepsilon \sigma T^4} \right\rangle \right)^2 \right.
$$
$$
\left. - \left[\left(\frac{\omega_\varepsilon}{\varepsilon} \right)^2 + \left(4 \frac{\omega_T}{T} \right)^2 \right] \left[1 - \left\langle \sum_{j=1}^{n} \left(\frac{T_j}{T} \right)^4 D' \right\rangle \right]^2 \right\}^{1/2}
$$
(15.46)

where now $\omega_{qi}/\varepsilon_i \sigma T_i^4$ and the summations have been replaced by their mean values. From Equation 15.46 we see that the fractional uncertainty in the distribution factors, ω'_D/D', depends on the number of surface elements n into which the enclosure has been divided. Specifically, it depends, through the summations, on the interplay between the temperature spread T_j/T_i, the values of the distribution factors D'_{ij}, and the number of surface elements n.

Step 1 then involves estimating the mean value of $(T_j/T_i)^4$; that is,

$$
\left\langle \left\langle \left(\frac{T_j}{T_i} \right)^4 \right\rangle \right\rangle = \frac{1}{n^2} \sum_{i=1}^{n} \sum_{j=1}^{n} \left(\frac{T_j}{T_i} \right)^4
$$
(15.47)

Bearing in mind that the temperature distribution on the surfaces of the enclosure is known, we proceed by computing the mean value of $(T_j/T_i)^4$ for increasing values of n until convergence is obtained. The value of n producing convergence is then accepted as adequate for estimating the unknown surface net heat fluxes, assuming that it is consistent with the required spatial resolution.*

---

* In many cases the value of n is fixed in advance by the spatial resolution of the surface temperature data. That is, the surface temperature distribution may have been provided to the user as a vector of surface element temperatures of length n.

Once a converged value of the mean of $(T_j/T_i)^4$ has been obtained using Equation 15.47, we are ready to proceed to step 2 of experimental design. The triangle inequality requires that

$$\left\langle \sum_{j=1}^{n} \left(\frac{T_j}{T_i}\right)^4 D_{ij}' \right\rangle \leq n \left\langle \left\langle \left(\frac{T_j}{T_i}\right)^4 \right\rangle \right\rangle \frac{1}{n} = \left\langle \left\langle \left(\frac{T_j}{T_i}\right)^4 \right\rangle \right\rangle \tag{15.48}$$

where the factor $1/n = \langle\langle D_{ij}' \rangle\rangle$ is the mean value of the elements of the distribution factor matrix. With the introduction of Equation 15.48, Equation 15.46 becomes

$$\left| \frac{\omega_{D'}}{D'} \right| \leq \left[\left\langle \left\langle \left(\frac{T_j}{T_i}\right)^4 \right\rangle \right\rangle \right]^{-1} \left\{ \left(\frac{\omega_q}{\varepsilon \sigma T^4}\right)^2 - \left[\left(\frac{\omega_\varepsilon}{\varepsilon}\right)^2 \right. \right.$$
$$\left. \left. + \left(4\frac{\omega_T}{T}\right)^2 \right] \left[1 - \left\langle \left\langle \left(\frac{T_j}{T_i}\right)^4 \right\rangle \right\rangle \right]^2 \right\}^{1/2} \tag{15.49}$$

Equation 15.49 has great theoretical importance, first because it provides an upper bound on the magnitude of the relative uncertainty of the distribution factors consistent with the desired uncertainty in the surface net heat fluxes. Moreover, it also demonstrates that once the number of surface elements n is sufficiently large to obtain reasonable estimates of the mean value of $(T_j/T_i)^4$, the maximum value of uncertainty in the distribution factors depends only on the temperature spread in the enclosure, the specified uncertainties of the temperatures and emissivities, and the desired mean uncertainty in surface net heat flux. More importantly, Equation 15.49 establishes a lower limit on the mean value of the uncertainty of the surface net heat flux; i.e., the quantity under the square root cannot be negative. This means that

$$\left(\frac{\omega_q}{\varepsilon \sigma T^4}\right)^2 \geq \left[\left(\frac{\omega_\varepsilon}{\varepsilon}\right)^2 + \left(4\frac{\omega_T}{T}\right)^2 \right] \left[1 - \left\langle \left\langle \left(\frac{T_j}{T_i}\right)^4 \right\rangle \right\rangle \right]^2 \tag{15.50}$$

Finally, note that equality of the two sides of Equation 15.50 can be achieved only when the uncertainty in the distribution factors is zero. Of course, this can occur only by tracing an infinite number of energy bundles.

After the mean value of the relative uncertainty of the distribution factors ω_D'/D' has been estimated using Equation 15.49, we are ready to move to step 3 of the experimental design process, determination of the number of energy bundles that must be traced per surface element. Referring to Equation 15.20, we recognize that the uncertainty in the distribution factors is related to their estimated values and to the related level of confidence according to

$$\frac{\omega_{D_{ij}'}}{D_{ij}} = \pm W_c \sqrt{\frac{(1 - D_{ij}')}{N D_{ij}'}} \tag{15.51}$$

Equation 15.51 is valid for a particular distribution factor between two specified surface elements i and j, but we need an expression for the mean value of relative uncertainty, ω_D'/D'. Introducing the mean value of all distribution factors, $\langle D_{ij}' \rangle = 1/n$, Equation 15.51 becomes

$$\left\langle \left| \frac{\omega_{D_{ij}'}}{D_{ij}'} \right| \right\rangle \equiv \frac{\omega_{D'}}{D'} = \pm W_c \sqrt{\frac{n-1}{N}} \qquad (15.52)$$

Example Problem 15.4

An MCRT model is to be used to estimate the surface net heat fluxes for an enclosure consisting of 100 surface elements, all of whose temperatures are known. How many energy bundles must be traced per surface element if the upper limit on the mean value of the relative uncertainty of the distribution factors needs to be 0.01 (1 percent) to achieve the desired limit on the mean uncertainty in the surface net heat fluxes? Assume that a 95-percent level of confidence is required.

SOLUTION

Applying Equation 15.52, we obtain

$$N = \frac{W_c^2 (n-1)}{(\omega_{D'}/D')^2} = \frac{(1.960)^2 (100-1)}{(0.01)^2} \approx 3.8 \times 10^6 \qquad (15.53)$$

That is, we would have to trace about four million energy bundles from each surface element to assure, to a 95-percent level of confidence, that the distribution factors were, on the average, obtained to within a 1-percent uncertainty.

The mean relative uncertainties of distribution factors, expressed as a percentage of the mean value of the distribution factors for the enclosure, are tabulated in Table 15.2 for the case of a 95-percent confidence interval. The cells corresponding to the 1- to 2-percent relative uncertainty range are shaded to highlight the general interplay between the number of surface elements and the number of energy bundles emitted.

15.11 VALIDATION OF THE THEORY

Figure 15.5 shows a nine-surface-element enclosure, first encountered in Chapter 11, intended to increase the apparent absorptivity of the sensing elements of a linear-array thermal radiation detector (see Problems P 11.12 through P 11.16). Collimated radiation enters the narrow slit at the top of the wedge-shaped cavity whose walls are highly specular reflectors. The radiation is directly incident to the linear array, whose absorptivity is high and whose small reflectivity is highly specular. Radiation not immediately absorbed is returned to the detector several times by specular reflections, and each time a large fraction of its remaining energy is absorbed. Design and per-

Table 15.2 Relative uncertainties of distribution factors as a function of the number of surface elements, n, and the number of energy bundles traced per surface element, N, for a 95-percent confidence interval

N (10^3)	Relative Uncertainty of Distribution Factors (Percent) Corresponding to a 95-Percent Confidence Interval — Number of Surface Elements, n											
	10	20	30	40	50	60	70	80	90	100	150	200
10	5.880	8.543	10.555	12.240	13.720	15.055	16.281	15.421	18.491	19.502	23.925	27.649
20	4.158	6.041	7.463	8.655	9.702	10.646	11.512	12.318	13.075	13.790	16.917	19.551
50	2.630	3.821	4.720	5.474	6.136	6.733	7.281	7.791	8.269	8.721	10.700	12.365
100	1.859	2.702	3.338	3.871	4.339	4.761	5.148	5.509	5.847	6.167	7.566	8.743
150	1.518	2.206	2.725	3.160	3.542	3.887	4.204	4.498	4.774	5.035	6.177	7.139
200	1.315	1.910	2.360	2.737	3.068	3.366	3.641	3.895	4.135	4.361	5.350	6.183
250	1.176	1.709	2.111	2.448	2.744	3.011	3.256	3.484	3.698	3.900	4.785	5.530
300	1.074	1.560	1.927	2.235	2.505	2.749	2.972	3.181	3.376	3.561	4.368	5.048
350	0.994	1.444	1.784	2.069	2.319	2.545	2.752	2.945	3.125	3.296	4.044	4.674
400	0.930	1.351	1.669	1.935	2.169	2.380	2.574	2.754	2.924	3.083	3.783	4.372
450	0.877	1.274	1.573	1.825	2.045	2.244	2.427	2.597	2.756	2.907	3.567	4.122
500	0.832	1.208	1.493	1.731	1.940	2.129	2.302	2.464	2.615	2.758	3.383	3.910
550	0.793	1.152	1.423	1.650	1.850	2.030	2.195	2.349	2.493	2.630	3.226	3.728
600	0.759	1.103	1.363	1.580	1.771	1.944	2.102	2.249	2.387	2.518	3.089	3.569
650	0.729	1.060	1.309	1.518	1.702	1.867	2.019	2.161	2.293	2.419	2.968	3.429
700	0.703	1.021	1.262	1.463	1.640	1.799	1.946	2.082	2.210	2.331	2.860	3.305
750	0.679	0.987	1.219	1.413	1.584	1.738	1.880	2.012	2.135	2.252	2.763	3.193
800	0.657	0.955	1.180	1.368	1.534	1.683	1.820	1.948	2.067	2.180	2.675	3.091
850	0.638	0.927	1.145	1.328	1.488	1.633	1.766	1.890	2.006	2.115	2.595	2.999
900	0.620	0.901	1.113	1.290	1.446	1.587	1.716	1.836	1.949	2.056	2.522	2.914
950	0.603	0.877	1.083	1.256	1.408	1.545	1.670	1.787	1.897	2.001	2.455	2.837
1000	0.588	0.854	1.055	1.224	1.372	1.506	1.628	1.742	1.849	1.950	2.392	2.765
1500	0.480	0.698	0.862	0.999	1.120	1.229	1.329	1.422	1.510	1.592	1.953	2.258
2000	0.416	0.604	0.476	0.866	0.970	1.065	1.151	1.232	1.307	1.379	1.692	1.955
2501	0.372	0.540	0.668	0.774	0.868	0.952	1.030	1.102	1.169	1.233	1.513	1.749
3000	0.339	0.493	0.609	0.707	0.792	0.869	0.940	1.006	1.068	1.126	1.381	1.596
3500	0.314	0.457	0.564	0.654	0.733	0.805	0.870	0.931	0.988	1.042	1.279	1.478
4000	0.294	0.427	0.528	0.612	0.686	0.753	0.814	0.871	0.925	0.975	1.196	1.382
4500	0.277	0.403	0.498	0.577	0.647	0.710	0.767	0.821	0.872	0.919	1.128	1.303
5000	0.263	0.382	0.472	0.547	0.614	0.673	0.728	0.779	0.827	0.872	1.070	1.237

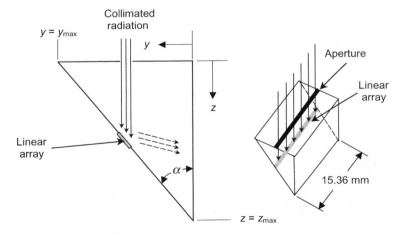

Figure 15.5 A linear-array thermal radiation detector concept

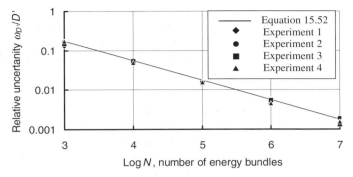

Figure 15.6 Comparison of predicted and "experimentally" determined global distribution factor uncertainties as a function of number of energy bundles emitted per surface element for the enclosure of Figure 15.5

formance details are given elsewhere [4–6]. Figure 15.6 [7] compares, as a function of the number of energy bundles emitted per surface element, the theoretical and "experimental" mean uncertainty of the elements of the distribution factor matrix used in the MCRT model of the detector. The theoretical line was computed using Equation 15.52, and the four sets of "experimental" data were obtained using FELIX, with different pairs of seeds used in the pseudo-random number generator for each experiment. Excellent agreement is obtained between theory and experiment, thereby providing limited validation of the theory.

Team Project

TP 15.1 Derive the expressions, equivalent to Equations 15.42 and 15.45, required to treat the case of specified surface net heat flux on the first N surface elements of an enclosure and specified temperature on the remaining $n - N$ surface elements (see Section 15.8). Then work out a strategy, similar to the one presented in Section 15.10, for the design of an MCRT experiment to obtain the unknown surface net heat fluxes and temperatures to a specified mean relative uncertainty and corresponding level of confidence.

Discussion Points

DP 15.1 Discuss the difference between a *probability* and a *proportion*. In what sense can it be said that the distribution factor is both a probability and a proportion?

DP 15.2 The *central limit theorem* of statistics states that as long as the sample size remains small compared to the population size, the distribution of an estimator involving sums of random variables tends toward a normal distribution with increased sample size, no matter what the underlying distribution of the population may be [1]. Investigate this theorem for the case of a uniformly distributed (between zero and unity) pseudo-random number generator. Specifically, show that the deviation of the mean from 0.5 follows a normal distribution.

DP 15.3   Discuss the difference between a *distribution function* (such as the Planck blackbody radiation distribution function or the normal probability distribution function) and a *cumulative function*. In your discussion pay particular attention to any difference in dimensions or units between the two. Include Equations 1.45, 2.86 and 15.3 in your discussion.

DP 15.4   Figure 15.4 indicates that the uncertainty in the estimate of the value of a distribution factor is directly related to the estimate of the value of the distribution factor itself, especially for small values of the distribution factor. Setting aside the mathematical development leading to Figure 15.4, why is this so? Why do you believe that the improvement achieved by increasing N is much greater for small values of the distribution factor than for large values?

DP 15.5   Show that Equations 15.32 imply

$$2\pi\sigma_{ik}'^2 p(\alpha_{ik,1}')p(\alpha_{ik,2}') = R_1 \tag{15.54}$$

and

$$\frac{\alpha_{ik,2}' - \mu_{ik}'}{\alpha_{ik,1}' - \mu_{ik}'} = \tan(2\pi R_2) \tag{15.55}$$

where $p(z)$ is the normal probability distribution function defined by Equation 15.5. Then discuss the physical significance of Equations 15.54 and 15.55. For example, Equation 15.54 states that the joint probability associated with randomly drawing a pair of specified values of a normally distributed variable (α_{ik}' in this case) is uniformly distributed, and Equation 15.55 states that the ratio of the difference between these two variables and their mean value is distributed as the tangent of a uniformly distributed random number. What is the underlying principle at work here?

DP 15.6   In the development of Sections 15.8 through 15.10 uncertainties are considered both in the emissivities and in the distribution factors, even though the distribution factors are themselves implicit functions of the emissivities. How would the analysis in these sections change if we took this into account? Do you think this is necessary? Please explain your answer.

DP 15.7   Verify Equations 15.38 through 15.42.

DP 15.8 Discuss the fact that, for the special case represented by Equation 15.44, the uncertainty in the surface net heat flux is independent of the uncertainty in the temperature itself.

DP 15.9 According to the book, Equation 15.44 represents the lower bound on the uncertainty of the surface net heat flux. Can you justify this? [*Hint*: Carefully consider the behavior of the summations in Equation 15.45 as the temperature spread T_j/T_i deviates from unity.]

DP 15.10 What is the minimum value of the quantity defined by Equation 15.47?

DP 15.11 What does the "triangle inequality" have to do with Equation 15.48? Can you derive the equation?

DP 15.12 Consider the sense of the inequality in Equation 15.49. Can you formally verify that the sense of the inequality is correct?

DP 15.13 Discuss the significance of Equation 15.50.

DP 15.14 What key assumption(s) relate(s) Equations 15.20 and 15.51?

DP 15.15 Consider the criticism of the result represented by Equation 15.44 (that it is "fictitious almost to a fault"). Criticize the defense that the result is still useful because the uncertainty represented is the one that would be produced by an MCRT model even though the answer ($q_i = 0$) is known a priori. Is this defense valid? [*Hint:* Just because the temperature distribution is known to be uniform does not mean that this is known without a specified uncertainty.]

DP 15.16 In what sense might it be argued that the estimate of the global mean relative uncertainty in the distribution factors given by Equation 15.52 is conservative?

DP 15.17 Considering Equation 15.45, suppose the uncertainties in emissivity and temperature are negligible in a hypothetical situation. Is it possible for the uncertainty in surface net heat flux, relative to the emitted heat flux, to be less than the relative uncertainty in the distribution factors? Under what circumstances might this be possible?

Problems

P 15.1 Rework Example Problem 15.1 using a sequence of 100 pseudo-random numbers from Excel or a similar source. What may you conclude about the pseudo-random number generator?

P 15.2 Assuming that when you worked Problem P 15.1 you obtained a result suggesting that your pseudo-random number generator may be biased toward either zero or unity, perform the appropriate hypothesis test to either verify or dispel this suggestion. [*Hint:* See Example Problem 15.2.]

P 15.3   Draw ten 100-member pseudo-random number sequences from Excel or a similar source. Then create a table similar to Table 15.1. What is the mean of the means of the ten sequences (see Equation 15.11)? What may you conclude from this experiment about your pseudo-random number generator?

P 15.4   Use the binomial probability distribution to compute the probability of rolling "boxcars" (two sixes) using a pair of fair dice.

P 15.5   Derive an expression for the sensitivity $\partial T_i / \partial q_j$, $1 \leq i \leq N$, where the dependence of T_i on q_j is as given by Equation 15.35.

P15.6   Show formally that the mean value of all of the distribution factors in an enclosure (the "global" mean) is $1/n$, where n is the number of surface elements defining the enclosure.

P 15.7   Randomly assign temperatures ranging between 1000 and 100 K to an enclosure consisting of 10 surface elements. Then compute the value of the quantity defined by Equation 15.47. Then repeat these calculations with the temperatures ranging randomly between 500 and 600 K. What may we conclude from these calculations?

P 15.8   Using the results from Problem P 15.7, what is the minimum possible value of $|\omega_q / \varepsilon \sigma T^4|$ if the uncertainties in emissivity and temperature are both 1 percent?

P 15.9   An MCRT model is to be used to estimate the surface net heat fluxes for an enclosure consisting of 25 surface elements, all of whose temperatures are known. How many energy bundles must be traced per surface element if the upper limit on the mean value of the relative uncertainty of the distribution factors needs to be 0.02 (2 percent) to achieve the desired limit on the mean uncertainty in the surface net heat fluxes? Assume that a 95-percent level of confidence is required.

REFERENCES

1. Koopmans, L. H., *Introduction to Contemporary Statistical Methods,* 2nd ed., Duxbury, Boston, 1987, pp. 206–207.
2. Press, W. H., S. A. Teukolsky, W. T. Vetterling, B. P. Flannery, *Numerical Recipes in C: The Art of Scientific Computing,* 2nd ed., Cambridge University Press, New York, 1992, pp. 288–290.
3. Kline, S. J., and F. A. McClintock, "Describing uncertainties in single-sample experiments," *Mechanical Engineering,* January 1953, p. 3.
4. Sánchez, M. C., J. R. Mahan, F. J. Nevárez, K. J. Priestley, "Uncertainty and confidence intervals in optical design using the Monte Carlo ray-trace method," *SPIE Proceedings, International Symposium in Remote Sensing,* Vol. 4169, Barcelona, 2000.
5. Mahan, J. R., S. Weckmann, M. C. Sánchez, I. J. Sorensen, K. L. Coffey, E. H. Kist, and E. L. Nelson, "Optical and electrothermal design of a linear-array thermopile detector for geo-

stationary earth radiation budget applications," *SPIE Proceedings, International Symposium in Remote Sensing,* Vol. 3498, Barcelona, 1998.

6. Sánchez, M. C., J. R. Mahan, E. A. Ayala, K. J. Priestley, "Tools for predicting uncertainty and confidence intervals in radiometric data products," EOS/SPIE Symposium on Remote Sensing (EUROPTO Series), University of Florence, Florence, Italy, 20–24, September, 1999 (published in *Proceedings of the Europto Series: Sensors, Systems, and Next-Generation Satellites III,* Vol. 3870, pp. 525–535).

7. Sánchez, M. C., J. R. Mahan, F. J. Nevárez, and I. J. Sorensen, "Bounding uncertainty in Monte Carlo ray-trace models," Paper No. 4540–50, 8th International Symposium on Remote Sensing, Conference 4540A, Sensors, Systems, and Next-Generation Satellites V, Toulouse, France, 17–21 September 2001.

APPENDIX A

RADIATION FROM AN ATOMIC DIPOLE

Maxwell's equations describing the propagation of electromagnetic radiation were introduced in Chapter 1, along with a model to explain how a thermally polarized atom can emit like a dipole antenna. In this appendix we use Maxwell's equations to derive an expression for the radiation of electromagnetic energy by an atomic dipole as a function of wavelength and the temperature of the atomic matrix.

A.1 MAXWELL'S EQUATIONS AND CONSERVATION OF ELECTRIC CHARGE

In Chapter 1 specialized versions of Maxwell's equations, Equations 1.16, were expressed in terms of the *magnetic field strength* $\mathbf{H}$ (A/m) and the *electric field strength* $\mathbf{E}$ (V/m). The validity of Equations 1.16 is limited to the case of a homogeneous, isotropic medium in which the *magnetic dipole moment density* $\mathbf{M}$ (A/m) and the *electric dipole moment density* $\mathbf{P}$ (C/m^2) are zero. The more general version of Maxwell's equations involves these latter quantities plus the *magnetic flux density*,

$$\mathbf{B} = \mu(\mathbf{H} + \mathbf{M}) \quad (\text{V·s/m}^2) \tag{A.1}$$

the *electric flux density,*

$$\mathbf{D} = \varepsilon\mathbf{E} + \mathbf{P} \quad (\text{C/m}^2) \tag{A.2}$$

440

and the *current density*,

$$\mathbf{J} = \sigma_e \mathbf{E} \quad (\text{A/m}^2) \tag{A.3}$$

In Equation A.3 the *electrical conductivity* σ_e ($1/\Omega \cdot$m) is the reciprocal of the electrical resistivity r_e. In terms of these variables Maxwell's equations can be written

$$\nabla \times \mathbf{H} = \frac{\partial \mathbf{D}}{\partial t} + \mathbf{J} \tag{A.4a}$$

$$\nabla \times \mathbf{E} = -\frac{\partial \mathbf{B}}{\partial t} \tag{A.4b}$$

$$\nabla \cdot \mathbf{D} = \rho_e \tag{A.4c}$$

and

$$\nabla \cdot \mathbf{B} = 0 \tag{A.4d}$$

In the absence of sources or sinks of electric charge, Equations A.4a and A.4c lead directly to the principle of *conservation of electric charge*,

$$\nabla \cdot \mathbf{J} = -\frac{\partial \rho_e}{\partial t} \tag{A.5}$$

A.2 MAXWELL'S EQUATIONS APPLIED IN FREE SPACE

We first consider the case of the propagation of electromagnetic radiation through *free space*, or in a vacuum; that is, where $\rho_e = 0$, $\sigma_e = 0$, $\mathbf{J} = 0$, $\varepsilon = \varepsilon_0$ and $\mu = \mu_0$. In this case Equations A.4 become

$$\nabla \times \mathbf{H} = \varepsilon_0 \frac{\partial \mathbf{E}}{\partial t} \tag{A.6a}$$

$$\nabla \times \mathbf{E} = -\mu_0 \frac{\partial \mathbf{H}}{\partial t} \tag{A.6b}$$

$$\nabla \cdot \mathbf{E} = 0 \tag{A.6c}$$

and

$$\nabla \cdot \mathbf{H} = 0 \tag{A.6d}$$

Applying the curl operator to Equation A.6a yields

$$\nabla \times \nabla \times \mathbf{H} = \varepsilon_0 \frac{\partial}{\partial t} (\nabla \times \mathbf{E}) \tag{A.7}$$

Then with the introduction of the vector identity

$$\nabla \times \nabla \times \mathbf{V} = \nabla(\nabla \cdot \mathbf{V}) - \nabla^2 \mathbf{V} \tag{A.8}$$

and Equations A.6b and A.6d, Equation A.7 becomes

$$\nabla^2 \mathbf{H} = \frac{1}{c_0^2} \frac{\partial^2 \mathbf{H}}{\partial t^2} \tag{A.9}$$

where

$$c_0 = \frac{1}{\sqrt{\mu_0 \varepsilon_0}} = 2.997925 \times 10^8 \text{ m/s} \tag{1.24}$$

is the speed of light in a vacuum, as demonstrated in Section 1.9. A similar process involving Equations A.6a, A.6b, and A.6c leads to

$$\nabla^2 \mathbf{E} = \frac{1}{c_0^2} \frac{\partial^2 \mathbf{E}}{\partial t^2} \tag{A.10}$$

Equations A.9 and A.10 are *wave equations* whose solution describes the propagation of an electromagnetic wave through free space (see Sections 1.8 and 1.9). It is interesting to note that Maxwell himself believed that his work supported the *ether* theory, which postulated that radiation could propagate only through an elastic (in the electromagnetic sense) medium [1]. Of course, we now know that the required "elasticity" is achieved by alternately storing the propagating energy in the electric and magnetic fields. A solution of Equations A.9 and A.10 for the special case of a y-polarized transverse electric wave propagating along the x axis is illustrated in Figure 1.7.

A.3   EMISSION FROM AN ELECTRIC DIPOLE RADIATOR

We now consider emission from an electric dipole radiator such as an atom in the crystal matrix of Figure 1.1. The development in this section follows that for a dipole antenna presented in Chapters 9 and 11 of Plonsey and Collin [2]. In this case the radiation propagates through a medium for which ρ_e, σ_e, and $\mathbf{J}$ are not zero and ε and μ are not necessarily equal to ε_0 and μ_0.

Application of the curl operator to Equation A.4b followed by use of Equations A.4a, A.4c, and A.8 yields

$$\nabla^2 \mathbf{E} - \frac{1}{c^2} \frac{\partial^2 \mathbf{E}}{\partial t^2} = \mu \frac{\partial \mathbf{J}}{\partial t} + \nabla \left(\frac{\rho}{\varepsilon} \right) \tag{A.11}$$

Similarly, application of the curl operator to Equation A.4a followed by use of Equations A.4b, A.4d, and A.8 yields

$$\nabla^2 \mathbf{H} - \frac{1}{c^2} \frac{\partial^2 \mathbf{H}}{\partial t^2} = -\nabla \times \mathbf{J} \tag{A.12}$$

Equations A.11 and A.12 are called the *inhomogeneous Helmholtz equations*. They are difficult to solve for the electric field strength $\mathbf{E}$ and the magnetic field strength $\mathbf{H}$ subject to prescribed distributions of ρ_e and $\mathbf{J}$. However, the degree of this mathematical difficulty can be reduced by introducing a vector potential function $\mathbf{A}$ and a scalar potential function Φ whose forms are themselves uniquely determined by the prescribed charge density and current density distributions.

The divergence of the curl of any vector is identically zero. Therefore, Equation A.4d permits the partial definition of an auxiliary vector $\mathbf{A}$ such that

$$\mathbf{B} = \nabla \times \mathbf{A} \tag{A.13}$$

Then we can rewrite Equation A.4b

$$\nabla \times \mathbf{E} = -\frac{\partial}{\partial t}(\nabla \times \mathbf{A}) = -\nabla \times \frac{\partial \mathbf{A}}{\partial t} \tag{A.14}$$

or

$$\nabla \times \left(\mathbf{E} + \frac{\partial \mathbf{A}}{\partial t}\right) = 0 \tag{A.15}$$

Equation A.15 can be interpreted to imply that the vector $\mathbf{E} + \partial \mathbf{A}/\partial t$ is *irrotational*, which means that there exists a scalar potential Φ such that

$$-\nabla\Phi = \mathbf{E} + \frac{\partial \mathbf{A}}{\partial t} \tag{A.16}$$

since the curl of the gradient of any scalar variable is identically equal to zero. Equations A.1, A.4a, A.8, A.13, and A.16 may now be combined to yield

$$\frac{1}{\mu}[\nabla(\nabla \cdot \mathbf{A}) - \nabla^2 \mathbf{A}] = -\varepsilon \frac{\partial(\nabla\Phi)}{\partial t} - \varepsilon \frac{\partial^2 \mathbf{A}}{\partial t^2} + \mathbf{J} \tag{A.17}$$

for the case of negligible magnetic polarization* ($\mathbf{M} = 0$).

Recall that a vector is completely specified by its divergence and its curl. It is convenient in the case at hand to define the divergence of the auxiliary vector $\mathbf{A}$

---

* Magnetic polarization is rarely an issue in materials involved in radiation heat transfer applications.

such that

$$\nabla \cdot \mathbf{A} = -\mu\varepsilon \frac{\partial \Phi}{\partial t} \tag{A.18}$$

This definition leads to the *Lorentz condition*

$$\nabla(\nabla \cdot \mathbf{A}) = -\mu\varepsilon \frac{\partial}{\partial t}(\nabla\Phi) \tag{A.19}$$

It may be demonstrated (see Problem P A.6) that the Lorentz condition is consistent with the conservation of electric charge principle, Equation A.5. With the introduction of the Lorentz condition, Equation A.18 becomes

$$\nabla^2\mathbf{A} - \frac{1}{c^2}\frac{\partial^2\mathbf{A}}{\partial t^2} = -\mu\mathbf{J} \tag{A.20}$$

Similarly, Equations A.4c, A.16, and A.17 lead directly to

$$\nabla^2\Phi - \frac{1}{c^2}\frac{\partial^2\Phi}{\partial t^2} = -\frac{\rho_e}{\varepsilon} \tag{A.21}$$

The vector potential $\mathbf{A}$ and the scalar potential Φ may now be determined as follows. Recalling Equation 1.41 for the periodic current of amplitude $I_0 = i\omega q$ carried by the electric dipole radiator of Chapter 1,

$$I = I_0 e^{i\omega t} \tag{A.22}$$

we are justified in writing for the current density $\mathbf{J}$*

$$\mathbf{J} = \mathbf{J}_0 e^{i\omega t} \tag{A.23}$$

Consistent with this assumption, the vector potential $\mathbf{A}$ must have the form

$$\mathbf{A} = \mathbf{A}_0 e^{i(\omega t + \varphi)} \tag{A.24}$$

where $\varphi = \omega r/c = kr$ represents the phase angle at a field point a distance r from the radiator consistent with the *wave number* $k \equiv \omega/c$. The distance r, illustrated in Figure A.1, is related to the source coordinates x', y', z' and the field coordinates x, y, z by

$$r = \sqrt{(x - x')^2 + (y - y')^2 + (z - z')^2} \tag{A.25}$$

---

* The current density in this case is the current divided by an effective cross-sectional area of the radiator and multiplied by the unit vector oriented along the axis of the radiator.

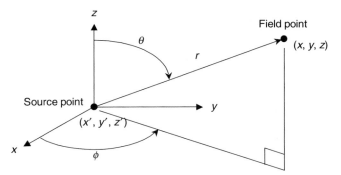

Figure A.1 Geometry showing the relationship between the field point and the source point

Equation A.24 allows Equation A.20 to be rewritten

$$(\nabla^2 + k^2)\mathbf{A} = -\mu\mathbf{J} \qquad (A.26)$$

It can be verified that a solution of Equation A.26 is

$$\mathbf{A} = \frac{\mu}{4\pi} \int_{V'} \frac{\mathbf{J}}{r} e^{-ikr} dV' \qquad (A.27)$$

where the integration is performed over the source coordinates, x', y', z'. The proof proceeds as follows. First, Equation A.27 is substituted into Equation A.26 to obtain

$$(\nabla^2 + k^2)\,\mathbf{A} = \frac{\mu}{4\pi} \int_{V'} \mathbf{J}(\nabla^2 + k^2)\left(\frac{e^{-ikr}}{r}\right) dV' \qquad (A.28)$$

In Equation A.28 it is emphasized that the Laplacian operator ∇^2 operates only on the field coordinates x, y, z while $\mathbf{J}$ is a function of the source coordinates x', y', z'. Viewed from its far field, the dipole radiator is essentially a point source located at the origin of a spherical coordinate system, as illustrated in Figure A.1. Therefore,

$$\nabla^2\left(\frac{e^{-ikr}}{r}\right) = \frac{1}{r^2}\frac{\partial}{\partial r}\left[r^2\frac{\partial}{\partial r}\left(\frac{e^{-ikr}}{r}\right)\right] = -\frac{k^2}{r}e^{-ikr}, \qquad r \neq 0 \qquad (A.29)$$

Thus, the integrand of Equation A.28 vanishes everywhere except at the singular point $r = 0$. The point $r = 0$ itself may be surrounded by an arbitrarily small sphere of radius Δ and volume V_0. Then if Δ is made sufficiently small,

$$\lim_{\Delta \to 0} e^{-ik\Delta} = 1 \qquad (A.30)$$

and $\mathbf{J}$ may be considered essentially constant within V_0. Under these circumstances the right-hand side of Equation A.28 becomes

$$\frac{\mu \mathbf{J}}{4\pi} \int_{V_0} (\nabla^2 + k^2) \frac{1}{r} \, dV' = \frac{\mu \mathbf{J}}{4\pi} (-4\pi + 2\pi k^2 \Delta^2) = -\mu \mathbf{J} \qquad (A.31)$$

since $2\pi k^2 \Delta^2$ is negligible compared to 4π. Therefore, Equation A.28 converges to Equation A.26, thereby verifying the assumed solution, Equation A.27. By a similar process it can be shown that Equation A.21 may be written

$$(\nabla^2 + k^2) \, \Phi = -\frac{\rho}{\varepsilon} \qquad (A.32)$$

and that its solution is

$$\Phi = \frac{1}{4\pi\varepsilon} \int_{V'} \frac{\rho}{r} e^{-ikr} dV' \qquad (A.33)$$

In the case of an atomic dipole oscillating along the z axis, as depicted in Figures 1.9b and 1.11a, Equation A.27 may be written

$$\mathbf{A}(r) = i \left(\frac{\mu \omega q \, \delta z_{max}}{2\pi r} \right) e^{-ikr} \mathbf{a}_z \qquad (A.34)$$

where the symbol $\mathbf{a}_z$ represents the unit vector oriented along the positive z axis. In the derivation of Equation A.34 we have exploited the fact that the atomic dipole is differentially sized so that

$$\mathbf{J} dV' = \left(\frac{I_0}{A_c} \right) \mathbf{a}_z \, (\delta z_{max} A_c) = i \omega q \delta z_{max} \mathbf{a}_z \qquad (A.35)$$

where $I_0 = i\omega q$ and A_c is the (fictitious) cross-sectional area of the atomic dipole normal to the z axis. From Figure A.2, we may deduce that

$$\mathbf{a}_z = \mathbf{a}_r \cos\theta - \mathbf{a}_\theta \sin\theta \qquad (A.36)$$

where $\mathbf{a}_r$ and $\mathbf{a}_\theta$ are the unit vectors directed along the r and θ directions indicated in the figure.

Considerations similar to those leading from Equation A.27 to Equation A.34, when applied to Equation A.33, yield

$$\Phi = \frac{q}{4\pi\varepsilon} \left(\frac{e^{-ikr_1}}{r_1} - \frac{e^{-ikr_2}}{r_2} \right) \qquad (A.37)$$

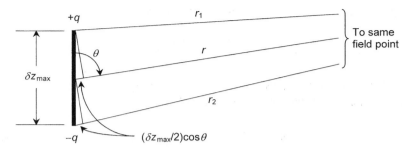

Figure A.2 Geometry associated with an atomic dipole radiator and a field point at a large distance r

where r_1 and r_2 are the respective distances from the point charges $+q$ and $-q$ to the same field point a distance r from their combined center of charge. Then from the geometry depicted in Figure A.2 we can write

$$r_1 \approx r - \frac{\delta}{2} \cos\theta \tag{A.38}$$

and

$$r_2 \approx r + \frac{\delta}{2} \cos\theta \tag{A.39}$$

where for simplicity the symbol δ has now replaced the symbol δz_{max}. The degree of approximation in Equations A.38 and A.39 improves as the distance r increases with respect to δ and the three lines represented by r_1, r_2, and r in Figure A.2 become more nearly parallel. A further approximation is achieved by introduction of the series

$$\frac{1}{1 \pm x} = 1 \mp x + x^2 \mp x^3 + \cdots \approx 1 \mp x \tag{A.40}$$

where the approximation holds for small x. Then

$$\frac{1}{r_1} \approx \frac{1}{r\left(1 - \dfrac{\delta}{2r}\cos\theta\right)} \approx \frac{1}{r}\left(1 + \frac{\delta}{2r}\cos\theta\right) \tag{A.41}$$

and

$$\frac{1}{r_2} \approx \frac{1}{r\left(1 + \dfrac{\delta}{2r}\cos\theta\right)} \approx \frac{1}{r}\left(1 - \frac{\delta}{2r}\cos\theta\right) \tag{A.42}$$

With the introduction of Equations A.41 and A.42, Equation A.37 becomes

$$\Phi = \frac{\mathcal{P}}{4\pi\varepsilon}\left(\frac{1}{r^2} + \frac{ik}{r}\right)\cos\theta \, e^{-ikr} \tag{A.43}$$

where $\mathcal{P} = q\delta$ is the *electric dipole moment* of the fully polarized atom. In deriving Equation A.43 use has been made of the fact that

$$e^{\pm ik(\delta/2)\cos\theta} \approx 1 \pm ik\left(\frac{\delta}{2}\right)\cos\theta \tag{A.44}$$

for small values of $k\delta/2$.

Now specializing Equation A.16 for the case of a periodic electromagnetic field we obtain

$$\mathbf{E} = -i\omega\mathbf{A} - \nabla\Phi$$

$$= \frac{\mu\omega^2\mathcal{P}}{4\pi r} e^{-ikr}(\mathbf{a}_r \cos\theta - \mathbf{a}_\theta \sin\theta) \tag{A.45}$$

$$-\frac{\mathcal{P}}{4\pi\varepsilon}\nabla\left(\frac{1}{r^2} + \frac{ik}{r}\right)\cos\theta \, e^{-ikr}$$

Also, Equations A.1 (with $\mathbf{M} = 0$), A.13, and A.34 may be combined to yield

$$\mathbf{H} = \frac{1}{\mu}\nabla\times\mathbf{A} = \frac{i\omega\mathcal{P}}{4\pi}\nabla\times\left[\frac{e^{-ikr}}{r}(\mathbf{a}_r\cos\theta - \mathbf{a}_\theta\sin\theta)\right] \tag{A.46}$$

After suitable manipulation, Equations A.45 and A.46 can be used with Equation 1.18 (the Poynting vector) to compute the power radiated to the far field of the atomic dipole. We begin by evaluating the gradient operation in the second term on the right-hand side of Equation A.45,

$$\nabla\left(\frac{1}{r^2} + \frac{ik}{r}\right)\cos\theta \, e^{-ikr} = -\left(\frac{2ik}{r} + \frac{2}{r^2} - k^2\right)\frac{\cos\theta}{r} e^{-ikr}\mathbf{a}_r$$

$$-\left(\frac{ik}{r} + \frac{1}{r^2}\right)\frac{\sin\theta}{r} e^{-ikr}\mathbf{a}_\theta \tag{A.47}$$

With the introduction of Equation A.47, Equation A.45 may be written

$$\mathbf{E} = \frac{\mathcal{P}}{4\pi\varepsilon r} e^{-ikr}\left[2\left(\frac{ik}{r} + \frac{1}{r^2}\right)\cos\theta \, \mathbf{a}_r + \left(\frac{ik}{r} + \frac{1}{r^2} - k^2\right)\sin\theta \, \mathbf{a}_\theta\right] \tag{A.48}$$

Carrying out the curl operation in Equation A.46 leads to

$$\mathbf{H} = \left(\frac{\omega\mathcal{P}}{4\pi r}\right)e^{-ikr}\left(\frac{i}{r} - k\right)\sin\theta \, \mathbf{a}_r \tag{A.49}$$

The total time-averaged power radiated by the atomic oscillator may be obtained by evaluating the surface integral [2]

$$P_r = \frac{1}{2} \operatorname{Re} \left[\int_S \mathbf{E} \times \mathbf{H}^* \cdot d\mathbf{S} \right] \tag{A.50}$$

where $\mathbf{H}^*$ is the complex conjugate of $\mathbf{H}$, and the surface S completely encloses the oscillator. The notation Re[§] indicates the real part of the integral §. The integrand of Equation A.50 is

$$\mathbf{E} \times \mathbf{H}^* \cdot d\mathbf{S} = \mathbf{E} \times \mathbf{H}^* \cdot \mathbf{a}_r dS = E_\theta H_\phi^* dS$$

$$= \left(\frac{\mathcal{P}}{4\pi} \sin\theta \right)^2 \frac{\omega}{\varepsilon r^2} \left(k^3 - i \frac{1}{r^2} \right) \tag{A.51}$$

where E_θ and H_ϕ are the relevant components of the electric and magnetic field strengths expressed in a spherical (r, θ, ϕ) coordinate system. Then, assuming that the atomic oscillator is located at the origin of the spherical coordinate system and that the surface surrounding the oscillator is a sphere of radius r, Equation A.50 becomes

$$P_r = \frac{1}{2} \left(\frac{\mathcal{P}}{4\pi} \right)^2 \left(\frac{k^3 \omega}{\varepsilon} \right) 2\pi \int_0^\pi \sin^3\theta \, d\theta \tag{A.52}$$

Finally, noting that

$$k^3 \omega = (2\pi)^4 \frac{c}{\lambda^4} \tag{A.53}$$

and

$$\int_0^\pi \sin^3\theta \, d\theta = 2 \tag{A.54}$$

Equation A.52 may be written

$$P_r = \left(\frac{\mathcal{P}}{4\pi} \right)^2 (2\pi)^5 \frac{c}{\varepsilon} \lambda^{-4} = \frac{\mathcal{P}^2}{\varepsilon} 2\pi^3 c \lambda^{-4} \tag{A.55}$$

This result represents the power in watts radiated by one of the three vibrational degrees of freedom of the atomic oscillator. Then Equation 1.42 in Chapter 1 is obtained by multiplying this result by three.*

---

* Equation 1.42 contains ε_0 and c_0 rather than ε and c. In Chapter 1 the atomic source is considered to be located in a vacuum where $\varepsilon = \varepsilon_0$ and $c = c_0$, whereas the treatment in this appendix is more general.

Problems

P A.1 Show that Equation A.5 follows directly from Equations A.4a and A.4c.

P A.2 Equation A.22 is an expression for the periodic electric current flowing along one axis of the atomic dipole illustrated in Figures 1.9b and 1.11a. Demonstrate that the conservation of charge principle, Equation A.5, leads to the expression $I_0 = i\omega q$ in this case. [*Hint:* Assume that the volume of the dipole is $A_c \delta z_{max}$, where A_c is the (fictitious) cross-sectional area of the dipole normal to the z axis.]

P A.3 Derive Equation A.11.

P A.4 Derive Equation A.12.

P A.5 Derive Equation A.17.

P A.6 Demonstrate that the Lorentz condition, Equation A.19, is consistent with the conservation of electric charge principle, Equation A.5.

P A.7 Derive Equation A.20.

P A.8 Derive Equation A.21.

P A.9 Verify Equation A.26.

P A.10 Derive the first equality in Equations A.31.

P A.11 Verify Equations A.32 and A.33.

P A.12 Derive Equation A.34 from Equation A.27.

P A.13 Draw and label a sketch that represents the r, θ, z coordinate system implied by Equation A.36. Clearly show the unit vectors $\mathbf{a}_r$, $\mathbf{a}_\theta$, $\mathbf{a}_z$.

P A.14 Derive Equation A.37 from Equation A.33.

P A.15 Derive Equation A.43.

P A.16 Verify Equation A.47.

P A.17 Verify Equation A.48. $\left[\textit{Hint: } \text{Note that } \mu\omega^2 \dfrac{\varepsilon}{\varepsilon} = \dfrac{\omega^2}{c^2} \dfrac{1}{\varepsilon} = \dfrac{k^2}{\varepsilon}. \right]$

P A.18 Verify Equation A.49.

P A.19 Verify Equation A.51.

P A.20 Supply the missing steps between Equation A.50 and Equation A.52.

P A.21 Verify Equation A.53.

P A.22 Verify Equation A.54.

REFERENCES

1. Maxwell, J. C., "A dynamical theory of the electromagnetic field," *Philosophical Transactions of the Royal Society of London,* 1865, pp. 459–512.
2. Plonsey, R., and R. E. Collin, *Principles and Applications of Electromagnetic Fields,* McGraw-Hill, New York, 1961.

APPENDIX B

MIE SCATTERING BY HOMOGENEOUS SPHERICAL PARTICLES: PROGRAM UNO

Mie scattering was briefly introduced in Chapter 6 as one of the two important models for describing the scattering of electromagnetic radiation by particles whose complex index of refraction differs from that of the propagating medium. This appendix presents a FORTRAN code, Program UNO, whose operation is based on the Mie scattering model. Program UNO permits prediction of the scattering phase function for scattering from homogeneous spherical particles when $2\pi r/\lambda < 250$. It is suitable for applications involving the scattering of visible and infrared radiation by atmospheric aerosols.

B.1 INTRODUCTION

Scattering centers in the atmosphere include individual molecules, aerosols, and precipitation, such as ice crystals and raindrops. While real scattering centers are generally inhomogeneous and irregular in shape, they are usually modeled as being homogeneous and regular in shape to permit a closed-form solution for the wavelength dependence of their scattering phase function. In 1908 Gustav Mie [1] published a seminal article in which he formulated the problem of scattering of electromagnetic

451

radiation by particles whose complex index of refraction differs from that of the propagating medium. The other important scattering model, attributed to Lord Rayleigh, is limited to scattering from particles, such as individual molecules, whose size is very small compared to the wavelength of the scattered radiation. The Mie scattering model on the other hand can be used to describe scattering for a wide range of values of the size parameter $a = 2\pi r/\lambda$, including molecular scattering. While a Mie model can be formulated to characterize scattering from irregularly shaped scattering volumes, Mie scattering theory is typically used to describe scattering from homogeneous spherical particles. Finally, the Mie scattering model can also be used to describe scattering from particles whose characteristic dimension is large compared to the wavelength of the scattering radiation, but this can also be accomplished—often more directly—using the MCRT approach developed in Part III of this book.

The Mie scattering model is based on solving Maxwell's equations in the interior of the scattering volume and in the surrounding medium, and then matching the two solutions at the interface. Even for homogeneous spherical scattering volumes the formulation is very tedious, involving as it does infinite series of products of periodic functions whose eigenvalues are roots of equations involving Bessel and Hankel functions. Program UNO, an executable version of which is provided on the compact disc packaged with this book, is based on a version of the Mie theory that incorporates certain simplifying assumptions that limit the validity of results obtained to values of the size parameter $a = 2\pi r/\lambda$ of less than 250. The program is valid for most

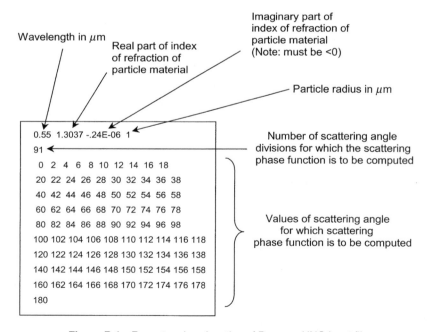

Figure B.1   Format and explanation of Program UNO input file

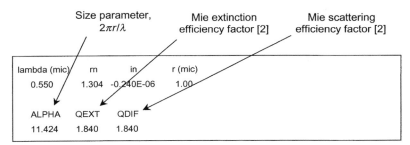

Figure B.2 Format and explanation of header of output file generated by Program UNO after execution using the input file of Figure B.1

atmospheric radiation applications of practical interest, including those involving the scattering of visible and infrared radiation by atmospheric aerosols and precipitation. Readers interested in details about formulating the Mie scattering model are referred to Lenoble's book on atmospheric radiation [2].

B.2 PROGRAM UNO

Program UNO was generously provided by Gérard Brogniez, an atmospheric physicist at the *Laboratoire d'Optique Atmosphérique* of the University of Lille I (France). Example input and output files, with explanations of variables, appear in Figures B.1, B.2, and B.3. The free-format input file for Program UNO, an example of which is shown in Figure B.1, must be created and named with a '.dat' extension before Program UNO can be executed. The output files shown in Figures B.2 and B.3 will result when Program UNO is executed with the input file of Figure B.1. The results for the scattering phase function are plotted in Figure 6.9 for this example. The phase function is normalized to the case of diffuse scattering. This means that when the scattering phase function is integrated over 4π-space and divided by 4π, a value of 1.0 is obtained.

Problem

P B.1 Use Program UNO to create plots of the scattering phase function versus scattering angle for values of the size parameter $2\pi r/\lambda$ of 0.01, 0.1, 1, 10, and 40 using the same value of the complex index of refraction $n - i\kappa$ as in the example case in this appendix. Discuss the results obtained. What trends with relative particle size do you see? Compare the result for the smallest particle size with the Rayleigh scattering phase function, Equation 6.59. What may we conclude from this comparison?

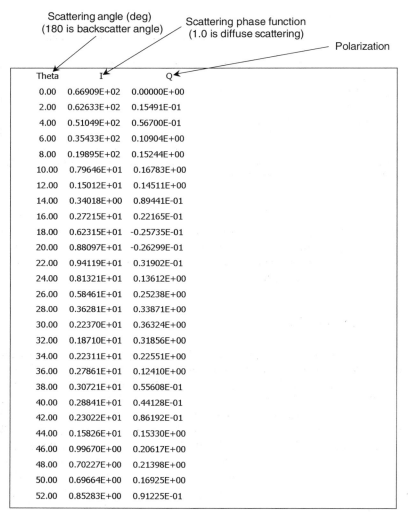

Figure B.3 Format and explanation of output file generated by Program UNO after execution using the input file of Figure B.1

54.00	0.10036E+01	0.15932E-01
56.00	0.10257E+01	-0.22330E-01
58.00	0.88822E+00	-0.86649E-02
60.00	0.64891E+00	0.44668E-01
62.00	0.41126E+00	0.10474E+00
64.00	0.26628E+00	0.13552E+00
66.00	0.24880E+00	0.11739E+00
68.00	0.32670E+00	0.57825E-01
70.00	0.42481E+00	-0.12639E-01
72.00	0.46847E+00	-0.57364E-01
74.00	0.42295E+00	-0.54129E-01
76.00	0.30837E+00	-0.76683E-02
78.00	0.18366E+00	0.53026E-01
80.00	0.11018E+00	0.91260E-01
82.00	0.11621E+00	0.83613E-01
84.00	0.18259E+00	0.33236E-01
86.00	0.25692E+00	-0.32231E-01
88.00	0.28666E+00	-0.76836E-01
90.00	0.25081E+00	-0.77047E-01
92.00	0.17091E+00	-0.34890E-01
94.00	0.95496E-01	0.24013E-01
96.00	0.68259E-01	0.65958E-01
98.00	0.10046E+00	0.68743E-01
100.00	0.16479E+00	0.33693E-01
102.00	0.21414E+00	-0.16533E-01
104.00	0.21285E+00	-0.52297E-01
106.00	0.16003E+00	-0.54274E-01
108.00	0.89847E-01	-0.23311E-01
110.00	0.48630E-01	0.22365E-01
112.00	0.64010E-01	0.59662E-01
114.00	0.12628E+00	0.74238E-01
116.00	0.19455E+00	0.66707E-01
118.00	0.22384E+00	0.48499E-01
120.00	0.19543E+00	0.31558E-01

Figure B.3 *(Continued)*

122.00	0.13021E+00	0.19824E-01
124.00	0.75596E-01	0.85667E-02
126.00	0.74064E-01	-0.85433E-02
128.00	0.13374E+00	-0.30697E-01
130.00	0.22105E+00	-0.47140E-01
132.00	0.28147E+00	-0.43699E-01
134.00	0.27586E+00	-0.14981E-01
136.00	0.20907E+00	0.26484E-01
138.00	0.13094E+00	0.52079E-01
140.00	0.10713E+00	0.31168E-01
142.00	0.17661E+00	-0.49718E-01
144.00	0.32260E+00	-0.17435E+00
146.00	0.47647E+00	-0.29995E+00
148.00	0.55490E+00	-0.37632E+00
150.00	0.50914E+00	-0.37125E+00
152.00	0.35691E+00	-0.28809E+00
154.00	0.17675E+00	-0.16497E+00
156.00	0.66358E-01	-0.55545E-01
158.00	0.88597E-01	-0.16161E-02
160.00	0.23627E+00	-0.13000E-01
162.00	0.43637E+00	-0.65504E-01
164.00	0.59107E+00	-0.11738E+00
166.00	0.63119E+00	-0.13390E+00
168.00	0.55112E+00	-0.10562E+00
170.00	0.40587E+00	-0.50102E-01
172.00	0.27408E+00	0.21874E-02
174.00	0.21104E+00	0.28080E-01
176.00	0.22053E+00	0.24832E-01
178.00	0.26147E+00	0.85308E-02
180.00	0.28265E+00	0.58781E-13

Figure B.3 *(Continued)*

REFERENCES

1. Mie, G., "Optics of turbid media," *Annalen der Physik,* Vol. 25, No. 3, 1908, pp. 377–445.
2. Lenoble, J., *Atmospheric Radiative Transfer,* A Deepak Publishing, Hampton, VA, 1993.

APPENDIX C

A FUNCTIONAL ENVIRONMENT FOR LONGWAVE INFRARED EXCHANGE (FELIX)

Much of the methodology developed in Part III of this book has been implemented as a Windows-based PC environment called FELIX (Functional Environment for Longwave Infrared eXchange). A student version of FELIX has been packaged with this book. The student version of FELIX is limited to surface-to-surface exchange where the surfaces are directional or diffuse emitters, diffuse absorbers, and diffuse–specular reflectors. FELIX is also useful for estimating, to a high degree of accuracy, the values of configuration factors or exchange factors used in the net exchange method.

C.1 INTRODUCTION TO FELIX

A student version of a functional environment for modeling infrared exchange, called FELIX, has been supplied with this book. FELIX is fully capable of providing highly accurate estimates of configuration factors and exchange factors for use with the net exchange method developed in Part II of the book. More to the point, however, it is also capable of computing the radiation distribution factors described in Part III of the book and using them to obtain accurate estimates of the net heat flux distribution on the walls of enclosures for which surface properties are known and surface tem-

457

perature distributions are specified. FELIX is a product of the doctoral research of Félix J. Nevárez-Ayala [1] carried out under the direction of the author. The compact disc packaged with the book refers to an extensive on-line users' manual from Dr. Nevárez's dissertation. All recent Virginia Tech doctoral dissertations are available on-line by visiting *www.vt.edu* and then following directions through the library link to "ETD" (Electronic Theses and Dissertations). The reader interested in more details about FELIX is encouraged to visit the Virginia Tech website and download Dr. Nevárez' dissertation.

C.2   WHAT THE STUDENT VERSION OF FELIX CANNOT DO

The student version of FELIX does not treat radiation through a participating medium, nor does it model the effects of directional absorptivity and bidirectional reflectivity. The student version of FELIX is not intended for use in the modeling of optical systems such as cameras, telescopes, and certain types of radiometric instruments. This is because all surface elements in the student version of FELIX are built up from plane triangular facets. As explained below, curvature can be simulated to almost any desired degree of accuracy by using a sufficiently large number of very small facets to represent a continuously curved surface, such as a sector of a spherical mirror; however, experience with FELIX has shown that the number of facets required for optical-quality design is prohibitively large.

C.3   WHAT THE STUDENT VERSION OF FELIX CAN DO

FELIX can accurately model thermal radiative exchange in any diffuse–specular enclosure. Its capabilities match and exceed the net exchange method developed in Part II. More to the point, it can do so with considerably less effort on the part of the user. While exchange factors are extremely difficult to obtain for all but the most regular of enclosures (unless, of course, FELIX is used to obtain them), the much more powerful distribution factors, whose values are estimated and used by FELIX, are only a CAD drawing away. FELIX uses the MCRT method developed in Part III of the book to estimate, to almost any desired degree of accuracy, the configuration factors, exchange factors, and distribution factors described in this book. It also permits export of the distribution factors to facilitate calculation of the net heat flux distribution on the walls of an enclosure given the surface properties and temperature distribution. Users of FELIX will no doubt appreciate the powerful graphical interface that produces high-quality false-color images of the distribution factor fields on the enclosure walls. Examples of these images appear in Figures 9.17 and 12.11.

C.4   HOW DOES FELIX WORK?

The internal workings of FELIX are generally based on the methodology developed in Part III of this book. However, features of the object-oriented programming lan-

guage at the heart of FELIX (C/C++) permit many programming shortcuts and effi-
ciencies not available in FORTRAN. Also, the need to import geometries as CAD
files limits the surface types to those that can be modeled as continuous meshes of triangular facets. This is not really a severe limitation since curved surfaces (for the purpose of ordinary heat transfer modeling) can always be adequately represented, as explained in the following subsection.

C.4.1 Simulation of Curvature

Figure C.1 shows facet 1, one of the generic triangular building blocks used to create
all surfaces in the student version of FELIX. The unit normal vectors $\mathbf{n}_A$, $\mathbf{n}_B$, and $\mathbf{n}_C$
are the area-weighted means of the unit normal vectors of the triangular facets whose
vertices meet at a point. For example, if a vertex of facet 1 shares point A with a vertex from each of facets 2, 3, 4, and 5 (the vertex angles at point A adding up to 360
deg), then the x component of the unit normal vector $\mathbf{n}_A$ is given by

$$u_{x,A} = \frac{u_{x,1}A_1 + u_{x,2}A_2 + u_{x,3}A_3 + u_{x,4}A_4 + u_{x,5}A_5}{A_1 + A_2 + A_3 + A_4 + A_5} \qquad (C.1)$$

and so forth. Once the area-weighted unit vector is known at each corner of facet 1,
the apparent local unit vector at any point x, y, z on the facet is approximated as the
trilinear interpolation of the three vertex values. Therefore, overall surface curvature
is well approximated by the mesh of planar facets, and, of course, the approximation
becomes more accurate as the mesh is refined. Still, experience has shown that exceedingly refined meshes must be used to approach the standard of precision required
for optical-quality design. For this reason the commercial version of FELIX has an
optics option that imports and uses actual curved surfaces from a library, as described
in Chapters 11 and 12.

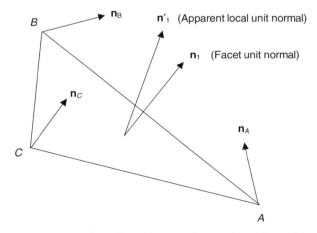

Figure C.1   A typical triangular facet illustrating pseudo-curvature using unit vector averaging

C.4.2   Definition of Triangular Facets and Random Location of Emission Sites

A typical triangular facet is defined by the location of its vertices, as shown in Figure C.2. Then if the vector from x_0, y_0, z_0 to x_1, y_1, z_1 is $\mathbf{V}_1$ and the vector from x_0, y_0, z_0 to x_2, y_2, z_2 is $\mathbf{V}_2$, the unit normal vector defining the front (radiatively active) side of the facet is

$$\mathbf{n} = \frac{\mathbf{V}_2 \times \mathbf{V}_1}{|\mathbf{V}_2 \times \mathbf{V}_1|} \tag{C.2}$$

The random emission sites are found by drawing two random numbers, R_α and R_β, and then testing their sum. If $R_\alpha + R_\beta \leq 1$, then they are redefined as

$$R_\alpha = 1.0 - R_\alpha \quad \text{and} \quad R_\beta = 1.0 - R_\beta \tag{C.3}$$

After a pair of independent random numbers whose sum exceeds unity has been thus obtained, the coordinates of the random emission site on the triangular facet are

$$\begin{aligned} x &= x_0 + R_\alpha (x_1 - x_0) + R_\beta (x_2 - x_0) \\ y &= y_0 + R_\alpha (y_1 - y_0) + R_\beta (y_2 - y_0) \end{aligned} \tag{C.4}$$

and

$$z = z_0 + R_\alpha (z_1 - z_0) + R_\beta (z_2 - z_0)$$

When the surface element is both plane and triangular, the process for finding the intersection of a ray and a surface element is much easier than that described in Chapters 11 and 12. The reader is referred to Reference 1 for the details.

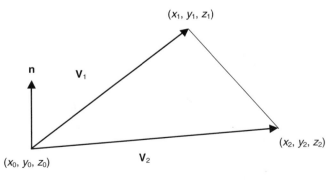

Figure C.2   Geometry and nomenclature for defining a triangular facet and its radiatively active side

Perhaps the most unique feature of FELIX is that it imports the enclosure geometry as a CAD file from any CAD engine that can export files as '.3ds' files. In addition, FELIX results can be exported either as ASCII files or bitmap files. For more information about the operation of FELIX, please install the environment on your PC and follow the directions to the tutorial and users' manual.

REFERENCE

1. Nevárez-Ayala, F. J., *A Monte Carlo Ray-Trace Environment for Optical and Thermal Radiative Design,* Ph.D. Dissertation, Department of Mechanical Engineering, Virginia Polytechnic Institute and State University, Blacksburg, VA, Spring 2002. (*Note:* Recent Virginia Tech dissertations are available by internet at *www.vt.edu,* then follow the indications through the library link to Electronic Theses and Dissertations, or "ETD.")

APPENDIX D

RANDOM NUMBER GENERATORS AND AUTOREGRESSION ANALYSIS

The Monte Carlo ray-trace approach to radiation heat transfer analysis is highly de-pendent on the availability of a fast, efficient random number generator. In this ap-pendix we learn that true random number generators do not exist, but rather that "random" numbers are computed using completely deterministic algorithms. In fact, the very success of the MCRT method requires that the same identical sequence of "random" numbers be available upon demand. The requirements for these pseudo-random number generators are established and an algorithm for computing them is identified. Finally, a brief introduction is provided to autoregression analysis, which is an important tool for testing the randomness of pseudo-random number genera-tors. Many readers will find autoregression analysis useful in other aspects of their work.

D.1 PSEUDO-RANDOM NUMBER GENERATORS

The accuracy of a Monte Carlo simulation depends on the quality of the random num-ber generator used. In the early days of computing, physical devices that supplied truly random numbers were attached to computers. Examples include devices for sensing Johnson ("shot") noise in an electronic circuit and for counting particles per unit time associated with radioactive decay. This approach was soon found to be impractical, mostly because of the inherently low rate at which these devices were capable of producing a sequence of random numbers.

Physical devices were eventually replaced with *pseudo-random* number generators (PRNGs), which use deterministic formulas to produce sequences of numbers whose statistical properties approach those of true random number sequences. Unfortunately, a correlation necessarily exists among members of a given sequence of numbers produced by a pseudo-random number generator; that is, knowledge of the value of a given member or members of the sequence is sufficient to predict the value of a later member in the sequence. Also, pseudo-random number generators are by their nature periodic; that is, the sequence they produce eventually repeats itself. If the periodicity of a given PRNG happens to match that of the sequence or sequences in which the random numbers are used in a simulation, biases may occur in the simulated result.

An alternative to the PRNG is the *quasi-random* number generator (QRNG), which produces sequences not intended to be random but rather intended to be distributed as uniformly as possible. In some Monte Carlo simulations QRNGs may be more appropriate than PRNGs. However, in the radiation heat transfer calculations developed in this book PRNGs usually are the appropriate choice.

D.2 PROPERTIES OF A "GOOD" PSEUDO-RANDOM NUMBER GENERATOR

A "good" pseudo-random number generator suitable for radiation heat transfer analysis will exhibit the following seven properties, adapted from James [1].

D.2.1 Property 1

The sequence of numbers generated will be uniformly distributed. Uniformity of distribution of a sequence of pseudo-random numbers may be tested using the standard χ^2 test from statistics, which compares the observed and expected frequencies of the pseudo-random numbers in the sequence. In this test a sequence of pseudo-random numbers of length N is sorted according to the range into which its individual members fall. For example, suppose $N = 10,000$ and a given member of the complete sequence has a value between $0.00 \ldots$ and $0.05 \ldots$ (the ellipses "$\ldots$" indicate an unspecified number of zeros, i.e., an unspecified precision). Then that number is placed in bin 1, while any number between $0.05 \ldots$ and $0.10 \ldots$ is placed in bin 2, and so forth. When all 10,000 numbers have been sorted into the 20 bins, the count of numbers in each bin is obtained. Of course, the expected count for each bin is $10,000/20 = 500$. We then compute

$$\chi^2 = \sum_{i=1}^{20} \left(\frac{O_i - E_i}{E_i} \right)^2 \tag{D.1}$$

where O_i is the observed count in bin i and E_i is the expected count in bin i. The value of χ^2 may then be interpreted as a measure of whether or not the random numbers are uniformly distributed. The probability that the random number distribution is uniform is read from a chart, available in standard texts such as Reference 2 (page 541). To enter the chart we need the value of χ^2 and the number of *degrees of freedom d* $= n - k$, where n is the number of observations (20 in the current example) and k is the number of conditions imposed on the distribution (one in this case: uniformity). For $d = 19$, the value of χ^2 would have to be about 30 to have a 95-percent confidence that this distribution was uniform.

D.2.2 Property 2

The sequence of numbers generated will "appear" random; that is, it will pass any statistical test for randomness. The correlation between each number generated and the numbers that precede it in the sequence should be low. This means that knowledge of the value of any one member or subsequence of members in the sequence cannot be used to predict, using any statistical method, the values of subsequent members of the sequence. (Clearly, the deterministic formula used to generate the pseudo-random number sequence can be used to predict values of the sequence. However, in a statistical test this formula would not be "known" a priori.

For any sequence s_1, s_2, s_3, . . ., s_{n+N+1} it is possible to find an *autoregression model of order* $n \leq N + 1$ of the form

$$s_i = a_i s_{i-1} + a_2 s_{i-2} + \cdots + a_n s_{i-n} + \varepsilon_i, \qquad i > n + N + 1 \qquad \text{(D.2)}$$

where a_1, a_2, . . . , a_n are *autoregression coefficients* and ε_i is the error in predicting s_i. The procedure for obtaining the values of the n autogression coefficients from $n + N + 1$ members of a sequence is given in Section D.4. Equation D.2 uses only a posteriori knowledge of a subsequence of $n + N + 1$ members of a sequence to predict, on the basis of statistics, the value of the ith member of the sequence, where $i \geq n + N + 2$. The accuracy with which Equation D.2 can be used to predict the ith member of the sequence also depends on the degree to which the sequence represents a deterministic process; that is, the degree to which the ith member is correlated with the previous $n + N + 1$ members. If the error ε_i is sufficiently large, the ith member can be said to be uncorrelated with the previous $n + N + 1$ members of the sequence. This, of course, is the desired property of a good pseudo-random number generator.

Equation D.2 can be used to generate a subsequence of numbers whose values can then be compared, using the standard χ^2 test, with the corresponding actual subsequence produced by the pseudo-random number generator. If the pseudo-random number sequence is uncorrelated, we should expect to obtain a large value for χ^2.

D.2.3 Property 3

The period of the sequence of random numbers should be sufficiently long to complete the simulation without repeating any part of the sequence. The period of a pseudo-random number generator is the count of numbers that can be generated before the sequence is repeated. Every pseudo-random number generator has a period.

D.2.4   Property 4

The PRNG should be repeatable. That is, given the same starting values, called *seeds,* the generator should generate the same sequence of numbers. This property is essential in debugging Monte Carlo algorithms.

D.2.5   Property 5

The PRNG should provide long, disjoint sequences. This means that different users performing subparts of a same overall simulation should be able to work simultaneously without danger of inadvertently using the same sequence of random numbers, thereby possibly introducing a bias to the overall simulation. If a PRNG does not have this property then each part of the simulation must be carried out serially. That is, the first part must be carried out and the final pseudo-random number recorded so that this number can be used as the seed to start the PRNG for the second part of the simulation, and so forth.

D.2.6   Property 6

The PRNG should be portable. Many "random" number generators written in FORTRAN use processor-specific programming shortcuts that do not work on all processors.

D.2.7   Property 7

The PRNG should be efficient because billions of random numbers may be needed in modern Monte Carlo simulations. Trade-offs will usually be required between efficiency and achievement of the other six properties of a "good" PRNG, however.

The statistical properties of a relatively short sequence of members of a long random number sequence are generally less important than the statistical properties of the larger sequence. However, if the MCRT method is to be used to model a system so large that only a few hundred energy bundles can be emitted per surface, the statistical properties of relatively short sequences will be important.

D.3   A "MINIMAL STANDARD" PSEUDO-RANDOM NUMBER GENERATOR

Park and Miller [3] have studied the manifold of available PRNGs and concluded that the family of *multiplicative linear congruential generator* (MLCG) algorithms first proposed by Lehmer [4] can form the basis for what they refer to as a "minimal standard" PRNG. At the center of the MLCG algorithm is the recursion formula

$$s_{i+1} = (as_i + c) \bmod m \tag{D.3}$$

where S_{i+1} and S_i are successive members of the pseudo-random sequence $S_0, S_1, S_2,$ . . .; the coefficient a is a well-chosen integer multiplier in the range $[2, m-1]$; the constant c is a number that can be (and usually is) equal to zero; and m is an integer usually equal to or slightly smaller than the largest integer that can be represented in one computer word. The notation "$n \bmod m$" means "the modulus of n/m," where the modulus operator produces the first remainder of the division of two integers. For example,

$$13 \bmod 15 = \text{first remainder of} \quad \begin{array}{r} 0 \\ 15\overline{)13} \\ \underline{0} \\ 13 \end{array} = 13 \tag{D.4}$$

and

$$13 \bmod 15 = \text{first remainder of} \quad \begin{array}{r} 1 \\ 15\overline{)18} \\ \underline{15} \\ 3 \end{array} = 3 \tag{D.5}$$

The initial value of the sequence, s_0, is called the *seed* and must be provided by the user.

The members of the sequence produced by Equation D.3 are integers whose values range between 1 and $m-1$. The s sequence is converted into a sequence of pseudo-random numbers R, uniformly distributed between zero and unity, by dividing by m. Note that neither zero nor unity can result from this and similar MLCG algorithms with the constant c equal to zero. However, if the resulting sequence contains billions of members, each expressed to 32-bit precision, the difference between zero and unity, respectively, and the smallest and largest members is arbitrarily small. A "good" PRNG is one for which each seed value will lead to a different random number sequence of the same (known) fixed period. As an example, consider the MLCG algorithm $S_{i+1} = 6s_i \bmod 13$. Then if the seed s_0 is taken to be 1, the resulting sequence is 1, 6, 10, 8, 9, 2, 12, 7, 3, 5, 4, 11, 1, . . .and the sequence repeats itself. It is clear in this example that the period is short and that the number of significant figures in the corresponding random number sequence ($R_i = s_i/13, i = 1, 2, 3,$ . . ., 12) is small. This suggests that in a practical PRNG, the quantities a and m should be as large as possible (although not all values of $a < m$ lead to full-period sequences, i.e., to sequences whose period is $m-1$).

Down through the years many MLCG algorithms have been proposed and implemented as "commercial release" PRNGs. Subsequent statistical studies have revealed that some are quite "good" as defined in Section D.2 while others are surprisingly "bad." Some combinations of values of a and m are known to be better than others in their ability to produce long-period sequences of statistically random numbers.

The largest *prime* integer that can be represented on a 32-bit machine is $2^{31} - 1$, and so this number has emerged, because of number-theory considerations beyond the scope of this book, as the favorite choice for *m* in 32-bit operating systems. Research into the best multipliers has led many to conclude that $a = 7^5 = 16,807$ is a good choice. The resulting MLCG algorithm,

$$s_{i+1} = 16,807 s_i \bmod (2^{31} - 1) \tag{D.6}$$

is a full-period algorithm. Of practical concern is that 46 bits would be required to store the largest possible value of the product $16,807 s_i$. The potential for integer overflow therefore requires careful programming if an MLCG algorithm based on these values of *a* and *m* is to be universally portable.

A double-precision FORTRAN version of an MLCG algorithm, RANMAR, by Marsaglia and Zaman [5] has been adapted for use in Virginia Tech's Thermal Radiation Group. Based on Equation D.6, it meets all of the requirements laid out in Section D.2. For example, it is capable of creating 900 million different random number sequences with each subsequence having a period of approximately 10^{30}!

RANMAR actually consists of two subroutines. Before the first call to RANMAR, the FORTRAN command CALL RMARIN(IJ, KL) must be executed, where the integer variables IJ and KL (the seeds) have already been defined such that $0 \le$ IJ $\le$ 31,328 and $0 \le$ KL $\le$ 30,081. The second subroutine, RANMAR, is the actual pseudo-random number generator. Then each time a random number is needed, the FORTRAN command R = RANMAR() is executed, where R contains the new random number after execution of the command.

RANMAR is structured so that a large number of teams working independently at different locations can perform parts of the same Monte Carlo simulation without danger of a given random number subsequence appearing in any two parts. This is accomplished by assigning each team a different (randomly assigned) value of the seed IJ and then allowing each team to pick its own value of the second seed KL.

D.4   AUTOREGRESSION ANALYSIS

A physical process can usually be represented by a deterministically based time series.* For example, the distance $x - x_0$ traveled in time *t* by an object dropped from rest in a vacuum is described by

$$x - x_0 = \frac{gt^2}{2} \tag{D.7}$$

---

* It is usual to refer to any equally spaced series of observations as a "time" series even if the independent variable is, say, space rather than time.

where g is the acceleration of gravity. Often the process in question is much more complex, being governed by a large system of coupled equations, some of which may be nonlinear. A well-known example is that of an aircraft maneuvering in a combat situation. In this case, while the aircraft trajectory is governed by a set of equations, a gunner tasked to shoot down the aircraft is neither privy to the equations nor does he have time to solve them. The pilot usually renders the gunner's job even more difficult by executing random twists and turns. How might the gunner succeed in shooting down the aircraft?

When first faced with the fire-control problem described above, early designers of radar-aimed anti-aircraft guns developed a technique called Kaman filtering, or *autoregression analysis.* The underlying idea is that the gunner does not need to know or solve the actual set of equations governing the trajectory of the aircraft if he can find a simpler model that represents the motion of the target over a short but sufficiently long period of time. The assumption in autoregression analysis is that information gained by observation of the trajectory over a limited period of time can be used to predict the trajectory for a brief period of time into the future.

Suppose the observed trajectory is known in terms of a *sampled time series,* as illustrated (in one dimension) in Figure D.1. In the figure the filled symbols represent the actual sequence of positions making up the trajectory, and the open symbols represent positions predicted using a fourth-order autoregression model. Then the open symbol at time increment 9 represents the position s'_9 predicted by a *fourth-order autoregression model* of the form

$$s'_9 = s_9 - \varepsilon_9 = a_1 s_8 + a_2 s_7 + a_3 s_6 + a_4 s_5 \tag{D.8}$$

In Equation D.8, ε_9 is the error between the observed and predicted positions at the ninth time increment,

$$\varepsilon_9 = s_9 - s'_9 \tag{D.9}$$

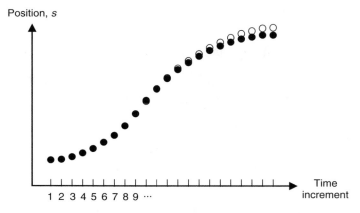

Figure D.1 A hypothetical sampled time series representing a segment of the trajectory followed by a maneuvering aircraft

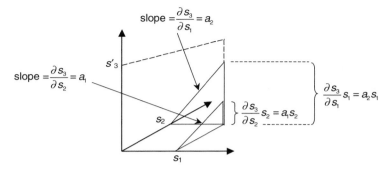

Figure D.2   Illustration of the interpretation of the autoregression coefficients as slopes for a second-order autoregression

and the coefficients a_1, a_2, a_3, and a_4 are the *autoregression coefficients* that correlate the position at the ninth time increment with the positions at the previous four time increments. We will learn in Section D.4.1 that the order n of the autoregression model is limited by the number of observations available for computing the n autoregression coefficients. The general form of Equation D.8 for an nth-order autoregression is given by Equation D.2.

The autoregression coefficients can be thought of as sensitivities, or slopes, in a linear regression, as illustrated in Figure D.2 for the case of a second-order autoregression model. In this simple example it is clear that

$$s_3' = s_3 - \varepsilon_3 = \frac{\partial s_3}{\partial s_2} s_2 + \frac{\partial s_3}{\partial s_1} s_1 = a_1 s_2 + a_2 s_1 \tag{D.10}$$

Since the regression is linear, the error $\varepsilon_3 = s_3 - s_3'$ will depend on the size of the sampling interval $\Delta s = s_3 - s_2 = s_2 - s_1$ and on the curvature of the true trajectory in the vicinity of the point s_3. Therefore, it is not always the case that an autoregression of the highest available order n will yield the smallest error. In the current example we may say that a_2 is the sensitivity of s_3 to s_1, all other variables being held constant. Then Figure D.2 illustrates a situation in which s_3 is more sensitive to s_1 than to s_2.

D.4.1   Calculation of the Autoregression Coefficients.

We consider a sampled time series such as the one illustrated in Figure D.1 and assume that there exist n autoregression coefficients a_1, a_2, . . ., a_n such that, for the range of discrete observations of s between $i - n$ and $i + N$,

$$\begin{bmatrix} s_i \\ s_{i+1} \\ \cdot \\ s_{i+N} \end{bmatrix} = \begin{bmatrix} s_{i-1} & s_{i-2} & \cdot & s_{i-n} \\ s_i & s_{i-1} & \cdot & s_{i+2-n} \\ \cdot & \cdot & \cdot & \cdot \\ s_{i+N-1} & s_{i+N-2} & \cdot & s_{i+N-n} \end{bmatrix} \begin{bmatrix} a_1 \\ a_2 \\ \cdot \\ a_n \end{bmatrix} \tag{D.11}$$

or

$$[S] = [[\bar{S}]] [A] \tag{D.11a}$$

The form of Equation D.11 assumes the availability of a sequence of $n + N + 1$ observations. Equation D.11 is solved to obtain the n coefficients of an nth-order autoregression model based on the $n + N + 1$ observations. The vector of unknown autoregression coefficients a is then used with Equation D.2 to estimate the value of s at the next $(i + N + 1\text{st})$ time step. Implicit in the method is the assumption that the values of the autoregression coefficients change slowly and so are essentially the same for the sequence between $i - n + 1$ and $i + N + 1$ as for the sequence between $i - n$ and $i + N$. The degree to which this is not true is reflected in the error ε. Note that a large error implies that the sequence is random, i.e., that knowledge of the past behavior of the sequence cannot be used to reliably predict its future behavior.

The vector $[S]$ of observed values of s on the left-hand side of Equation D.11a has at least one more element than the vector $[A]$ of unknown autoregression coefficients on the right-hand side. As a consequence of this, the matrix $[[\bar{S}]]$ is not square. Therefore, to solve for the vector of unknown autoregression coefficients, we must begin by multiplying Equation D.11a by the transpose of $[[\bar{S}]]$,

$$[[\bar{S}]]^t [S] = [[\bar{S}]]^t [[\bar{S}]] [A] \tag{D.12}$$

or

$$
\begin{bmatrix}
\sum_{j=i}^{i+N} s_j s_{j-1} \\
\sum_{j=i}^{i+N} s_j s_{j-2} \\
\cdots \\
\sum_{j=i}^{i+N} s_j s_{j-n}
\end{bmatrix}
=
\begin{bmatrix}
\sum_{j=i}^{i+N} s_{j-1} s_{j-1} & \sum_{j=i}^{i+N} s_{j-1} s_{j-2} & \cdots & \sum_{j=i}^{i+N} s_{j-1} s_{j-n} \\
\sum_{j=i}^{i+N} s_{j-2} s_{j-1} & \sum_{j=i}^{i+N} s_{j-2} s_{j-2} & \cdots & \sum_{j=i}^{i+N} s_{j-2} s_{j-n} \\
\cdots & \cdots & \cdots & \cdots \\
\sum_{j=i}^{i+N} s_{j-n} s_{j-1} & \sum_{j=i}^{i+N} s_{j-n} s_{j-2} & \cdots & \sum_{j=i}^{i+N} s_{j-n} s_{j-n}
\end{bmatrix}
\begin{bmatrix}
a_1 \\
a_2 \\
\cdots \\
a_n
\end{bmatrix}
\tag{D.13}
$$

We are now able to solve Equation D.13 directly for the vector of autoregression coefficients,

$$[A] = [[\bar{\Sigma}]]^{-1} [\Sigma] \tag{D.14}$$

where $[[\bar{\Sigma}]]^{-1}$ is the inverse of the matrix in Equation D.13 and $[\Sigma]$ is the vector on the left-hand side.

D.4.2   Example

We now consider Equation D.7, the model for the distance an object falls from rest in a vacuum. Table D.1 lists the "observed" vertical displacement of the object, based

Table D.1 Vertical displacement observed and predicted from a third-order autoregression model based on $n + N + 1 = 7$ observations ($a_1 = 2.9885$, $a_2 = -2.9591$, $a_3 = 0.9553$)

Time (s)	Observed Displacement (m)	Predicted Displacement (m)	$\dfrac{\text{Observed} - \text{Predicted}}{\text{Observed}}$
0	0	—	—
1	4.90	—	—
2	19.6	—	—
3	44.1	—	—
4	78.4	—	—
5	122.6	—	—
6	176.5	—	—
7	240.3	239.6	0.0029
8	313.8	310.8	0.0096
9	397.1	388.6	0.0214
10	490.3	470.3	0.0408

on Equation D.7, at eleven equally spaced time steps beginning at $t = 0$ s. We wish to predict the displacement 7, 8, 9, and 10 s after the object is released using a third-order autoregression model with $N = 3$. Thus, we will need to use the first seven observed displacements to compute the three autoregression coefficients. The results are given in Table D.1 and in Figure D.3.

Inspection of Table D.1 reveals that the error in the predicted displacement is less than 1 percent at the 8th and 9th time steps ($t = 7$ and 8 s), but then doubles to 2 and 4 percent at the 9th and 10th time steps. From this we may conclude that the model performance degrades as the model is applied farther from the time interval used to

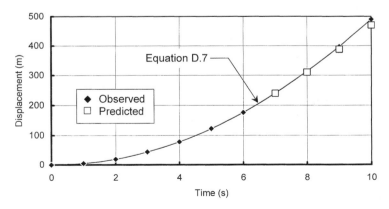

Figure D.3 Observed displacement (Equation D.7) and displacement predicted using a third-order autoregression model for an object dropped from rest falling under the influence of gravity in a vacuum

compute the autoregression coefficients. In fact, this is a general property of autoregression models.

Team Project

TP D.1   Test the random number generator provided with your Windows-based spreadsheet against as many of the seven properties identified in the chapter as possible. How could you test its periodicity? Do you have any control over the sequence produced? For example, can you force it to give you the same sequence of random numbers twice? What limitations of this random number generator were you able to identify?

Discussion Points

DP D.1   What "naturally occurring" sources of random numbers can you think of? Go to the library and do some research on this topic. Is there really any such thing as a true random number sequence or is this concept a product of human imagination?

DP D.2   Give an example of an algorithm that produces a sequence of quasi-random numbers.

DP D.3   What kinds of biases could arise in an MCRT analysis if a nonrandom sequence of "random" numbers were to be used?

DP D.4   In the discussion of property 4 (repeatability) it is stated that this property is necessary for debugging MCRT algorithms. Please explain.

DP D.5   Referring to property 5, what precautions could a research team, running parts of the same MCRT study at different remote sites, use to ensure that different members of the team used different sequences of the same random number generator?

DP D.6   Is the "modulus" operator generally available in FORTRAN? In other programming languages such as C or C++? In your spreadsheet? Write a brief algorithm that computes c mod m.

Problems

P D.1   A PRNG algorithm has been proposed in which an n-digit integer, where n is an even number, is squared and then the middle n digits of the $2n$-digit result are isolated and squared, and so forth. For example, if $n = 4$ and the integer $s_0 = 1234$ is selected as the seed, then the first few members of the resulting sequence are 1234, 5227, 3215, 3362, 3030, 1809, 2724, 4201, 6484, 422, 1780, 6840, ... Dividing each of these by the maximum possible value, 9999, yields the pseudo-random number sequence (rounded to eight significant figures) 0.12341234, 0.52275227, 0.32153215, 0.33623362, 0.18091809, 0.27242724, 0.42014201, 0.64846484, 0.04220422, 0.17801780, 0.68406840. Write a computer program to create a sequence of pseudo-random numbers based on this algorithm. Test the program for the case of $n = 8$ and $s_0 = 12345678$.

P D.2 Explore the capabilities of the PRNG algorithm created in Problem P D.1. Specifically, determine how the length of the sequences generated depends on the values of n and s_0. Discuss the behavior of the PRNG for $n = 2, 4, 6$, and 8 with at least three values of the seed for each value of n.

P D.3 Run the PRNG in Problem P D.1 with $n = 8$ and $s_0 = 12,345,678$ until a sequence of 100 (or the largest number of random numbers available if less than 100) random numbers has been obtained. Remember to scale the sequence by dividing by 99,999,999 so that the values vary between zero and unity.

  (a) Use the χ^2 test to evaluate the uniformity of distribution of the sequence obtained. (Use ten bins.)

  (b) Use a fourth-order autoregression model to test the degree of correlation of the sequence.

P D.4 Write a computer program that implements an MLCG algorithm based on

$$s_{i+1} = 5s_i \bmod 101 \qquad\qquad (D.15)$$

to produce a sequence of pseudo-random numbers, and execute the program with $s_0 = 7$. What is the period of the random number sequence produced by this algorithm? Is the resulting sequence a full-period sequence?

P D.5 For the pseudo-random number sequence produced by the algorithm in Problem P D.4, use the χ^2 test to evaluate the uniformity of distribution of the sequence obtained. (Use 10 bins.)

P D.6 Use a fifth-order autoregression model to test the degree of correlation of the pseudo-random number sequence produced by the algorithm in Problem P D.4.

P D.7 Verify the autoregression coefficients (given in the title of Table D.1) for a third-order autoregression model of a body falling under the influence of gravity (Equation D.7) using the first seven "observed" displacements in Table D.1. Note that $i = n = N = 3$ in this case. Then use the autoregression model to verify the predicted displacements for $t = 7, 8, 9$ and 10 s given in Table D.1. Also compute the percent difference between the observed and predicted displacements at these times.

P D.8   Compute the four autoregression coefficients for a fourth-order autoregression model of a body falling under the influence of gravity (Equation D.7) using the first nine "observed" displacements in Table D.1. Note that $i = n = N = 4$ in this case. Then use the autoregression model to verify the predicted displacements for $t = 9$ and 10 s given in Table D.1. Also compute the percent difference between the observed and predicted displacements at these times. Are the percent differences smaller for the fourth-order autoregression model than they were for the third-order model considered in Section D.4.2?

REFERENCES

1. James, F., "A review of pseudo-random number generators," *Computer Physics Communications,* Vol. 60, 1990, pp. 329–344.
2. Dally, J. W., W. F. Riley, and K. G. McConnell, *Instrumentation for Engineering Measurements,* Wiley, New York, 1984, p. 541.
3. Park, S. K., and K. W. Miller, "Random number generators: good ones are hard to find," *Computing Practices,* Vol. 31, No. 10, October 1988, pp. 1192–1201.
4. Lehmer, D. H., "Mathematical methods in large-scale computing units," *Annu. Comput. Lab. Harvard Univ.,* Vol. 26, 1951, pp. 141–146.
5. Marsaglia, G., and A. Zaman, Florida State University Report FSU-SCRI-87-50, 1987.

INDEX